Transforming Climate Finance and Green Investment with Blockchains

Transforming Climate Finance and Green Investment with Blockchains

Blockchain Climate Institute

Managing Editor

Alastair Marke

Associate Editors

Bianca Sylvester
Justin Macinante
Stefan Klauser

ELSEVIER

ACADEMIC PRESS

An imprint of Elsevier

Academic Press is an imprint of Elsevier
125 London Wall, London EC2Y 5AS, United Kingdom
525 B Street, Suite 1650, San Diego, CA 92101, United States
50 Hampshire Street, 5th Floor, Cambridge, MA 02139, United States
The Boulevard, Langford Lane, Kidlington, Oxford OX5 1GB, United Kingdom

British Library Cataloguing-in-Publication Data
A catalogue record for this book is available from the British Library

Library of Congress Cataloging-in-Publication Data
A catalog record for this book is available from the Library of Congress

ISBN: 978-0-12-814447-3

For Information on all Academic Press publications
visit our website at https://www.elsevier.com/books-and-journals

Publisher: Candice Janco
Acquisition Editor: J. Scott Bentley
Editorial Project Manager: Hilary Carr
Production Project Manager: Priya Kumaraguruparan
Cover Designer: Mark Rogers

Typeset by MPS Limited, Chennai, India

Contents

List of Contributors

Mira Ahmad Abt Associates, Princeton, NJ, United States

Jack Aldane Development Finance magazine, London, United Kingdom

Matias Aranguiz PhD, Shanghai Jiaotong University, Shanghai, China

Gustavo Arciniegas López Geo-Col GIS and Collaborative Planning, Amsterdam, Netherlands

Julien Bouteloupe Doshup Ltd, London, United Kingdom

Paul R. Brody Principal & Global Innovation Leader, Ernst & Young, San Francisco, CA, United States

Pascale Bronder Abt Associates, Washington, DC, United States

Delton B. Chen Center for Regenerative Community Solutions 501(c)(3), Brisbane, QLD, Australia

Edward Dodge Independent, Technology Consultant, Ithaca, NY, United States

Katrina Donaghy Civic Ledger, Brisbane, QLD, Australia

James Duchenne Managing Partner, Sutton Stone, Houston, TX, United States

Steven Dunkel Rochester Institute of Technology, Rochester, NY, United States

Paul Ellis Electron, London, United Kingdom

Anton Galenovich DAO Integral Platform for Climate Initiatives, Moscow, Russian Federation

Cristián Retamal González Amanhã Ventures, Santiago, Chile

Alexander Harris Independent Journalist, London, United Kingdom

Rob Hitchens CTO & Co-founder, Volt Markets, Houston, TX, United States

Edward Hoyt Abt Associates, Washington, DC, United States

Joanna Hubbard Electron, London, United Kingdom

Markus Hüwener Independent Scholar, Frankfurt am Main, Germany

David Huysman Neocapita, Bucharest, Romania

Adrian Jackson Edinburgh Parallel Computing Centre, University of Edinburgh, Edinburgh, United Kingdom

Ben Joakim Disberse, Swansea, United Kingdom

Loretta Joseph Australian Digital Currency Commerce Association, Sydney, NSW, Australia

Jenya Khvatsky CEO of CleanTek Market, Melbourne, VIC, Australia

Ashley Lloyd Edinburgh Business School, University of Edinburgh, Edinburgh, United Kingdom

Sergey Lonshakov Airalab, Togliatti, Russian Federation

Justin Macinante Law School, University of Edinburgh, Edinburgh, United Kingdom

Alastair Marke Secretary-General, Blockchain Climate Institute, London, United Kingdom

David Martin Power Ledger, Perth, WA, Australia

Sebastien Meunier Director, Chappuis Halder, New York, NY, United States

Mihail Mihaylov Enervalis, Houthalen-Helchteren, Belgium

Ann Nowé Department of Computer Science, Vrije Universiteit Brussel, Brussels, Belgium

Anna Poberezhna Smart4Tech, London, United Kingdom

Steven Pocock Abt Associates, Port Moresby, Papua New Guinea

Iván Razo-Zapata Luxembourg Institute of Science & Technology, Esch-sur-Alzette, Luxembourg

Tim Reutemann Managing Director RedMorpheus GmbH, Zurich, Switzerland

Adam Richard CTO & Co-founder, Volt Markets, Houston, TX, United States

Neil Salisbury Founder & Director of CleanTek Market, Melbourne, VIC, Australia

Owen Sanderson GlobalCapital, London, United Kingdom

Alexey Shadrin Russian Carbon Fund, Moscow, Russian Federation

David Shrier University of Oxford, Oxford, United Kingdom

Bianca Sylvester Associate Director, Clean Energy Finance Corporation, Melbourne, VIC, Australia

Jane Thomason Abt Associates, Brisbane, QLD, Australia

Tony Willenberg Neocapita, Vienna, Austria

Adam Woodhall Advisor, Activator, Author: Inspiring Sustainability, London, United Kingdom

Duoqi Xu Professor, Shanghai Jiaotong University, Shanghai, China

Xinran Xu Fellow, FinTech4Good, Washington, DC, United States

Xiaochen Zhang President, FinTech4Good, Washington, DC, United States

Xing Zhang Operations Officer, IFC, Washington, DC, United States

About the Editors

Managing Editor

Alastair Marke, FRSA, FRGS is the Founder of the *International Core Group on Blockchain Climate Finance*, which is being upgraded as an NGO—"*Blockchain Climate Institute.*" He is a sustainability and climate finance policy advisor to a variety of institutional clients. Currently the Head of Blockchain Innovations (Climate Change) at Abt Britain, Alastair is also the Vertical Lead on climate change with the British Blockchain Association and an expert member (Greenhouse Gas Management Institute) of the ISO 14097 Climate Finance Standards Committee. During his tenure with several think tanks, including the Overseas Development Institute, Alastair has authored papers and reports covering a wide range of policy issues, including digital revolution in global governance, food and energy security, climate finance, low-carbon investment planning, emissions trading, and associated green growth issues internationally. At the Global Solution Networks, where he worked for Don Tapscott, Alastair was a key faculty member of the Climate & Planet Stewardship Hub. He started his policy career as a civil servant in Hong Kong.

Alastair holds an MSc in Environmental Strategy (Distinction) from the University of Surrey, United Kingdom. He is qualified as certified expert in renewable energy, climate, and adaptation finance from the Frankfurt School of Finance & Management, as well as an Oxford University-certified FinTech entrepreneur. Alastair is now a candidate for MSt in Sustainability Leadership at the University of Cambridge, United Kingdom; and LLM Climate Change Law & Policy at the University of Strathclyde, United Kingdom.

Associate Editors

Bianca Sylvester is an associate director at the Clean Energy Finance Corporation in Australia. Bianca was formerly a Climate Change Specialist in the World Bank's Carbon Finance Unit, based in Washington DC. She led the World Bank's Networked Carbon Markets initiative which develops the tools, services, and institutions needed to link the next generation of carbon markets. Bianca has also worked in the World Bank's Latin America and Caribbean region, implementing a range of renewable energy and carbon finance initiatives. Previously, Bianca was a trader in J.P. Morgan's commodities sales and trading team based in Singapore and Sydney.

Bianca holds a Master of Government from Johns Hopkins University in Washington DC, United States.

Justin Macinante has 35 years of experience as a lawyer in banking and finance, environmental regulation, and policy as well as in commercial practice in Europe, Australia, and the Pacific. He was a legal counsel for the First Climate Group, a carbon trading firm based in Germany and Switzerland for 8 years and more recently has been consulting to the World Bank's Networked Carbon Markets initiative. Justin is part of the "DLT4NCM project" team at Edinburgh University, United Kingdom.

Justin holds BSc and LLB degrees from the University of New South Wales, Australia and a Master of Environmental Law (First Class Honors) from the University of Sydney, Australia.

Stefan Klauser is a political scientist (MA, University of Zurich, Switzerland) and an MIT-certified FinTech specialist. He leads the subject areas "Digital Society" and "Finance 4.0" in Professor Helbing's Chair for Computational Social Science at ETH Zurich. Stefan is a Co-Founder of the Blockchain & IoT School BIOTS and the BlockchainX Network. Before his engagement at ETH, he worked for the Swiss Department of Foreign Affairs and the Swiss State Secretariat for Education, Research, and Innovation, where he served as Project Leader Innovation Initiatives and Swiss Representative to the OECD Global Science Forum.

About the Lead Contributors

Jack Aldane is currently an Editor of the Development Finance magazine. He holds MA Financial Journalism from the City University London, plus 5 years' experience as a reporter and an editor. He has the strongest aptitude as a reporter being his style of interviewing. This style gives the illusion of natural progression to structured conversation, and which adapts to off-piste dialogue without losing focus. He has interviewed high-profile interviewees in the past and has succeeded in drawing out unexpected information from such conversations.

Alexander Harris is a journalist from the United Kingdom. He holds a Bachelor's degree in Journalism from the City University London. He is a passionate proponent of the value of Blockchain technology in global governance. He is an active member of the International Core Group on Blockchain Climate Finance (now Blockchain Climate Institute). With keen interest in languages, he can speak Chinese and German fluently, alongside his native English. He currently resides in Beijing, China, where he teaches oral English and Journalism at the University of Science and Technology Beijing.

Sebastien Meunier is ranked as a top Blockchain influencer by "Richtopia." Currently the Director at Chappuis Halder & Co., he is an expert in financial technology, who has published articles and delivered keynotes about innovation, Blockchain, and artificial intelligence on various occasions. As a strategy consultant, Sébastien has a successful track record of defining innovative business strategies and leading global transformations in financial institutions.

Adam Woodhall is the founder of Inspiring Sustainability. He is a storyteller who works with rapid growth small enterprises creating sustainable solutions. He helps them create and execute exponential growth plans. He has over 20 years of commercial experience, working in sustainability since 2006, and having worked with over 100 clients, large and small. He is the author of the *Empower Change* guide and a regular contributor to the *Ethical Performance* magazine.

Paul R. Brody is the Principal and Global Innovation Leader at Ernst & Young. He has spent more than 15 years in the electronics industry doing extensive consulting work across supply chain, operations, and business strategy. He currently leads technology strategy for Ernst & Young with a focus on the implementation and deployment of Blockchains across industries. His client experience in the past has included companies like IBM, HP, Samsung, LG, Huawei, Apple, Canon, Toshiba, Sony, Panasonic, Ericsson, Nokia, Alcatel, FIAT, Honeywell, and Raytheon. Prior to joining Ernst & Young, he did two stints at IBM, including running the global electronics industry organization, as well as working at five different start-up companies, including co-founding an online movie company. He started his career as an analyst at McKinsey & Co. in Los Angeles. He holds a BA in Economics and a Certificate in African studies from Princeton University.

Adam Richard is a co-founder and the CEO of Volt Markets, a company that is bridging energy and capital markets to Blockchain technology in order to streamline the distribution, tracking and trading of energy and energy attributes. He is a serial entrepreneur who has been a Blockchain consultant since 2014 and helps entrepreneurs build successful companies. He is a three-time Rice University start-up competition winner, Surge accelerator alumnus, Boost VC accelerator alumnus, Plug and Play accelerator "Energy and Sustainability" alumnus, and founded the Houston Bitcoin and Ethereum meetups.

Joanna Hubbard had 5 years' experience in renewable energy finance with 3 years in cleantech, fintech, and digital consulting. They are now co-founders and on the executive team of Electron, a company that builds Blockchain platforms for the energy sector. She holds a BA in English Literature from the University of Oxford.

Mihail Mihaylov is a senior research engineer at Enervalis and a research affiliate at Vrije Universiteit Brussel, Belgium. He holds a PhD in artificial intelligence with over 10 years of research experience in decentralized systems within both private and public organizations. He has been involved in Blockchain technology since 2011, giving lectures and invited talks on this subject to international audiences worldwide. He founded and co-organized the Barcelona Bitcoin Community and is currently co-organizing the P2P.Gent Bitcoin community in Belgium. He started developing the NRGcoin concept in 2013. His work has been featured in academic journals and popular science magazines alike. Mihail's main interests and work focus on renewable energy, smart grids, and Blockchain technology.

Ann Nowé finished her PhD in collaboration with Queen Mary and Westfield College, University of London. Her PhD thesis studied the intersection of computer science (artificial intelligence), control theory (fuzzy control), and mathematics (numerical analysis, stochastic approximation). With years of scientific research experience at higher institutions, she is currently a professor both in the Computer Science Department of the Faculty of Sciences and in the Computer Science group of the Engineering Faculty at Vrije Universiteit Brussel.

Jane Thomason holds a PhD in public health policy and has 30 years of experience in international development. As the CEO of Abt Associates Australia, she is currently managing over $400 million in development programs for the poor. Over the past 18 months, She has studied Blockchain and its applications for the poor. Being a frequent speaker and blogger on Blockchain and social impact, she is currently working on several Blockchain pilots in developing countries. Abt Associates has a strong track record in climate change and multidisciplinary development approaches.

Tim Reutemann is an independent climate finance expert and science fiction author. He is working on the forefront of new financial mechanisms under the Paris Agreement, while exploring future opportunities for more democratic and efficient global governance mechanisms in his writing. Tim is currently advising the Peruvian and Swiss Governments on structuring a first pilot for the new market-based approaches under the Paris Agreement, while helping UNEP to improve further their training program on the technicalities of accessing finance from the Green Climate Fund. He holds a PhD in Environmental Economics from ETH Zurich and has worked on climate change projects for the private sector, academia, and the public sector over the last 10 years.

Cristián Retamal González has over 10 years of experience in the international climate regime. He has been involved with the UNFCCC process since 2006 supporting the commitment of developing countries under the convention. The scope of his work covers both spheres: climate mitigation and adaptation. He has proved experience on a broad range of topics such as climate finance, carbon pricing, market-based mechanisms, and climate-induced migration, among others. He holds a civil engineering degree with environmental minor from Pontificia Universidad Catolica in Chile, and an MSc in Environment and Resources Management from VU University Amsterdam in the Netherlands. He is currently affiliated with Amanhã Ventures focusing his work on the transformations and mobilization of efforts that climate change entails. His most recent work is oriented to supporting developing countries in formulating climate change initiatives, including the mobilization of climate finance to achieve the goals of the Paris Agreement.

Iván Razo-Zapata is a computer scientist with more than 10 years of experience in research within both private and public organizations. He holds a PhD in Science from VU University Amsterdam. He has been involved in EU projects dealing with business modeling, service innovation, integration of renewable energy, and artificial intelligence. Between 2014 and 2015, he was involved in the Scanergy project that designed the NRGcoin concept. He is affiliated to the Luxembourg Institute of Science and Technology (LIST) as Research & Technology Associate. His current research interests encompass smart cities, business modeling, and service science.

Gustavo Arciniegas López received a PhD in "Map-based decision support tools for collaborative land use planning" from IVM, VU University Amsterdam in the Netherlands.

He also holds an MSc in Geoinformatics from the University of Twente, ITC faculty, the Netherlands, and a BSc in civil engineering from the UIS, Colombia. He specializes in the integration of GIS, SDSS, PSS, and geo-instruments for collaborative spatial decision-making and geodesign, focusing on sustainable planning and climate adaptation. He is a certified CommunityViz authorized GOLD consultant. His most recent work includes a role as consortium partner of an EU-funded project called REPAiR, which focuses on providing geodesign decision support tools to support the transition to circular economy of six European metropolitan areas.

Neil Salisbury is an entrepreneur and sustainability leader with over 20 years of experience. He is the Managing Director and co-founder of Point Advisory a leading integrated sustainability consultancy and the founder of CleanTek Market Pty Ltd, a clean technology marketplace looking to accelerate climate finance in both developing and developed countries. He is the Vic-Chair and board member of US-based NGO, The Efficiency Valuation Organization (EVO) and on the advisory board of the World Intellectual Property Organization (WIPO Green).

Jenya Khvatsky is currently the CEO at CleanTek Market. He is a career clean energy professional, with 20 years' experience in energy and low-carbon technologies in Australia, Asia, and Europe in senior technical, commercial, and investment roles, including CEO of Green Terrafirma (Hong Kong, bio-products) and Future Blends (UK, advanced biofuels), Head of Biofuels Programmes at the Carbon Trust, and as investment manager at Sindicatum Sustainable Resources and Nord Engine Capital. Jenya Khvatsky specializes in interfacing between technology and investors through the judicious application of strategic, financial, and risk frameworks.

Anna Poberezhna is a strategic and innovation expert with a focus on the impact business, Global Goals for Sustainable Development, Blockchain, and green finance. An author and speaker on Blockchain for climate change and sustainability, she is the founder of Smart4tech. She is an advisor on Blockchain and strategy for a number of start-ups, while she researches and develops the circular model for water trading on Blockchain, which is closely interconnected with Waste, Energy, Food, and Carbon sectors. She also advised clients on a strategic vision for enhancement of the UK–Ukraine business relations with a particular focus on digital economy and sustainability. Prior to founding Smart4tech, she headed the Strategy and Development in SCG International. Her industry experience includes strategy, international relations, branding, sustainability, and construction sectors.

Delton B. Chen is an Australian civil engineer with over 14 years of experience as a professional engineer. He also holds a PhD from the University of Queensland for his environmental study of Heron Island in the Great Barrier Reef. He is a thought leader in monetary science, having developed a new market theory for addressing climate systemic risk with a Central Bank Digital Currency (CBDC) and decentralized ledgers. Delton is the lead author and co-founder of the "Global 4C Risk Mitigation" policy (global4c.org). He is a recognized expert in developing computer-based models of natural and engineered systems, such as groundwater flow, "hot rock" geothermal energy, and greenhouse gas mitigation. He is a research fellow with Project Drawdown, and is a contributor to the book "Drawdown—The Most Comprehensive Plan Ever Proposed to Reverse Global Warming" (drawdown.org).

Edward Dodge is an independent technology consultant and writer with 20 years of experience in information technology and renewable energy, plus experience with federal energy and carbon policy in the United States. His proposal for a new carbon pricing model is drawn from his personal technology and policy experience and represents a unique idea that, to his knowledge, has never been presented elsewhere. He holds MBA and bachelor's degrees from Cornell University.

Anton Galenovich has 25 years of experience in international relations and external liaison, including government and public relations. Dr. Galenovich has profound theoretic and practical expertise in the field of international diplomacy, economic, corporate, business intelligence, planning, structuring, and implementing of top-level negotiation process. For the last 10 years, he has been dealing with the complexities of diplomatic, international, economic, environmental, emissions trading issues relating to the UNFCCC and Kyoto Protocol. He has worked with major Russian Corporations and NGOs, and other organizations interested in emissions credits and quota trading, energy efficiency, and capacity building in Russia for facilitating negotiations and cooperation among state authorities, major business companies, and NGOs.

Sergey Lonshakov is a project architect at Airalab. Since 2011, he has been engaged in research and development of bitcoin, altcoin, cryptocurrency, smart contract, and Ethereum platform projects to provide solutions for DAO and Internet of Things.

Alexey Shadrin is the Founder and CEO of the Russian Carbon Fund and has co-developed the Blockchain Ecosystem platform for environmental assets trading at IPCI. He holds a Master's degree in strategic management from the Admiral Makarov State University of Maritime and Inland Shipping, and a Bachelor's degree in international economics from the International Institute of Energy Policy and Diplomacy of the Moscow State Institute of International Relations.

Steven Dunkel is a submarine veteran of the US Navy and graduate of the Rochester Institute of Technology. He is a founding member of a bitcoin wallet company that offers a service capable of broadcasting Blockchain transactions to over 100 countries. He communicates with customers living in remote areas around the world and addresses their regional issues maintaining connections to the bitcoin Blockchain. The bitcoin company now utilizes the same technology to issue credentials through a wearable ring. As a technology innovator, he has conceptualized improvements in the Clean Development Mechanism to increase the success rate of Intended Nationally Determined Contributions that he is eager to share and discuss with readers of this book.

Adrian Jackson is a research architect with two decades of computing experience at The University of Edinburgh. He focuses on developing highly scalable software, from large parallel programs, to large-scale data analytics. He is part of the DLT4NCM consortium, developing distributed ledger solutions for networking carbon markets, and carbon trading, with particular focus on low-cost consensus mechanisms.

Ashley Lloyd has researched widely in the emerging technologies arena, with publications ranging from the *IEEE Journal of Quantum Electronics* to the *International Journal of Innovation Management*. He led the project that built the first Global Computing Grid to connect three continents and demonstrated the first connection of China's Ministry of Education and Ministry of Science and Technology networks in a global "Big Data" project. His environment expertise includes formulating new advice for diffusion correction of NO2 sensors (now a policy adopted by the British government), and leading a successful case at the United Nations, under the Aarhus Convention, setting a precedent for public access to environmental data. Now his work focuses on the DLT4NCM project to improve the efficiency of global carbon markets using distributed ledger and relevant technologies. Ashley holds a BSc (Hons), PhD, MBA, and CPhys CSci from the University of Edinburgh Business School.

Markus Hüwener has lectured at Frankfurt School of Finance & Management. A former Managing Director of Dresdner Kleinwort investment bank, he is a founder and long-term CEO of the First Climate AG, a carbon asset management, trading, and project development firm. He is a seasoned expert who has contributed to several books on climate change, renewable energy finance, and infrastructure. He holds an MBA from the Technical University of Dortmund, Germany.

Owen Sanderson has spent 8 years writing about the global capital markets, largely for EuroWeek, GlobalCapital, International Financing Review and Reuters, covering investment banking, financial technology, regulation, and strategy. He has worked on the daily newspapers of the World Bank and IMF annual meetings, interviewing finance ministers, many of the largest green bond issuers and structurers, and senior investment bankers about their sustainable finance plans. He has extensive knowledge of the structures, practices, and approaches in modern bond markets and has been closely watching the rise of green finance and blockchain initiatives in banking and markets.

Xiaochen Zhang leads the design and implementation of FinTech4Good and Blockchain Frontier Group's strategy, which aims to introduce cutting-edge fintech and Blockchain solutions to emerging markets through incubation, acceleration, and investment. He brings thought leadership with more than 16 years of global experience to build impactful solutions and scale up innovative ideas in North America, Africa, Latin America, Europe, and Asia. He serves on the Board of Directors of the Crowdfunding Professional Association, the United Nations Economic and Social Commission for Asia and the Pacific's Digital Economy Task Force, and as a senior adviser of the China Social Entrepreneurs Foundation and the Inter-American Development Bank. He is also the co-chair of the Global Digital Economy Council and several Blockchain labs. He holds an MA in Global Studies from Leipzig University.

Matias Aranguiz is a partner at a Chinese consulting firm, Mantian Investment, focusing on cross-border finance between China and Latin America, greentech, fintech, and social innovation. In 2018, he taught at Massachusetts Institute of Technology: a new course, Artificial Intelligence and Global Risks (6.S092). He also works as an advisor to Fintech4Good. Prior to joining Mantian Investment, he worked in several private and public institutions as a lawyer, investor, and advisor. He is a Chilean lawyer (PUC) who holds an MSc in Finance (SUFE). He is currently a PhD candidate at Shanghai Jiao Tong University, where he studies law, artificial intelligence, sustainable development goals, and fintech.

Duoqi Xu is a Professor of tax and financial law at Shanghai Jiao Tong University, where she is also the Director of Shanghai Jiao Tong University Tax Law and Policy research center, and Co-director of Internet Financial Law Innovation Research Center. Professor Xu is a Shanghai Shuguang Scholar (2011), the Fifth Top Ten Young and Middle-aged Jurist (2012), as well as the founder and the Editor-in-Chief of the Internet Financial Law Review since August 2013. She was a Fulbright Senior Visiting Scholar at Harvard University in the United States from 2016 to 2017, and a Visiting Scholar of the New York University Hauser Global Program from 2008 to 2009. Her main research areas include fiscal and taxation law, finance law of science and technology, and the Internet security.

Xing Zhang has more than 10 years of experience in climate market and international standard setting for both the public and private sectors. He advises international financial institutions on tracking climate finance, creating opportunities for climate business, and exploring how climate actions could benefit from disruptive technologies such as the Blockchain. He is also active in investigating use cases of Blockchain and other emerging digital technologies in facilitating post-2020 climate markets under the Paris Agreement. Before that, he worked with the UNFCCC Secretariat in supporting the operationalization and negotiation of international carbon market. He holds a PhD in Environmental Engineering.

Xinran Xu is a fellow at FinTech4Good, the world's leading blockchain and fintech platform agency. The organization identifies and scales up the most promising fintech and Blockchain technology solutions worldwide and brings them to frontier markets through incubation, acceleration, and investment. She supports various FinTech4Good projects, including research, material development, and operational support. She holds a Master of Finance degree from Johns Hopkins University Carey Business School and a Bachelor of Science degree in Business Administration from Boston University.

James Duchenne has been involved in the Blockchain (including bitcoin) business as an advisor and venture builder since 2013. He is the Chief Operating Officer of Crowd Machine, Managing Partner at Sutton Stone, and the Honorary Representative of the Board of Investment of Mauritius in the United States assisting and advising the government on innovation and Blockchain initiatives. In 2010, he was a carbon credit assessor (voluntary market) in Australia. Today, with Volt Markets, his team is designing the future of RECs (renewable energy credits) and Carbon Credit trading using Blockchain technology. He holds advanced degrees in engineering, law, and finance. He is also a qualified lawyer of the Supreme Court of New South Wales and the High Court of Australia.

Guest Foreword

Prof Anne-Marie Slaughter
Director of Policy Planning, U.S. Department of State, 2009-11

The new technologies of the 19th and 20th centuries have helped create climate change: endless engines burning fossil fuels—coal, oil, and gas—and emitting the carbon stored into the atmosphere for millennia. The new technologies of the 21st century will provide the means of mitigating and adapting to the "change" we humans have created. The critical and urgent means of saving our planet are certainly new energy technologies: cost-efficient and reliable ways of harnessing the ever renewable sources of sun, wind, and water; safer nuclear energy; new ways of tapping the heat deep in the earth; new micro-power grids that can distribute and share the energy we do not need. Equally important, however, even if far less noticed, are *breakthroughs in financial technology*.

One of the biggest obstacles to fighting climate change effectively is *lack of trust*. Trust among nations is necessary to conclude international agreements; and it is even more essential to implement them, as late US President Ronald Reagan so memorably said in the course of concluding arms control agreements with the Soviet Union, "*Trust, but verify*." In other words, trust is bolstered by the actual ability to see whether or not your negotiating partners are in fact living up to the commitments they have made.

The international acronym for this process is MRV: Measurement, Reporting, and Verification measures. The Paris Agreement on climate change commits its 195 signatories to reducing carbon emissions to levels that will "hold the increase in the climate's average temperature to well below 2 degrees centigrade above pre-industrial levels" and to working toward limiting that increase to 1.5°C.

Those commitments, in turn, depend on the willingness of the developed countries—all of which burned carbon with abandon as they developed—to finance the adoption and use of cleaner energies by developing countries. Indeed, Article 2 of the Paris Agreement further commits to "making finance flows consistent with a pathway toward low greenhouse gas emissions and climate-resilient development."

The developed countries have pledged to "mobilize" US$100 billion a year in climate finance through 2025. These funds are supposed to enable developing country governments and subnational governments—provincial governors and mayors—to convert or install or adopt sustainable energy systems that will lower their emissions. Making sure that developed country governments actually provide the funds they promise—which are only a fraction of the actual need—is hard. But equally hard is ensuring that the funds once committed actually get spent.

Think about the process of getting a loan from your bank. The bank needs to know that who you say you are, that you will actually spend the money the way you say you will spend, and that you actually own the assets you put up as collateral. The bank will not take your word for it; you have to fill out lots of documents and get them verified by third parties—your employer, your local government, credit agencies, etc. Now think of all the verification that has to take place when the government of a small, poor, and risky country—or the mayor of a badly functioning and corrupt city—seeks millions or hundreds of millions of dollars to implement a clean energy project. How can the lender be sure that the funds will actually be used as the borrower says they will be used, rather than being siphoned off for individual gain or for some very different projects?

Enter *Blockchain*! Here is the simplest explanation I can think of. Imagine that you are sharing a house with three housemates. You know each other, but you are not family. You each agree to contribute US$25 a week for household expenses into a common pot. You could each spend your US$25 every week on food and other household items to be shared and keep a ledger of your expenses. At the end of the week you could meet with your housemates and compare ledgers and the receipts that verify that you did in fact spend the money the way you say you did. That process is cumbersome and complex enough that no one will want to meet; the entire arrangement will quickly break down.

Alternatively you could each agree to put your US$25 into a glass jar on the kitchen counter; anyone who is going to the grocery store could take an amount from the jar and replace it with a receipt showing what was purchased with the money. But suppose one of your housemates just takes money from the jar for his/her own purposes and does not admit it? Or loses a receipt for food that is consumed before the end of the week? Or forges a receipt?

Now imagine that every time one of you takes money from the jar and spends it on a household expense, and that set of transactions—the taking and spending—is recorded, time-stamped, and sent to each of you via your email. All of you see the transaction in real time on one set of books that is collectively stored and monitored. You can see when money was put into the jar, when it was taken out, how it was spent. And if one of you pays a third party to go shopping or to repair a household item and the third party has permissioned access to the Blockchain set of accounts, you can all see the transfer to the third party and whether the third party meets its commitments. *Trust but verify*—easily, quickly, and cheaply.

I will not explain all the wonderful ways that Blockchain can create trust and thereby unblock chains of financial commitments. That is what this book is for. Not only financial transactions, a Blockchain network can also track and trace virtually anything of value, thereby minimizing risk and g costs for all stakeholders. At the age of limiting and adapting to climate change, those things of value include the flow of power itself from household to household on distributed micro-grids, the exchange of carbon credits, and community and crowd funding for local renewable energy permits, to name a few.

The contributions in this book are optimistic and exciting. They offer hope and genuine progress on the most important global issue facing our generation. Equally important to me, however, is how this book came about.

The Paris Agreement was a classic example of statecraft: governments coming together to negotiate and reach an agreement that each was then responsible for implementing. President Donald Trump's decision to pull the United States out of the Paris Agreement was also an example of statecraft, even if it was bad and shortsighted statecraft. But the day after Trump's decision, 30 mayors, 3 governors, over 80 university presidents, and 100 business leaders began negotiating with the UN to have their submissions of commitments to reduce carbon emissions accepted alongside other countries. Their action was not statecraft, but what I have called "*webcraft*"—the ability to contribute to solving global problem and to shape global outcomes through the design, creation, and mobilization of many different types of networks.[1]

The emergence of the International Core Group on Blockchain Climate Finance, founded by Alastair Marke, as a not-for-profit web-based international network of individuals who are concerned about the current global climate finance gap and who share a vision of how the deployment of Blockchain technology could help bridge that gap is another example of webcraft. In moving from a LinkedIn group to a more formal Global Blockchain for Climate Network capable of undertaking projects like writing and publishing this book, the Network is an inspiring example of citizen power in global affairs. It brings together academics, development professionals, technologists, economists, journalists, lawyers, and

[1] Anne-Marie Slaughter, *The Chessboard and the Web: Strategies of Connection in a Networked World* (Princeton, NJ: Princeton University Press, 2017).

scientists, aligning expertise and interests in ways that can do things which governments often cannot do.

Blockchain is just the beginning!

Anne-Marie Slaughter

President and CEO, New America.

Bert G. Kerstetter '66 University Professor Emerita of Politics and International Affairs, Princeton University.

Director of Policy Planning, U.S. Department of State, 2009–11.

Guest Biography

Anne-Marie Slaughter is the President and CEO of New America[1], a think-and-action tank dedicated to renewing America in the Digital Age. She is also the Bert G. Kerstetter '66 University Professor Emerita of Politics and International Affairs at Princeton University. She has been a Director at Abt Associates Inc. since November 7, 2011. From 2009 to 2011, Dr. Slaughter served as Director of Policy Planning for the United States Department of State—the first woman to hold that position. Prior to her government service, Dr. Slaughter was the Dean of Princeton's Woodrow Wilson School of Public and International Affairs from 2002 to 2009 and the J. Sinclair Armstrong Professor of International, Foreign, and Comparative Law at Harvard Law School from 1994 to 2002.

Dr. Slaughter has written or edited eight books, including *The Chessboard and the Web: Strategies of Connection in a Networked World* (2017), *Unfinished Business: Women, Men, Work, Family* (2015), *The Idea That Is America: Keeping Faith with Our Values in a Dangerous World* (2007), and *A New World Order* (2004), as well as over 100 scholarly articles. Dr. Slaughter is also a contributing editor to the *Financial Times* and writes a bi-monthly column for *Project Syndicate*. She provides frequent commentary for both mainstream and new media and curates foreign policy news for over 140,000 followers on Twitter. *Foreign Policy* magazine named her to their annual list of the Top 100 Global Thinkers in 2009, 2010, 2011, and 2012. She received a BA from Princeton, an M. Phil. and D. Phil. in international relations from Oxford, where she was a Daniel M. Sachs Scholar, and a J.D. from Harvard.

[1] https://www.newamerica.org/

Editor's Prologue: Blockchain Movement for Global Climate Actions

You have picked up the world's first book marrying Blockchain and climate change, probably because you are at least intrigued by the potential capabilities of Blockchain technology to ameliorate and accelerate inexcusable climate actions worldwide.

I was first intrigued on how digital technologies can give new solutions to solve some very old problems in global governance when I worked for Don Tapscott, one of the world's leading authorities on the socioeconomic impact of technological innovations, at his Global Solutions Network a couple of years ago. As a climate finance policy adviser, it is my mission to enhance, in particular, least developed countries' readiness to access international climate funding for their survival. I share the desperation of Anote Tong, former President of Kiribati, over the access to climate finance being difficult to his country. "It's a paradox. We need [the funds] the most but we don't have the capacity to get it because we're not accredited... and the accreditation process involves so much bureaucracy," he said at a public lecture in the London School of Economics right before the end of COP21 in December 2015. Some recently accredited national entities, from India to Namibia, have uttered that the accreditation process was "excruciatingly painful."

It is absolutely reasonable for beneficiaries to meet certain fiduciary standards before they receive public funding coming out of taxpayers' pockets. That said, at that moment, I was pondering: "Can we find ways to go through the procedures and smooth the implementation of the Paris Agreement in a 'smarter', 'faster' way?" Thanks to the experience I had gained with Don's research program and my exposure to Blockchain discussions in the financial services industry, I was sure that it was time to bring these Blockchain talks into the climate change policy community in hopes of playing a part to ratchet up much-needed public and private climate finance for the climate-vulnerable. A decent starting point was to curate a series of innovative ideas from some "rare species" at the time—a few people working at the intersection of Blockchain and climate change. This is how I started the **International Core Group on Blockchain Climate Finance** on LinkedIn in late 2016.

I am always incredibly grateful for the overwhelming enthusiasm of intellectuals from all walks of the international community upon this ground-breaking project—the first piece of hard work by over 40 experts from over 20 countries working selflessly together.

Indeed, Anote Tong's comment epitomizes the warlike challenges facing many least developed countries with low-resourced governments and/or not-yet-mature green capital markets. The need for accreditation process represents a fundamental lack of trust between donors and recipients, for example. Such lack of trust may have caused three daunting "gaps" in global climate finance that we have to urgently address without delay.

The first one is "funding gap." To keep the global temperature rise to within 2°C above preindustrial level, the colossal need for green investment would require US$ 300 billion a year by 2020 and up to US$ 500 billion a year by 2030, according to the World Resources Institute. The US$ 100 billion a year pledged by developed countries to be transferred through the Green Climate Fund is far from sufficient. The second one is "transparency gap." In 2016, just over half of donor countries complied with the UNFCCC climate finance transparency requirements for their climate funding to be trackable. The third one is even more daunting—the "efficiency gap." Like Kiribati, many least developed countries simply cannot meet the stringent accreditation requirements of many international climate funds. Even when money is available, it often takes years for a climate change mitigation or adaptation project to go through the funding process with various intermediaries. Lengthy delays in the disbursement of climate finance have increasingly become a norm. I would imagine that by the time these aggrieved receive the much-needed funding, the water level will have already got their eyebrows soaked. We have spent no less than 25 years on climate negotiations, but, the progress cannot catch up with how fast the climate changes. We simply do not have the luxury to wait as climate change never stops as we need to finish our procedures!

Blockchain technology, as a breakthrough of fintech, may be giving us a silver lining to address these gaps in a "smarter" and "faster" way. As Don Tapscott describes one of its defining features, "The Blockchain is an incorruptible digital ledger of economic transactions that can be programmed to record not just financial transactions but virtually everything of value." Not only that, it also introduces a fresh platform on which to build transactions or interactions without a privileged central operator or authority. It can be co-owned and co-operated by platform users themselves. And Smart Contracts can automate and hence, accelerate much of green investment and global climate finance flows. By moving our existing procedures onto some tailor-made Blockchain networks, it would hopefully revolutionize most, if not the entirety, of the global climate change governance.

But how can we do that? This is the core question this book will "try" to answer. Through the **International Core Group on Blockchain Climate Finance**, we gathered rare experts working in this niche area—they are scholars, economists, financial analysts, entrepreneurs, technologists, journalists, and lawyers—on a common platform to seek to break the climate finance logjam. In this book, we have provided some inspiring use cases or insights from multiple perspectives. Apart from the scene-setting team, the Core Group is composed of four subgroups of members contributing respectively to the four major sections of this book, i.e., how Blockchain can lead to:

- smarter renewable energy deployment;
- smoother international climate finance transfers;
- fraud-free emissions management; and
- better green finance law enforcement.

They will be introduced further in the Interludes to precede each section after the scene-setting part. Yet, this book is *not* to give readers any model or ultimate answers to such a profound question but rather to stimulate further discussions among the climate change policy community in search of more thorough answers. Many concepts presented in this book are at the forefront of Blockchain innovations; however, most of them are so new that they are either untested or recently tested. To translate our vision into reality, we have huge amount of work to do and absolutely welcome like-minded readers to join our movement.

I hope you enjoy reading this book!

Alastair Marke, FRSA FRGS
Founder and Convenor,
International Core Group on Blockchain Climate Finance

December 2017

INTERLUDE I

How to Read This Book

I appreciate that not all readers would have days or weeks to read every page of this book. Therefore, I suggest a couple of ways to read this book, depending on your needs or priorities.

1. **If you do not have any prior knowledge about Blockchain technology, it is very important that you start with:** Chapters 1 and 3.
2. **If you have had some knowledge about Blockchain and have an hour to learn about its applications in climate change in general, you may read:** Chapters 2 and 4, and all interludes preceding each section. It would be helpful if you had a clear grasp on the state of global climate finance reported in Chapter 4.
3. **If you are intrigued by Blockchain's capacity of stimulating peer-to-peer renewable energy trading, you may read:** Section 2, including Chapters 5–9. Chapter 5 summarizes the potential role of Blockchain in shaping the energy sector. Chapters 6 and 7 discuss the changing energy sector from the supply-side perspective while Chapters 8 and 9 discuss from the demand-side perspective.
4. **If you are particularly interested in how Blockchain technology can accelerate international climate finance transfers, you may read:** Section 3, including Chapters 10–14. Chapter 10 represents a kaleidoscope of innovative Blockchain solutions for developing countries. Chapters 11 and 12 discuss on possible streamlining of procedures in international climate funds, while Chapters 13 and 14 focus on the smart mobilization of private climate finance.
5. **If you are concerned about the way in which Blockchain could increase the efficiency of emissions trading schemes, you may read:** Section 4, including Chapters 15–19. Chapters 15 and 16 introduce new approaches to price carbon credits that underpin emissions trading schemes, while Chapters 17 and 18 suggest new, administratively efficient systems to transfer or trade carbon credits on a Blockchain. Chapter 19 goes further by proposing the "networking" of carbon markets with Blockchains.
6. **If you would like some food for thoughts as you are pondering the interaction between the law and disruptive technologies, you may read:** Section 5, including Chapters 20–22. Chapters 20 and 21 present how Blockchain can facilitate the enforcement of green finance-related law, while Chapter 22 envisages new legal frameworks to suit the "Blockchain era."

SECTION 1

Scene-Setting — What's Going On?

CHAPTER 1

A Conversation with Dr. Kelce Wilson on the Foundations of the Blockchain

Jack Aldane

Journalist and Editor of Development Finance, London, United Kingdom

Born in New Jersey on October 14, 1966, Kelce Wilson,[1,2] 50, lives in a single-family residence near Dallas Texas with his wife Robyn. He has three daughters, a dog called Cassie, and a cat named Chanel. In 2009 he started PEDDal, an online service that protects patents and trade secrets against legal dispute. He is an attorney in patent law and cybersecurity at Grable Martin Fulton PLLC, a US law firm. Wilson's resume includes his service in the US Air Force, a Bachelor of Business Administration, and a doctorate in electrical engineering. It doesn't mention that he may also have patented the first Blockchain.

JA: Were you drawn to technology growing up?

KW: Yes, I was.

JA: What form did this take? Judging by your age, I should think you were really becoming cognisant of the influence of technology in the early to mid-eighties.

KW: Yes, when I was about in the neighborhood of 13 or 14 years old, my father bought one of those home computer game consoles, named Odyssey 2, and it actually had a game cartridge that was an assembly language programming module. Now, I don't know who it was that thought that teenage kids would want to, instead of playing PacMan, program an assembly language. However, I did, and even before I learned Basic, I was programming that game cartridge in assembly language and one of the first games I programmed was a little shooter game with two people. I made the bullets from my gun travel faster than the bullets from the other person's gun and then I would play it with my sister. Of course, she figured it out quickly and refused to play it with me anymore. I took an interest in electronics when my sister and I would fight about what television show to watch after school. My mother would tell me "You can watch today and then your sister watches tomorrow and then you can have the day

1 Interview recording–https://www.dropbox.com/s/coetodk7i2of44o/Jack%20Aldane%20interview%20Kelce%20Wilson.mp3?dl=0

2 Patent–https://www.dropbox.com/sh/1luya9bzjixut55/AACLBJl0GeQyeMip6gGlIP3Ha?dl=0

Transforming Climate Finance and Green Investment with Blockchains.
DOI: https://doi.org/10.1016/B978-0-12-814447-3.00001-X

after." I used to love watching Star Trek with Captain Kirk, but you know, some episodes are better than others. So when my day to watch Star Trek fell on a lousy episode, and my sister's day to watch her TV show fell on a good episode for Star Trek, I would want to change with her. But of course, she would not want to change because she was my sister and she would try to fight with me just to fight with me. So I mailed away for a TV jammer kit—this was in the days of broadcast TVs where people would have aerials to pick up the local TV stations instead of cable. It arrived and I assembled it in the garage and I didn't let my parents know about it. I hid it in the sock drawer of my dresser and then when my sister would not change days with me, I would go into my room and I would turn on the TV jammer and her TV show would of course be blanked out. She'd get up, turn off the TV and I'd turn off the TV jammer, go out to the TV room, turn on the TV and change it to Star Trek.

JA: Let's talk about your early career. You were in the US military and eventually went on to become a patent lawyer. At what point along the trajectory of a military career did you first start to think about ways to create, protect and verify the authenticity and accuracy of documentation?

KW: When I worked in a SCIF (Sensitive Compartmented Information Facility), we were concerned about the possibility of eavesdropping. We didn't have Internet connectivity, so we would communicate via couriers. And we had to worry that couriers could be careless, corrupt or blackmailed. Instead of providing us with an optical disc that had been written by another SCIF and given to us, they would actually be delivering us a forged disc that had malicious logic, such as spyware, on it. You see, we would generate software and then we would give the software to other, different, facilities that would use it. And once in a while they would see some anomalous behaviour or they might have a special use case and they'd want us to do some special investigations for them. So they'd put down the results or some other type of information on an optical disc and send it to us. So these things are coming in and out of our facility and when they would come to me what I would do is hash the disc image and I would get the number from it. Then I would call the people who'd sent it to us and ask them what number I should see. They would tell me the number, and I would have some degree of confidence that the person I spoke to was the person who had been involved in writing the disc. Now, I didn't just use any phone, I used a special kind of phone called an STU, a secure telephone unit. And it had a cryptographic verification scheme in which I could have certainty where or what other STU I was connected to. The telephone units would do some kind of cryptographic handshaking where there was a third-party identification entity that would, if you hadn't spoken to that phone before, vouch for the phones on either end and tell you that this is the phone you're talking to. And so if the phone I was talking to correlated with the other SCIF that sent us the disc, then I believed I was talking to somebody that was involved with sending us the original disc. So during this, I began to think of two things. One was that if they had told me when they first wrote the

disc "Hey we are sending you a disc and when you get it, the hash value is X.[3]" They didn't have to wait until after I got the disc. They could've told me that ahead of time. That was the first thing I realised, that I didn't have to wait.
Sometimes, the disc would come in and I'd call and nobody was there. And then I couldn't use the disc until I could get hold of somebody, and that sometimes is inconvenient. But if they had called before they gave it to the courier, then as soon as I got the disc I could hash it and use it right away. And so that inconvenience of that person not being there on the phone started me thinking that this number could be communicated at any time, which then gave me a derivative realization, which is "Hey, you know what? If they happened to call me and then they don't get around to sending it or forget to send it until a year later, guess what, I know not only that it's from them, but also that it's a year old." And so I started thinking that the hash value could function as a dating mechanism. And then there's this other realisation I had. We would use those STUs only for the point of verifying with whom it was that we were speaking. The hash values are one-way, but that is because if you know the data, then you can create the hash value. If you know the hash value, you cannot recreate the original data. So this got me thinking that the only reason we have to use these special telephone units is to authenticate who the party is, because the actual value could very well be published in the newspaper.
That is a critical part of my PEDDaL system that popped up later. Typically, the CDs would show up not just by themselves but they'd show up with a letter by the people who wrote them. The letter was a formality, a nicety, we could throw it away, and we didn't even really need to read it. But because we were human beings interacting, they wanted to be nice so they sent us a letter. So one day I commented: "Isn't it odd that these computer files have only been around for a matter of decades and we have a way to mathematically test them for accuracy and lack of forgery? This letter's got written language which has been around for thousands of years and we have no way of knowing that the courier didn't just stop and reprint one with an extra word in it!" One of the people in the office heard me saying this and gave me a lecture about how, if I were to scan that letter and do an optical character recognition on it and then try to hash it, there is no way I could try to recreate the original hash value, because the programme I used would not necessarily know if gaps between letters were tabs or spaces or margins or columns of anything along those lines. The original data file could never be recreated, and the hash could never be recreated, so therefore I couldn't do it. He told me this in a rather condescending manner, and I was already aware of it. So I became rather steamed and went home and I told my wife about it because I was still, you know, simmering. She said "Well then, prove him wrong,

[3] A *hash value* is a numeric value of a fixed length that uniquely identifies (secret) data or information. Hash values represent large amounts of data as much smaller numeric values, so they are used with digital signatures.

come up with a way to do it." And I said "I can't. Written language has been around for thousands of years and nobody has ever figured out a way to perform math on written-down words in a way that can be mathematically processed to prove lack of forgery. Nobody's been able to do it and written language has been around for thousands of years". And she said "Oh, I'm sure you could do it." I laughed.

JA: So where did you go from there?

KW: Okay, so I figured out a way to reliably recreate a hash value when all you have is a paper document, which you have to scan and do optical character recognition on with some program that you don't know what it's going to do with margins, tabs, spaces, whatever. Just so long as you have a font with which you can accurately collect the letters so you don't mistake 1 with L and zero with O and lower case R followed by a lower case N for an M. Just so long as you can get the characters correct, you can reliably recreate the hash value. The issue I had with the phone system is that I had to trust the central arbiter. What if I eliminated that so I don't have to trust anybody? And this issue of "Hey, I can use this to establish dates," I have all three of those things in my head the same morning after my wife told me that I was going to solve it. I must have figured it out in my sleep, because as I was showering I'm thinking 'Well, okay, so you process the scanned words this way, then you hash it' and, oh well, just as I'd been thinking with those optical discs, they can hash it before they send it out the door. So they had it right away and then those hash values could be printed by a newspaper. That's kind of expensive if everybody's doing it, so how about instead everybody just produces this common block of information? I called it a database edition at the time. Then you hash that block of information, and then you put that in the newspaper and then everybody knows the date that that block of information existed, and that block of information had to have existed after the letters were done. So there you get your date-proof for anybody in the world because it's in the newspaper. And it goes back. They just have to do a two-tier verification on the hash values.

> "I figured out a way to not only reliably recreate a hash value when all you have is a paper document."
>
> "Just so long as you can get the characters correct, you can reliably recreate the hash value. The issue I had with the phone system is that I had to trust the central arbiter. What if I eliminated that so I don't have to trust anybody?"

JA: So this essentially eliminates the problem of a forged or inauthentic document written by hand, in ink?

KW: Correct, so my thinking of it was for correspondence between military officers, but I originally also thought about wills being challenged. Imagine somebody were to write a five-page will and they were to leave their nephew Ben a dollar, and their nephew Charlie a million dollars, then Ben breaks into the computer and swaps Ben and Charlie around, then reprints that page and forges the initials and somehow sneaks that in where the will is kept. Twenty years later, when the person dies and the will is being read, how is anybody supposed to know that there was originally Ben and Charlie the other way? That's what I was thinking and I came up with that in the fall of 2000.

JA: Is that when you knew you had something unique on your hands? Is this the moment you consider the invention of Blockchain?

KW: It would be, although I wasn't motivated yet to go to law school. It was in January 2001 when I added another piece. I had a way to prove dates of documents in the fall of 2000, but I considered it to be boring. Not sexy, not interesting, not the type of thing that would get anyone's attention. But law firms might use for documents and contracts maybe, someday. I wasn't motivated; I didn't have a sense of urgency.

> "I had a way to prove dates of documents in the fall of 2000, but I considered it to be boring. Not sexy, not interesting, not the type of thing that would get anyone's attention."

KW: I'm not sure if you have much trust in websites, but you know, anybody can put up a website and when you go to it with your internet browser you have no idea how long it's been since there was a modification to it.

JA: I don't trust a large number of websites, but that has less to do with the date of their creation than the integrity of their content. What it is about the date on which a website is created that strikes you as particularly significant?

KW: Well, it goes back to when I was in high school and we had to do research papers and my high school teacher said that you need to footnote everything and I asked why. She said: "Well, because if it's published and it's been around for a while, then it's authoritative." And I said: "Why is that? Anybody can publish garbage." And she said "Yes, but if something has been around for, say, five years and nobody has criticised it, even though there has been plenty of opportunity and people have the motive to criticise it, you have at least some slight basis to trust the content." And I've seen that, throughout academics, if there has been an academic article and it has been published in kind of a journal and then five years later other people are still citing it then you can guess, well there's probably some degree of credibility I can give to this. However, if five years later it has received nothing but criticism for being inaccurate and untrue, then you probably shouldn't trust it. But if there isn't criticism of it then maybe it might be something that you can at least give some credence to. So I'd been thinking at least since high school that a little bit of older information that hasn't been criticised has a higher degree of trustability than information where you have no idea whatsoever how old it is. It could very well be that you're the first person to look at it and so the reason why there's no criticism is because there hasn't been time for the criticism. So, in January 2001, while I was surfing the internet and it was occurring to me "Oh wow this website that I'm looking at, I can't trust this, it could've been put up like two minutes ago." And then I thought "Oh wait a minute, I have this thing for the paper documents but if you're just starting out with digital files, you don't need to do the whole processing of the OCR text, you can just do the hash of the website. Hey, we can now prove dates and integrity for websites." That's when I realised that it could be huge because it could very well change the way that the internet is used. Right now, everybody has to have scepticism of it, but if you have basically established dates for material, now you can start putting in this tentative level of trust with information. And then I realized "Okay, somebody else is going to figure that out," and so in February I applied to law school.

JA: OK, supposing you pick a complete stranger to explain your invention to. What if they then ask you "How does this invention in any way alter the world as I know it?"

KW: Well, with using the Internet, there's either using the internet or just plain documents. For documents such as legal documents: wills, estate planning documents, or contracts, what happens is that if there's a dispute down the road, that dispute is going to be a lot cheaper if you don't have to pay so much money for lawyers. With regard to the Internet usage, there're two ways that can be done. First of all, you can never get away from the fact that there might be hackers. You can never get away from the fact that there're idiots posting stuff online that's just not true, or maybe there're malicious people posting untrue stuff online. But let's say you're trying to look for a vacation spot, okay? And so you go to some hotel websites and you know that hotel has got some customer reviews. I personally have been in the situation where something on the Internet looked great, and when I show up there, it is absolutely horrible. As I tell people, when you're planning travel, anything can look good on the Internet, okay? But suppose you have this hotel website that has been up with reviews and it's two years old, and you do a search and you cannot find any other information about the hotel except for the stuff that's two years old. And then you have this other hotel and it hasn't been registered on a block chain. So that website, what you find on it, is data that can't be date-proven. I would say you're probably better going to the hotel that has had positive reviews on its website after two years and you can't find negative information that's newer, than going to some hotels that might've just changed its name yesterday and so you have no idea that there's going to be negative reviews, you just haven't seen them yet.

> "We can now prove dates and integrity for websites. That's when I realised that it could be huge because it could very well change the way that the Internet is used."
>
> "Right now, everybody has to have scepticism of [the Internet], but if you have basically established dates for material, now you can start putting in this tentative level of trust with information."

JA: There's some serious discussion right now about how Blockchain could become a way for people to economize on the energy used in their homes by allowing them to pay for and trade surplus energy independently of large energy firms. Do you consider the potential ripple effects of a decentralized system on consumer behavior to be economically or politically appealing?

KW: Yes, I'll tell you I hadn't thought of that, but I see how Blockchain will go that way because you have all these people that are trying to make money or cut their bills on utility usage and homeowners, some of them, have a motivation to cheat. So, that introduces a trust problem. How do you verify everybody's transactions with the smart grid? That verification can be really expensive if everybody has to be trusted and this one person has to trust all these other homeowners. So, anyway there're some types of communal activity in which there's value transferred around. There's going to be a motivation to cheat, so if there's an opportunity to cheat, then you have a trust problem and then you can come in and say "Well, Blockchain is a specific way of handling the trust problem, maybe we can set up a system in which we can trust without there being a central trusted entity." So, within this neighborhood you might have a few bad actors, but guess what, those bad actors can't cheat everybody else because if they attempt it, it'll be discoverable. You see the whole thing about Blockchain is that it doesn't prevent tampering and forgery, instead it makes tampering and forgery readily detectable, so that something else can come along and fix it. For example, as long as a preimage attack is not feasible, if someone inserts a forged or tampered block, an independently-generated hash value for that block won't match the one that had been used to link the chain with the subsequent blocks. Anyone who can locate the proper block, from whatever source may hold it, can repair the chain merely by identifying which hash values do match the proper sequence. Trusting the source of the proper block is unnecessary, as long as a preimage attack is not feasible.

So, I believe we were discussing this idea that if somebody was trying to spend some digital currency with you and you say "Well, how do I know you're the person that still owns it?" You go back and look through the ledger history until the last time that particular currency was mentioned and since that time that currency hasn't been mentioned yet and so you have to know that none of those ledgers have been tampered with. Otherwise you can't trust that that person actually still owns the digital currency. Well, it's the same thing in any type of a communal effort that you have all of these people putting in information and that information has got some time aspect to it, so there's going to be certain events at a certain time. So for example, maybe somebody's meter readings, what's one thing that always happens with electric meter readings? They're monotonically non-decreasing, the meter reading only goes up. So if somebody has been putting in their meter readings and then one day their meter reading is lower than it was before, then you know that they're cheating, okay? But if they're trying to cheat, then they might want to go back and forge the earlier blocks of information to hide the fact that they cut the numbers but if the forgery attempt is instantly detectable, then guess what? They won't be able to do it, so now they're stuck with being honest, at least honest in regard to their meters and not decrease it.

> Blockchain ... doesn't prevent tampering and forgery, instead it makes tampering and forgery readily detectable, so that something else can come along and fix it.

JA: It's been almost a decade since the financial crisis of 2008, which dealt a pretty brutal blow to people's trust in the notion of free, self-governing, rational markets. Now, with Blockchain, people are once again being asked to embrace another supposedly infallible, autonomous system in which we may all have a stake sooner rather than later. Why should people accept the idea that humanity takes its hands off the wheel and let an automated ledger regulate some of our most important institutions?

KW: That's a good question. I can tell you that the bank crisis was a real shaker because with the bank what you do is you take your money to a building and give it to somebody and you walk away; and you just have to hope that when you go back it's going to be there. And you do that because you trust that whoever owns the bank isn't going to rip you off, but of course we have seen through at least the savings and loans scandal in the United States. I believe that was in the 1980s. That was huge too—sometimes those people let you down. And with the 2008 crash, a lot of it was because banks were maybe not just outright stealing funds—they were misrepresenting risks. And it was just because the system was so complex and they were telling people "Trust us, we know what we're doing," when, in reality, they were introducing all sorts of risks but not being honest about it. So I don't know if you're ever going to get away from this idea that if you put your trust in people, there's a chance you're going to be disappointed. I don't know if you're going to get away from that but then, the alternative might be this issue of antiseptic trust, which is you trust the numbers independent of people. That is kind of dehumanizing.

> "How do you verify everybody's transactions with the smart grid? That verification can be really expensive if everybody has to be trusted and this one person has to trust all these other homeowners."
>
> "I don't know if you're ever going to get away from this idea that if you put your trust in people there's a chance you're going to be disappointed. The alternative might be this issue of antiseptic trust, which is you trust the numbers independent of people."

JA: The appeal of Blockchain appears symptomatic of a world in which our ideas about progress increasingly follow our ideas about efficiency, the aim of which is to eliminate human doubt and error. At the helm of all this stands artificial intelligence and the rise of automated systems. Has the will to improve technology replaced the will to improve society?

KW: I believe that probably some people view those as synonymous, not everybody but I don't. Certainly some advances in technology can improve some aspects of human civilization. But not all advances in technology necessarily improve; some are probably agnostic and some might be detrimental. So I think that, leveraged properly, the first rule should be "do no harm", so if somebody comes up with an improvement and says "Okay, hey guys we've got some trust issues here that are constraining commerce. People aren't willing to do certain things because they don't trust that they're going to get a return on their investment." Okay, but you know wealth is created by efficient economic activity and so societies have an incentive to facilitate efficient economic activity. And if lack of trust is an impediment, then removing that can improve things. However it can also, if used the wrong way, be a problem—for example, the ransomware on people's computers. You get this notice that your computer is locked, everything's encrypted and it's going to be deleted in five days unless you give certain people some value and then they provide you with the decryption key. Now guess what they don't do? They don't give you a mailing address and a name, to whom to address the cheque. You know why? Because they'll get arrested. What do they do? They give you a bitcoin account because it's anonymous and they can get away with it. They can collect from you today and somebody else tomorrow with impunity. So this idea of trust without a person to trust can be good when it facilitates positive economic activity that improves people's lives. But it can also be used to facilitate harmful, destructive criminal behavior.

JA: I want to talk a little more about accountability. It strikes me that, no matter how little a hash value knows or cares about the individual user, any hack, any glitch, however slight, inevitably raises questions in the user which, if they can't be answered by the Blockchain, surely must be answered by whoever creates or sells them that technology. In other words, if Blockchain were to fail our expectations, wouldn't this march us straight back to the same questions of who we can ultimately trust?

KW: Yeah, it's sort of like a company that doesn't have a customer service hotline. I guess one way of thinking about the question you just asked is Blockchains don't come with a customer service hotline either. It either works for you 100 per cent when the hash value is the same, but if you've got a special request or you fell off the conveyor belt, then you're just out of luck. The Blockchain just has a simple one-size-fits-all solution. That is, you put things in and then later on you can't tweak it. If you made a mistake, that's too bad, it's there forever. For everyone else to see and you're not ever going to fix it. And later on maybe you could put in the fix, but forever your first mistake will be there. You can't reconcile it. There's no do-overs, no resets, no rewinds. Although Bitcoin did one a while ago when they had some pornography put into the chain. But that was a huge effort for them. But if it's not caught in time then it's too bad, you're stuck.

> "Blockchains don't come with a customer service hotline either. It either works for you 100 per cent when the hash value is the same, but if you've got a special request or you fell off the conveyor belt, then you're just out of luck."
>
> "This idea of trust without a person to trust can be good when it facilitates positive economic activity that improves people's lives. But it can also be used to facilitate harmful, destructive criminal behavior."

JA: In what state of mind are you about the fact that you have not yet received mainstream recognition for this invention?

KW: I've never been a publicity seeker. I don't care if nobody ever knows my name. I just want to get my PEDDaL system up and running to help register documents and keep costs down for somebody who has a dispute. And then they don't have to spend all sorts of money on lawyers, so then they can resolve their dispute. Because I personally, although I'm a lawyer, I'm like sickened by how much it costs for the average person to settle a dispute in the legal system. Even if I had to hire a lawyer, that would be expensive and difficult on me, and I am a lawyer. I've been doing well financially, but I certainly wouldn't want to have to hire a lawyer to sue somebody or defend me from a lawsuit because lawyers are so expensive. So I would feel good knowing that I was helping people avoid spending money in a courtroom. And just so long as my PEDDaL business is running along and moving, I'm happy. Nobody needs to know my name, actually. I mean, I'm not going to hide it. You're going to put my name in your book and I'm fine with that. But if I don't ever get any formal recognition or anything like that, I'll be fine with that too.

Endnote:

There is no widely held consensus on the exact moment at which Blockchain first came into being, much less a consensus on who takes credit for its invention. Wilson's claims along with his patents are a unique and as yet little-known addition to the subject among enthusiasts. The author holds that Wilson's story is germane to readers' understanding of how Blockchain functions, as well as its potential future influence. The contents of this chapter in no way seek to challenge or refute similar claims or patents that have or may be recorded.

CHAPTER 2

A Conversation with Masterminds in Blockchain and Climate Change

Alexander Harris

Independent Journalist, London, United Kingdom

Every paradigm shift starts with terminology of which many make use yet only a few can make sense. Sometimes the shift is enough on its own to give the terminology meaning. Climate change, having left a high watermark on our collective conscience, is an example of a term constantly undergoing this process of redefinition. Edging up the modern lexicon to meet it is 'Blockchain'. This tonne-weighing word almost designed to drop on the sceptic's foot feels, like climate change, strangely realer and closer to home every time it gets mentioned. An indispensible link between the two is being drawn. Blockchain technology was the topic of heated discussion across the climate change policy community at the UNFCCC Bonn Conference in May 2017.

Blockchain is a distributed ledger technology that stores a decentralized record of activity across a digital network. Secured by cryptography and constantly verified, the system removes the need for trusted third parties such as banks or central authorities. The UNFCCC has recognized that Blockchain's model of decentralization could prove crucial to scaling and accelerating climate finance transfers, which is often weighed by bureaucracy and obfuscation involved in huge volume of procedures (UNFCCC, 2017).

Stefan Klauser of ETH Zurich and Sven Braden of the LIFE Climate Foundation Liechtenstein are both seasoned experts in the respective fields of blockchain technology and climate change, appearing at the UN conference under the banner of their newly founded CarbonBC initiative, later known as the Climate Ledger Initiative[1,2] (CLI). CLI is an international, multi-stakeholder initiative with the mission of aligning Blockchain-supported climate actions with the goals of the Paris Agreement. Supported by public and private donors, the initiative offers collaboration and regulatory support for innovation in climate action.

1 https://www.climateledger.org/

2 Interview recording – https://www.dropbox.com/s/hr3xmjkd9zuyvxx/Alex%20Harris%20Blockchain%20x%20Climate%20Change.mp3?dl=0

Transforming Climate Finance and Green Investment with Blockchains.
DOI: https://doi.org/10.1016/B978-0-12-814447-3.00002-1

2.1 Unleashing the Potential of Blockchain in Addressing Climate Change

Blockchain started gaining its reputation in the wake of the 2008 financial crisis for being a core infrastructure for the cryptocurrency—Bitcoin, although individual components of it had existed long before that. Only in recent years has the term really found a place in intellectual conversations about the future of secure transactions without state involvement. At its heart, Blockchain represents a promising antidote to trust issues faced in a multitude of different sectors. Sven and Stefan are both confident supporters of Blockchain being applied in climate change actions worldwide and doubtless that its use will bring the change badly needed to save the planet from environmental catastrophes.

> *"Corruption certainly plays a role in climate finance or in development aid, but I think it's the bureaucracy that hinders development and leads to a decrease in efficiency," Sven Braden, LIFE Climate Foundation Lichtenstein.*

"It is pretty easy for everybody to see that our current financial system is not sustainable," Stefan says. "Part of the reason can clearly be found when looking at how intermediaries such as banks, loan providers and other actors behave. Blockchain is a very attractive solution [in] that what we need is not more centralization, but more diversity in economic and financial markets."

Sven describes the CLI as a provider of a sound knowledge base on how blockchain's potential can be fully unlocked to address climate change. "If we don't use blockchain's full potential," Sven says, "we will end up with just a nice little feature here and a little bit of efficiency increased there and, in terms of Paris Agreement implementation, we just don't have the time."

2.2 Blockchain's Role in Peer-to-peer Energy Trading

Blockchain has been identified as an essential component in a range of different climate-related sectors, including carbon pricing and the tracking and monitoring of emissions reduction. Peer-to-peer energy trading is a widely foreseen hotspot materialized by Blockchain, which offers a cutting-edge solution to the double-spend and accountability problems facing various trading platforms. With the price of solar photovoltaic modules, a key type of solar panel, falling on an almost yearly basis—down by 33.8% since the first half of 2016 alone (Hill, 2016)—the ability of households to generate, sell and trade energy, as a "prosumer," from one neighbor to another is becoming a not-so-distant prospect.

Although still at its early stage of development, a New York-based start-up Brooklyn Microgrid becomes the world's first ever peer-to-peer energy sale system via its blockchain-verified network. Two neighbors in the Park Slope neighborhood in Brooklyn conducted their first energy transaction through the TransActive Grid (TAG), a platform jointly built by Lo3 Energy and Consensys Systems (Nguyen, 2016). Stefan believes that the way in which energy companies manage the market's response to peer-to-peer trading will be critical to the future of energy blockchain innovation.

> *"One of the main achievements I see in the short or mid-term for Blockchain is in getting people to take more control," Sven says.*

"What [the energy companies] usually do is try to marginalize new technology," he says.

"Then, in the second step, incumbents actively fight against the new technology and mark it as the evil. New markets players usually see the potential of new technologies. They try to adapt and embrace it. I can definitely see the potential for this, although it is at the moment still only a very small project which tries to experiment with how you could produce, sell and trade energy in a distributed way."

2.3 Closing the Private Finance Gap

Private sector investment constitutes the lion's share—62%—of the global climate finance flows, according to the Climate Policy Initiative (CPI, 2015). This gives extra weight to the importance of maintaining and perhaps even upscaling private sector contributions. One of the enduring challenges of climate finance, however, is to lure much-needed international private investment, particularly in developing countries where the perceived risks are higher.

Barbara Buchner, senior director of the Climate Policy Initiative, characterized the problem in 2013, remarking that: "Currently, climate finance is mostly a domestic game." She iterated that addressing the perception of additional risk to overseas investment would be critical to increasing climate finance on a global level (CPI, 2013).

Convinced that the application of Blockchain in mobilizing climate finance could secure greater public confidence and close the finance gap, Sven pointed out that "Corruption certainly plays a role in climate finance or in development aid, but I think it's the bureaucracy that hinders development and leads to a decrease in efficiency. Blockchain eliminates the need for trusted third parties and, in many cases; these parties are the cause for slowing down development and efficiency. What I think at the end is that any project developer, investor or grand recipient will enjoy a much higher degree of credibility from his stakeholders just by using blockchain-based processes. They can be designed to be permission-less, they provide for a great transparency and that is exactly the kind of message that we want to spread with our Climate Ledger Initiative[1] that we started in Marrakech in 2016."

Sven is, nevertheless, aware that the adoption of Blockchain as a digital trading infrastructure would mean an overhaul to the well-entrenched status quo in the climate change governance regime. "There are processes that have been planted there for years. There are people involved that know their lines and know how money flows go. They don't necessarily have a big interest to change that. We can only work with the argument that we can increase transparency, accessibility and accountability along every line of climate finance flows, hoping that it leads to an automatic adoption over time. But, for that to happen, we need to constantly come up with new use cases and new research. We need to keep that development at the attention level of our colleagues under the UN climate change negotiations constantly," Sven said.

2.4 Restoring Faith in Climate Negotiations

Sven has lots of faith for blockchain applications to be capable of stretching far beyond facilitating international finance flows and have serious ramifications for society itself, especially in emboldening the average person to become more proactive in climate change matters.

"One of the main achievements I see in the short or mid-term for blockchain is in getting people to take more control," Sven says. He is also doubtful about any lack of technical understanding being a barrier to extensive adoption of this revolutionary technology. "It's complex, but it's the same as with the Internet, right? Who knows the protocols that make us use Google, Amazon, Facebook, Airbnb, Uber and so on?"

With over 10 years of experience representing Liechtenstein at international climate negotiations, Sven understands how important the participation of the population at large can be. "I've been involved with the [UNFCCC] negotiations for a decade now and I can confirm that the politicians mostly act within the mandates that they received back home. In this case, blockchain has the potential to get people directly involved with impact-driven climate actions."

Sven gave a hypothetical example of a village community in a country investing in the renewable energy project of another village abroad. "Every step of that project involvement is visible and accessible to every stockholder involved at any time. So, maybe it's not just one village, maybe it's a hundred villages or a thousand villages. I think it would be smart for any politician to align with such initiatives. I do therefore think that the decentralised ledger technology will lead to a greater decentralisation of powers at the end and to [greater] private sector involvement."

2.5 Finance 4.0.—Incentivizing a Sharing Economy

This concept of blockchain-based community mobilization is taken a step further by Stefan Klauser's recent work on a pioneering incentivization scheme called "Finance 4.0". Run by a European consortium called FuturICT 2.0[3] and led by the ETH Zurich Chair of

[3] FuturICT 2.0 aims at promoting disruptive innovation to tackle the main challenges of modern society toward a more resilient and adaptive society. It harnesses emerging knowledge in order to address urgent global challenges around sustainability.

Computational Social Science, the Finance 4.0 system offers insight into community-based solutions which could be unlocked with the use of Blockchain technology.

"The idea is that whenever you produce an externality, there is a price or compensation attached to it," Stefan explains. "Let's take the positive ones as an example. Like doing ... a service for your community—promoting the sharing economy or recycling, you can measure these actions and then you can translate these actions into digital currencies or coins that are going to be automatically paid out when you can prove that you did one of these activities. When you use basic resources, create a noise or pollute the air, these negative externalities can be measured too; and you can then get negative coins through your actions that you have done."

Stefan sees this system as one that incentivizes people to contribute to positive externalities by getting the good coins paid out to them and thereby leading to a society of individuals who act on a day-to-day basis in a way which supports sustainability. This system would not be centralized, but would instead be governed by a hierarchy of high-level smart contracts and involve a local-level subsidiarity engaged in collective decision-making activities.

"You could start from the highest level, which could be a Paris-style Agreement with hundreds of millions in a smart contract and, whenever regions or countries show that they contributed in a good or positive way, they would then get positive coins paid out. Then you could go down from level to level, until you reach an individual or a community level. This would then lead to a situation in which people compete for the good externalities."

Although Stefan admits that this system is still far from being implemented today, it is a step in the right direction. "We want to create a system that combines a high-level incentive system with very local subsidiary, community-based decision making. We envisage a future that could in principle be more sustainable, fair and efficient."

> *"We envisage a future that could in principle be more sustainable, fair and efficient." — Stefan Klauser, ETH Zurich*

2.6 The Future of Blockchain

Neither Sven nor Stefan are in denial about the challenges facing innovators in adopting Blockchain and the issues that still stand in the way.

"There are still a few things that we have to work on over the next few months, or even years," Stefan says. "Mainly, there are issues around scalability and these issues are usually also related to power consumption."

Sven offers unregulated Initial Coin Offerings (ICOs) as another potential problem that needs to be confronted in the near future. He stresses, however, that both cryptocurrencies and ICOs are already "out of the box and here to stay," adding that the downsides should be confronted but that the tools should be embraced as vital to unleashing the full potential of Blockchain.

Both are highly optimistic that many of the pitfalls are just small bumps in the road for this new technology, citing a vibrant and lively community that is working together to improve the security of Blockchain solutions.

"Now we have this climate blockchain initiative, where we really try to push on three levels," Stefan explains. "One level is dialogue, one level is research and one level is use cases. I think it's really important to push on all of these levels simultaneously to make people aware that the technology is ripe enough to actually be used at this point, that it also makes sense to use it and that we can prove that it makes sense."

Sven is under no illusions about the important role that cooperation must play in Blockchain's future. "It's just not enough to say that the technology is not ready yet for broad application, so we don't do anything," he iterated. "The idea is that we work on test cases and experiment with the technology so that we can make it more stable and able to be applied in a broader sense. We really want to call out to everybody that has the possibility to support such experiments, that they take it serious and that they really help blockchain to develop its full potential."

Through the publication of this book, the Blockchain Climate Institute (BCI) is working to highlight the multitude of use cases for Blockchain in the critical fight against climate disasters. The success of this battle is entirely dependent upon the ingenuity and innovation of those involved in the technology's applications and the speed of their adoption by those at the top. Any failure to recognize the critical importance of Blockchain or act swiftly for accelerating global efforts to combat climate change would leave our children suffering from dire consequences.

References

CPI. *Climate change investment totals USD$359 billion worldwide.* (2013). <https://climatepolicyinitiative.org/press-release/climate-change-investment-totals-usd-359-billion-worldwide/> Accessed 23.07.17.

CPI. *Climate finance landscape.* (2015). <http://www.climatefinancelandscape.org/> Accessed 24.07.17.

Hill, J. *"Dramatic" solar price reductions will only continue.* (2016). <https://cleantechnica.com/2016/12/02/dramatic-solar-price-reductions-will-continue-fall/> Accessed 24.07.17.

Nguyen, C. *Brookyln's 'Microgrid' did its first energy sale.* (2016). <https://motherboard.vice.com/en_us/article/d7y7n7/transactive-grid-ethereum-brooklyn-microgrid> Accessed 24.07.17.

UNFCCC. *Blockchain technology can boost climate action.* (2017). <http://newsroom.unfccc.int/climate-action/how-blockchain-technology-could-boost-climate-action/> Accessed 24.07.17.

CHAPTER 3

Blockchain 101: What is Blockchain and How Does This Revolutionary Technology Work?

Sebastien Meunier
Director, Chappuis Halder[1], New York, United States

3.1 Introduction to Distributed Ledgers

3.1.1 It All Started With Bitcoin

After it was launched early 2009, Bitcoin lived a rather underground life for almost 5 years. It attracted a passionate community of developers and was used on the dark corners of the Internet, but it was not well known by the public.

Starting 2014, a larger community began to get interested in Bitcoin cryptocurrency and the underlying technology used to securely record Bitcoin transactions: "Blockchain."

2015 and 2016 were two years of intense hype for Blockchain. It made the cover of mainstream publications including *The Economist* in October 2015; attracted more than $500 million both in 2015 and 2016 from venture capitalists (Crunchbase, https://cointelegraph.com/news/vc-investments-in-blockchain-companies-on-track-to-exceed-2017s-numbers) and gave birth to hundreds of start-ups and an entire new fintech industry.

Blockchain systems are sometimes referred to as Distributed Ledgers Technology (DLT) because they allow users to record and share a common view of a system's states across a distributed network. There are some slight differences between the two terms; but for the sake of simplicity, we will use the generic terms "Blockchain technology" and "distributed ledger technology" without differentiation in this book.

[1] www.chappuishalder.com/

Transforming Climate Finance and Green Investment with Blockchains.
DOI: https://doi.org/10.1016/B978 0 12 814447 3.00003-3

3.1.2 The Essence of Distributed Ledgers

Distributed database management systems (DDBMS) have existed since the 1990s. To understand the potential of Blockchain technology, it is essential to understand the value it added over preexisting systems.

Distributed databases store data across a common network rather than at a centralized location. With the development of the Internet in the nineties, businesses needed solutions that could process huge amounts of structured and unstructured data which could scale across networks. DDBMS solve this problem through consensus mechanisms such as Paxos or Raft which control read/write permissions and establish secure communication channels among participants. Common applications of this technology include NoSQL, NewSQL, and Hadoop databases. But these protocols assume that each participant cooperates in good faith which limits their application to private networks under a centralized authority where participants can be trusted.

Distributed Ledgers (DL) are like DDBMS protocols in that they maintain a consensus about the existence and status of a shared set of facts but they do not rely on this assumption of good faith. They achieve this by leveraging strong cryptography to decentralize authority. They are different from generic distributed databases in two fundamental ways:

1. The control of the read/write access is truly decentralized (whereas, it remains logically centralized for distributed databases); and
2. The integrity of the data can be assured in adversarial environments, without employing trusted third parties (whereas, distributed databases rely on trusted administrators).

3.1.3 Various Degrees of (De)centralization

Public distributed ledgers such as Bitcoin allow anyone to create and validate transactions on the network. Fully decentralized, they combine a system of economic incentives with resistance to censorship. Economic incentives are provided through the remuneration (in the form of cryptocurrency) of the provision of computational power to the network. Resistance to censorship is achieved through the decentralized and immutable character of the ledger.

Distributed ledgers implemented in business environments are by nature less decentralized because they already include some levels of organization with some built-in trust. For instance, government authorities regulate financial institutions by requiring them to collect identifying information on their clients through formal "know your customer" (KYC) processes. Regulators therefore act as a centralized authority who define the rules for participation.

When used in the context of regulated environments, distributed ledgers function like traditional distributed databases, but can still benefit from the technology's strong cryptography to enforce auditing, accountability, and automation of existing business processes.

In small local networks where trust can be reasonably assured, in small organizations for example, the advantages of distributed ledgers and databases are muted. In these cases, a centralized database is the most efficient data management solution.

Even if the technical solution, the cost distribution and the economic incentives can be very different among fully decentralized, semi-decentralized, and centralized environments described above, in each you need operators to run and maintain the platform.

3.2 Examples of Distributed Ledgers

3.2.1 Bitcoin

The Bitcoin Blockchain was designed to solve one problem: the double spending of digital money. In the modern era, it has become difficult to prevent users from duplicating digital products. The movie and music industries have been fighting illegal downloads for years yet without huge success. Obviously, digital money is not viable if users can forge copies at will. Enabling peer-to-peer exchanges of a digital currency in a secure way is the *tour de force* accomplished by "Satoshi Nakamoto"—the pseudonym of the inventor(s) of Bitcoin. Through cryptographic mechanisms and economic incentives, the Bitcoin system creates trust among strangers in a fully decentralized fashion.

This is how a Bitcoin transaction occurs:

1. Before the transaction takes place, all the network nodes share an identical copy of the Bitcoin ledger. Like all ledgers, this shows the Bitcoin addresses with their associated balances;
2. Each participant holds a pair of mathematically related keys. One key is public and one is private. The sender initiates a transaction by sending some Bitcoins to the address of the beneficiary. The sender signs the transaction with both his private key and the public key of the beneficiary;
3. The transaction is then broadcasted to the global network. Anyone can authenticate the origin of the transaction using the public key of the sender but, importantly, the transaction can only be deciphered with the beneficiary's private key;
4. A specific class of participants called "Miners" validate transaction and assemble it with all the other transactions into a "block"; and
5. Balances on the global ledger are updated. The beneficiary now owns the transferred Bitcoins.

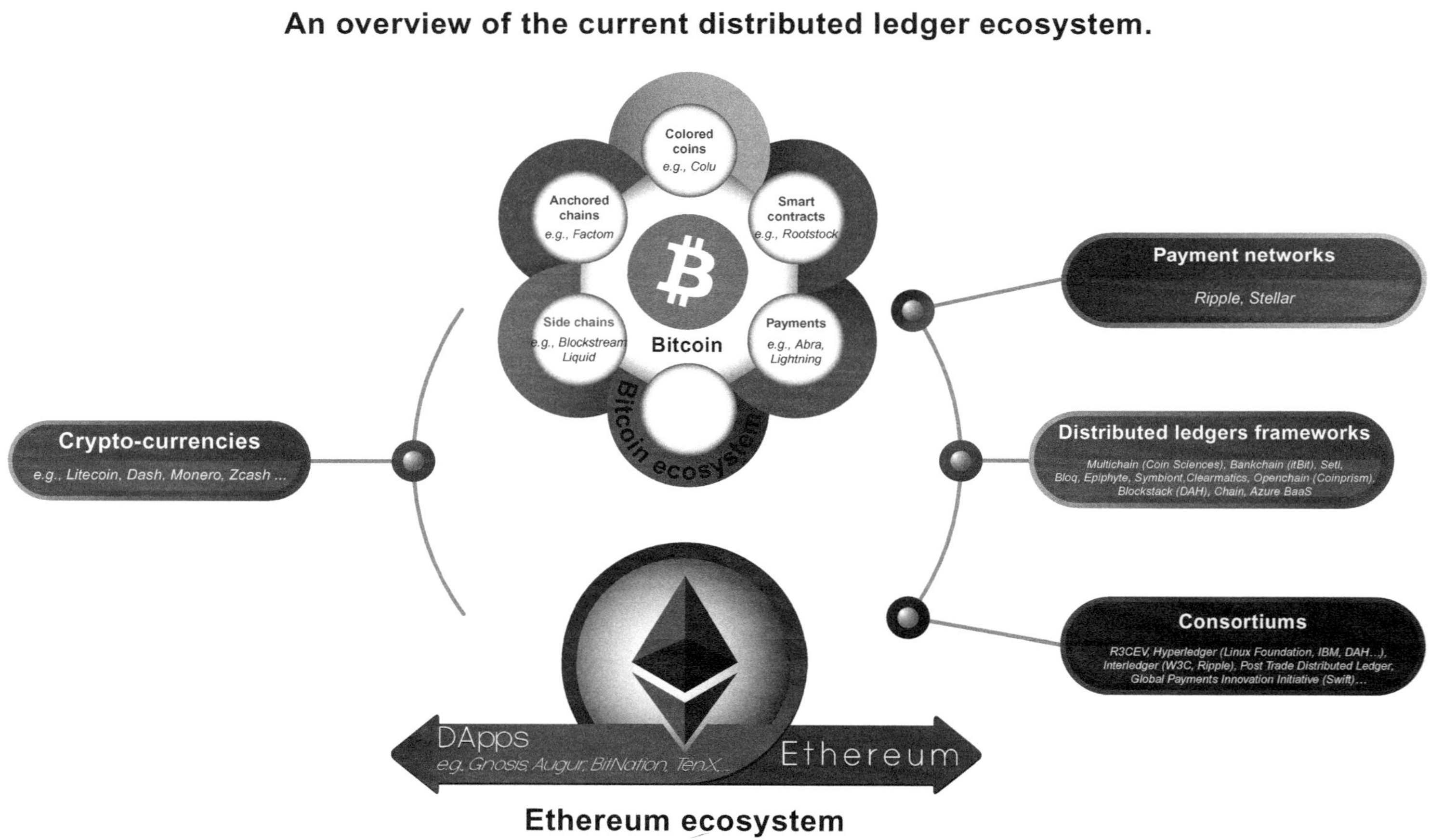

Figure 3.1
Overview of the current distributed ledger ecosystem.

The consensus protocol used by Bitcoin miners to build and validate blocks is called "proof of work." To create a block, miners must solve a mathematical problem: the hash of the block including an arbitrary number called the "nonce" must be under a given target. A hash is the result of a mathematical function that converts an input into a unique fixed length alphanumeric string. The problem can only be solved by trying random numbers until the target is reached. The first miner to reach the target broadcasts the block to the network. Other miners validate that the target was reached, and the new block is added to the database. The difficulty of reaching the mathematical target is adjusted by the software to ensure that blocks are mined every 10 min on average. Mining is a costly process; it requires a lot of computational power and consumes a significant amount of electricity. In exchange, miners are rewarded with new "minted" bitcoins and transaction fees for their work.

Because of the intrinsic difficulty to mine blocks and because blocks are chained together since the first block was mined, it is practically impossible for any attacker to modify the history of transactions and the balances. It would be too costly in terms of computational resources to recreate a chain and force it onto other miners. Bitcoin is censorship-resistant: the system provides auditability, immutability, accountability (time-stamping), and nonrepudiation (digital signature) at transaction level.

Bitcoin system itself is extremely secure. There are, however, two main ways to attack it:

1. Getting control of 51% of the network to "force" certain transactions onto the rest of the network. This is theoretically possible but practically very difficult because the control of the Bitcoin network is split between many actors across different continents;
2. Stealing the private key of someone to get access to its wallet and withdraw its funds. This is the reason why it is important to protect one's private keys.

The Bitcoin Blockchain system is public (everyone has access to the history of transactions), permissionless (anyone can join the network), pseudonymous (a secure address is used to identify users), and Byzantine-fault tolerant (it uses a "proof-of-work" consensus protocol to overcome Byzantine failures and maintain a coherent view of the system's state across the network).

The network effect is vital for Bitcoin and promotes a virtuous cycle. To wit, the security of the network is directly proportional to it size. The market value has increased accordingly. As of July, 29 2017, Bitcoin is supported by a network of around 8,600 active nodes and its market capitalization has grown to reach more than US$45 billion.

3.2.2 Ethereum

Launched in 2015, Ethereum shares many features with Bitcoin but is designed to execute programmable transactions called **smart contracts**. A smart contract is a computer program that can automatically execute the terms of a contract in a deterministic manner, without interference of any kind. Smart contracts are programed using Turing's complete scripting language and stored on the network, although any Ethereum node can independently verify their inputs and execution.

These are the general steps in the execution of a smart contract:

1. A program is registered on the platform with clear conditions for execution (i.e., if A, then B.);
2. Certain events trigger the execution of the program (Events can refer to the initiation of transactions or to information received from "Oracles" although an Oracle is a software agent that finds and verifies real world events and can push external data onto a Blockchain); and
3. The terms of the contract dictate the movement of value. For digital assets on the chain, such as cryptocurrencies, accounts are settled. For assets represented off the chain, such as stocks and fiat, changes to accounts on the ledger will match off-chain settlement instructions.

Two other concepts often relate to smart contracts:

- A **'DAO'** is a Decentralized Autonomous Organization is an organization that is run purely by rules written into smart contracts on a Blockchain; and
- A **'DApp'** is a Decentralized Application that includes a front-end user interface with a decentralized backend which typically leverages Blockchain technology and smart contracts.

As of July 2017, Ethereum is the second largest *permissionless* network with approximately US$18 billion capitalization and 20,600 nodes.

3.2.3 Alternative Cryptocurrencies

Alternative cryptocurrencies have been developed to address some limitations of Bitcoin. Some alternatives aim to change economic incentives and energy use through developing different cryptographic algorithms, different inflationary models or different consensus protocols such as "proof of stake" where block creation chances are proportional to the stake participants have in the cryptocurrency.

Some alternatives aim to add features such as Namecoin which stores data for use as a decentralized Domain Name System; or Emercoin which offers a distributed trusted storage service to name just two.

As of July 2017, there are around 830 cryptocurrencies listed on Coinmarketcap.com but, so far, none of them has surpassed Bitcoin's market capitalization.

3.2.4 Enterprise Blockchain Frameworks

Despite Blockchain's early reputation for failing to meet performance expectations and regulatory requirements, firms started to experiment with Blockchain frameworks specifically designed for their own needs. Some individual firms have developed internal frameworks such as JP Morgan's "Quorum" although others rely on the Blockchain services offered by large technology providers such as IBM and Microsoft.

Additionally, industry consortiums have been launched with the objective of establishing a set of Blockchain standards. Today, there are two frameworks dominating in this field:

- R3 Corda is designed to operate in regulated environments with a limited number of known participants (e.g., Financial Institutions, Regulators). Auditability is based on the "need to know" and consensus about a transaction is basically reduced to its validation by the two contracting parties.
- Hyperledger is an open source effort to create cross-industry Business Blockchain standards (primarily Healthcare and Finance). It was built with contributions from IBM, Intel, Soramitsu, and Monax. As of July 2017, it is supported by more than 100 companies.

Finally, many start-ups have been launched with the objective to sell Blockchain frameworks to other firms. There are 775 firms that apply Bitcoin's Blockchain architectural principles *outside* of the Bitcoin ecosystem, according to the AngelList as of July 2017.

3.3 The Four Main Use Cases of Distributed Ledgers

3.3.1 Peer-to-peer Value Transfer

As demonstrated by Bitcoin, Blockchain can be used to clear and settle peer-to-peer transactions without a central third-party. Beyond cryptocurrencies and digital payments, Blockchain can be used to trade any digital token of value such as:

- Loyalty tokens: Many start-ups such as Loyyal are working on building platforms to trade loyalty points;
- Private securities: Nasdaq and Hong Kong Stock exchange are actively experimenting such use cases. This use case also relates to Initial Coin Offerings (ICO) which have become a very popular way for start-ups to raise money; and
- Energy units: Several experiments around the world have already allowed neighbors to trade solar energy leveraging Blockchains. This use case will be delineated in great details later in the book.

3.3.2 Shared Ledgers

As explained earlier in this chapter, the core feature of Blockchain technology is to allow participants to share the same view of a system's state at any point of time. This is particularly relevant for multiparty accounting: organizations can literally share the same ledger without fear of tampering or fraud.

Blockchain technology can also be used to build trusted registries. Countries such as Georgia and Sweden are actively experimenting with Blockchains to register land titles. Blockchain could also be applied to share registries of fraudulent insurance claimants and corporate financial behavior, for instance.

3.3.3 Immutable Source of Truth ("Proof of")

Blockchain allows information to be time-stamped, authenticated, and immutably stored. It can be applied to notarize information without a central third-party. Possible applications include the following:

- Notarization of documents – *Factom* is an early example of Blockchain solutions providing a collaborative unalterable record-keeping platform for businesses.
- Cross-entity client accreditation – Customers on-boarded by one financial institution could share their credentials with other institutions ('KYC Passporting').
- Provenance / IP / Supply chain management – Blockchain could be used to track the provenance of goods or certify intellectual property.
- Sovereign identity – This problem has many aspects beyond technology; however, Blockchain might be employed to manage digital identities. Several start-ups and organizations are focusing on this issue, including Civic.io and the "World Identity Network" initiative that was recently launched in July 2017.

3.3.4 Enforceable Agreements

By using smart contracts, two strangers can execute agreements in a secure manner. Smart contracts can be applicable to a multitude of use cases including but not limited to:

- Escrow services – Money can be stored on the platform and sent to the beneficiary only when a service is delivered;
- Parametric insurance – A payment can be agreed in advance in case of catastrophic events;
- Royalty distribution – Payments can be triggered when customers purchase a copyrighted service or product; and
- Internet of Things (IoT) / Mark-to-Market contracts – Machines could consume services between each other on an open market in real time.

3.4 Challenges and Road Ahead

3.4.1 Real World Enforcement

Representing real world "off-chain" assets on a Blockchain remains a challenge. Smart contracts are simply software that can only enforce the state of data to which they have access on the Blockchain. In a fully decentralized network, there should be consensus on the state of the ledger among all the nodes at all times. Nodes cannot individually connect to external systems without compromising the consensus. The external information needs to be trustworthy. When introducing an external trusted third party such as an Oracle to a Blockchain, it challenges one of the principal benefits of using a decentralized system. Oracles themselves need to be strictly secured to avoid compromising the Blockchain with false or inaccurate information.

Smart contracts entail another issue in terms of immutability. Printed agreements, once signed, cannot simply be revoked. However, we have means to renegotiate, amend, or cancel them through governance bodies that can arbitrate issues and appeal the decisions. Standard software systems are fixed and enhanced all the times. However, once registered there is no way to fix a bug in a Blockchain smart contract, because it is unmodifiable! The only way out would be to create a new contract to reverse and cancel the operations done by the first contract. To minimize the risks, the business logic should be simple and the code thoroughly verified.

Finally, smart contracts are still executed within the regular legal frameworks of the jurisdiction in which they are operating. They are not yet recognized as legal contracts until the legal framework evolves and lawyers become quasi-programmers. Noteworthy is the *Ricardian Contract* providing a solution by placing the defining elements of a legal agreement as a contract at law, and linking it securely to other systems such as accounting, in a format that can be expressed and executed in software. It is robust through the use of identification by cryptographic hash function; transparent through the use of readable text for legal prose; and efficient through markup language to extract essential information. Later in this book, there will be chapters dedicated to the implications of Blockchain roll-out on traditional legal frameworks.

3.4.2 Performance

Bitcoin Blockchain can theoretically process seven transactions per second and Ethereum Blockchain 25 transactions per second. These transaction throughputs have yet to meet the standards required in industries such as Finance and Retail. In private environments, the throughput performance can be improved because they do not require complex consensus mechanisms. But this is still a challenge for the firms currently implementing Blockchain solutions.

A solution called BigChainDB aims to provide a scalable distributed data storage by adding Blockchain characteristics (i.e., decentralized control, immutability, plus creation, and movement of digital assets) to a standard distributed database. BigchainDB inherits characteristics of modern distributed databases: linear scaling in throughput and capacity with the number of nodes, a full-featured NoSQL query language, efficient querying, and permissioning. BigchainDB could be one of the technical bricks of a future high-performance Blockchain infrastructure.

3.4.3 Standardization

As we have seen in the first section, there are many different implementations of distributed ledgers. As the technology matures, businesses will need to build sustainable and interoperable solutions based on industry standards. Private consortiums such as Hyperledger have launched with the objective to build industry standards. International Standard bodies have also launched Blockchain initiatives: W3C and IEEE created community work groups and ISO has launched a formal work group with its "TC 307 Blockchain and electronic distributed ledger technologies."

But standardization challenges remain because, again, Blockchain technology does not "fit" into the current regulatory frameworks. Regulators and market participants need to work together to ensure the compliance of future standards and define the associated supervisory models.

3.4.4 Governance

The governance of a private Blockchain is straightforward: the one company or the consortium of companies that created it will rule the platform. However, the governance of fully decentralized public systems is complex. By design, public Blockchain systems are governed by rules written in their source code. Interestingly, the decision to update the code is not entirely at the hand of the core developers, even when they all agree on the path forward (which is rarely the case). This creates a tension among the ecosystem stakeholders: developers who produce the software, miners who install and run the software, front-end systems (e.g., exchanges) bringing users and traffic to the network, and investors who invest in the system.

The decision process in the Bitcoin community is as passionate as it is messy. This process is both a blessing and a curse. It is positive in the sense that the decision process itself is decentralized, ensuring that no one is going to take over the system. However, it slows down the development and the adoption of the system—international companies and governments are highly reticent to endorsing something that they cannot control. This is the paradox of Bitcoin: it was designed to run independently from nation states, but it needs them to be massively adopted across the world. Governments probably cannot ban Bitcoin itself but they can implement policies to restrict its use.

3.5 Conclusion

Fundamentally, Blockchain is composed of three major pieces of technology: cryptography, peer-to-peer protocols, and data storage. Individually, these technologies existed before. But when these technologies combined together with smart economic rules, they created revolutionary decentralized markets that never existed before. That is the power of Blockchain technology.

This technology is fascinating because it questions how human activities have been conducted for centuries. All human interactions in society rely on trust. In most cases, trust between strangers is enforced through centralized third parties who manage every aspect of society from policies (governments) to finance (banks) and business activities (corporate lawyers). Blockchain invites us to rethink these critical questions: **What is trust? What is a contract? What is ownership?** Blockchain technology is a new way to create trust with no middleman. Because it enables decentralization, it is also about ownership: with Blockchain you can truly own your identity, own the content you produce and own your financial assets.

Another fascinating fact about Blockchain is the hiatus between the concept and the practice. In effect, the concept is simple and appealing: "sending digital assets should be like sending emails, let's just exchange everything (e.g., securities, solar electricity, loans, etc.) the same way we exchange Bitcoins." Some people are quick to jump to conclusions and prophesize the advent of the universal "*internet of value,*" though, in reality, implementing a Blockchain solution is not straightforward. The Bitcoin Blockchain was a solution designed to solve only one specific problem: the exchange of units of value in a peer-to-peer and trustless environment. In contrast, businesses operate in business-to-business or business-to-consumer, trusted, and regulated environments. It should be obvious that for a different problem, you need a different solution!

Blockchain is not a one-size-fits-all solution for everything, it should be considered as only an "enabler" to solving specific business problems. It has the power to create disruptive decentralized business models. It also has the power to develop incremental innovation in more centralized environments. In both cases, extensive adoption of Blockchain is not only about technology, but also a serious challenge to human's traditional models of regulatory compliance, organization, governance, and business operations.

How can Blockchain be applied in the context of Climate Finance? Can it bring efficiency to or even disrupt the renewable energy markets, carbon trading markets, and the order of international development finance transfers? The rest of the book will answer these questions.

Recommended YouTube videos on Blockchain basics:

- What is blockchain? (by World Economic Forum)
- What is blockchain? CNBC explains
- 19 industries the Blockchain will disrupt

CHAPTER 4

Decoding the Current Global Climate Finance Architecture

Alastair Marke[1] and Bianca Sylvester[2]

[1]Secretary-General, Blockchain Climate Institute, London/Hong Kong [2]Associate Director, Clean Energy Finance Corporation, Melbourne, Australia

4.1 Introduction

Last autumn, Vestine Mukeshimana, a Rwandan villager, purchased electric lights from an off-grid solar company, BBOXX. The product helped her spot snakes in her garden and stopped thieves stealing her cow. She and her neighbors can now cook and their children study in the evenings. The neighborhood had never heard of the Green Climate Fund (GCF)—a UN initiative to bring climate finance to developing countries. But autumn 2016 is when the GCF invested in such household solar schemes as part of its first projects (Economist, 2016). Earlier last year, "excruciatingly painful" was used by Indian and Namibian governments to describe the GCF's accreditation process. "We need [the funds] the most but we don't have the capacity to get it because we're not accredited," grumbled Anote Tong, the then Kiribati president in December 2015 (Lo, 2016). These epitomize the prevalent problems in the entire global climate finance architecture—lack of **transparency** to recipient communities (local involvement) and **trust** among stakeholders of fund management, resulting in unnecessary barriers in raising funds plus delays in approving and implementing climate change projects urgently needed in the most vulnerable countries.

In this scene-setting chapter, let us take you through the climate finance story—where it comes from, where we are, and where we should be heading in the Blockchain era.

4.2 Scale of Climate Investment Required

The story comes from some numbers by the economists. By 2030, climate change and environmental degradation will be eroding the world's gross domestic product (GDP) by 1.6–3.2% (at least US$1.2 trillion) a year, with the least developed countries bearing the brunt as they could lose up to 11% a year (DARA, 2012). Climate change is costly because it has profound implications for the environment in which economic activity takes place.

Transforming Climate Finance and Green Investment with Blockchains.
DOI: https://doi.org/10.1016/B978-0-12-814447-3.00004-5

Published in 2007, the famous "*Stern Review on the Economics of Climate Change*" explains that climate change is the world's greatest market failure and ignoring it would profoundly threaten economic growth. If the messages to reduce emissions are ignored, Stern indicates that this could lead to, "major disruption to economic and social activity, later in this century and in the next, on a scale similar to those associated with the great wars and the economic depression of the first half of the 20th century." To this end, addressing climate change is positive for economic growth in the long term without hindering the short-term growth aspirations of rich and poor countries (Stern, 2007).

The Stern Review also highlights the cost of delaying action on climate change, and concludes that the benefits of strong, early action considerably outweigh the economic costs. The International Energy Agency (IEA) estimates that "for every year of delay before moving to a 450 parts per million (ppm) trajectory the world pursues, an extra US$500 billion is added to the global bill of US$10.5 trillion for mitigating climate change" (Stern, 2007; IEA, 2009).

Put economic costs aside, the uneven distribution of climate change impact signifies social injustice and ethical consequences, which explains the divide between developed and developing countries in international climate change negotiations. Stern indicates that poor countries and their populations will be at the highest risk, even though their contribution to the causes of climate change is minimal. The uneven distribution of the social consequences of global warming could have significant implications for migration flows.[1] The environmental degradation caused by climate change, such as the shrinking of lakes, drying up of rivers and the disappearance of forest and nature reserves, could lead to resource competition with local and/or regional-level political and security ramifications.[2] Much of the scientific literature indicates that climate change is a major factor that can exacerbate the condition of fertile land and water through the heat and weather disruptions it generates. For example, Stern indicates that "melting glaciers will initially increase flood risk and then strongly reduce water supplies, eventually threatening one-sixth of the world's population, predominantly in the Indian subcontinent, parts of China, and the Andes in South America" (Gilding, 2011; Stern, 2007) In the middle east, for example, a source of conflicts stems from the upstream construction of dams and seizing of river waters which can completely externalize the costs of water use

1 For example, climate change could lead to mass migration, with sub-Saharan populations being drawn toward the Mediterranean, Europe, and the Middle East. In Southern Asia, coastal inundation, environmental pressure on land and acute economic competition could affect large populations in Bangladesh and on the East coast of India. Similar effects may be felt in the major East Asian archipelagos, while low-lying islands may become inhabitable (DCDC, 2006).

2 Climate change has the potential to trigger military conflict because access to fertile land and water are considered vital resources and adequate supplies are therefore perceived by States as an issue of "national security" and survival.

by downstream users. Iraq, Syria, and Turkey could foreseeably become trapped in an "escalating struggle" over Turkey's control of the headwaters of the Tigris and Euphrates rivers (Dyer, 2011). Pressure on fertile land and water supplies would be further exacerbated by population increases and rising affluence which will increase consumption levels. Indeed, avoiding these dire socioeconomic consequences justifies the enormous scale of climate investment required.

Following his landmark report, Lord Stern in 2008 suggested that investing just 2% of the global GDP annually into key sectors, including energy and agriculture, could keep the global temperature rise within 2°C above the preindustrial level—the threshold above which the world will embrace itself for climate catastrophe (Jowit and Wintour, 2008). Experts at the World Resources Institute estimate that developing countries alone will require US$300 billion annually by 2020—growing up to US$500 billion annually by 2030—to adequately mitigate their carbon emissions, plus several hundred billion additional dollars to adapt themselves to the impact of climate change. These figures are just part of the US$5.7 trillion needed for green infrastructure investment in developing countries (World Resources Institute, 2017).

Over the past decade, public and private climate finance has been on the rise, as both sectors increasingly recognize that climate actions are in their economic interest and hence, conducive to achieving their goals (UNFCCC, 2016). These low-carbon investments deliver profits, reduce emissions, and build resilience to energy and climate shocks. At Conference of Parties (COP)15 in 2009, for the Green Climate Fund, developed countries pledged to mobilize US$100 billion annually by 2020, with US$30 billion delivered as "fast-start finance" for 2010–12.

Figures from the Climate Policy Initiative show that in 2014 global climate finance increased by 18% to US$391 billion. Apparently, most of this capital comes from the private sector, with government policies playing a facilitating role (Buchner, Trabacchi, Mazza, Abramskiehn, & Wang, 2015). When the scale of needed investment is considered, even a total of US$391 billion was far short of what is required to keep the globe on the 2°C trajectory.

4.3 The State of Climate Finance Post-Paris Agreement

Although, by the end of 2012, contributor countries self-reported that they had exceeded the fast-start finance targets, the Paris Agreement reiterates that developed countries should play leading roles in mobilizing climate finance "from a wide variety of sources, instruments and channels" in a "progression beyond previous efforts," with the accompanying COP decision agreeing to set a new collective goal with a floor of US$100 billion by 2025. Many (developing) countries have underlined the necessity to upscale international support in financing the implementation of their National Adaptation Plans (NAPs) and Nationally Determined Conditions (NDCs). Meanwhile, developing countries have also made the case for finance to address loss and damage attributable to human-induced climate change (arguably caused by industrialized countries over the past centuries). But neither the Warsaw Mechanism on Loss and Damage, nor the new Paris Agreement, commits any states to additional climate finance for this domain (Nakhooda, Watson, & Schalatek, 2016).

The Paris Agreement has been ratified by nearly 150 nations to date, including the US and, upon entering into force on November 4, 2016, it has become one of the fastest ratified international agreements; receiving more first-day signatures than any other UN agreement. However, in June 2017, US President Donald Trump announced the US' withdrawal from the Paris Agreement. This means that the US joins Syria and Nicaragua as the only three nations in the world that are not signatory to the Paris Agreement. Despite this, in the weeks and months leading up to and following this announcement, an overwhelming number of countries, states, and cities have reiterated or even scaled up their commitments to Paris Agreement. Businesses have also followed suit. A stronger global momentum is seen in many businesses which have already priced-in and internalized the impact of the Paris Agreement pledges and recognize its potential to create new business models, investment opportunities, new markets and jobs, and mega-trends toward clean energy technologies and renewable energy.

4.3.1 The "Gap" between Pledged and Disbursed Finance

With regard to public climate finance, in fact, there is a huge gap between the pledged and disbursed climate finance from governments worldwide. As of June 2, 2017, despite the US $100 billion a year pledged by countries in 2009, the Green Climate Fund has raised only US$10.3 billion equivalent from 43 state governments. The major contributors are the USA, Japan, the UK, France, and Germany, accounting for over 76% of the funding signed (Green Climate Fund, 2017).

4.3.2 Global Climate Finance Architecture

Today, there are more than 99 international climate funds or initiatives set up by various actors. Public actors include governments, bilateral aid agencies, climate funds, multilateral, bilateral, and national development finance institutions (DFIs). By funding source, the mix of instruments used to channel climate finance differs (Green Climate Fund, 2017). Approximately, 35% of the bilateral, regional, and other finance reported to the UNFCCC in Biennial Reports is spent as grants, 20% as concessional loans, 10% as nonconcessional loans, and the remainder through equity and other instruments. About 38% of reported finance is channelled through multilateral institutions, many of which are multilateral development banks (MDBs) that utilize capital contributions and commitments from member countries to raise low-cost capital from other sources of funding, including for donor contributions (UNFCCC, 2016).

4.3.2.1 Bilateral and Regional Channels for Climate Finance

On one hand, some developed countries have set up climate finance initiatives at various levels or channel climate finance through their bilateral development assistance institutions. Many developing countries, in parallel, have established regional and national funds or channels to receive climate finance in the form of grants, concessional loans, guarantees, or private equity. By estimate, around US$15 billion had come from direct public contributions from governments, ministries, and bilateral agencies to climate projects in low- and middle-income countries in 2014. In late 2015, in the lead up to COP21 in Paris, climate funds had received US$6.4 million in pledges from the subnational governments of Quebec, Paris, and Wallonia. Approximately, US$26 billion of Overseas Development Assistance (ODA) was climate-related, according to the Organisation for Economic Co-operation and Development's Development Assistance Committee (OECD DAC) (Buchner et al., 2015). The Table below summarizes the state of major initiatives (Table 4.1).

On the other hand, some developing countries have founded national channels and funds resourced through international finance and/or domestic budget and private sector. For instance, to receive Norway's US$1 billion support by late 2015, Brazil has established its Amazon Fund, administered by Brazilian National Development Bank. The Indonesian Climate Change Trust Fund is another example. Mostly with United Nations Development Program (UNDP) as the initial administrator to increase donor trust and meet good fiduciary standards, more national climate change funds have been set up in Bangladesh, Benin, Cambodia, Ethiopia, Guyana, the Maldives, Mali, Mexico, the Philippines, Rwanda, South Africa, etc. Common to these national climate change funds are independent governance structures that met high-level transparency and inclusiveness, and their ability to transfer finance to projects aligned with national priorities.

As regards regional channels, the Caribbean Catastrophic Risk Insurance Facility is an example. Established in 2007 with World Bank's support, it is a 16 member-country risk pool and offers parametric insurance financed with developing countries premiums. Another example is the African Risk Capacity Centre which offers a similar model as a specialized agency of the African Union (Nakhooda et al., 2016).

4.3.2.2 Multilateral Channels for Climate Finance

A unique characteristic of multilateral climate finance initiatives like MDBs is the absence of governance structures dominated by donor countries. Rather, developing country governments are given greater voice and representation in decision-making processes. Nongovernmental observers, with different degrees of participation, are also present in some fund meetings to

Table 4.1: Summary of major bilateral climate funds.

Country/ Initiative	Amount of Aid Approved as of 2014	Source of Finance	Remarks
Germany's International Climate Initiative (IKI)	US$1.1 billion (additional to existing development finance commitments) for 377 mitigation, adaptation, REDD+ (Reducing emissions from deforestation and forest degradation-plus) projects	Funded partly through the sales of national tradable emission certificates (additional to existing ODA)	• The UK, Germany, Denmark and the European Commission jointly contribute to the NAMA Facility (supporting nationally appropriate mitigation actions (NAMAs) in developing and emerging economies) • The UK and Denmark support the Global Climate Partnership Fund, managed by Germany's environment ministry, focusing on renewable energy and energy efficiency through public-private partnership • Germany and the UK also support the US$141 million REDD+ Early Movers Programme
UK's International Climate Fund	US$5.95 billion channelled through multilateral funds (including Climate Investment Funds and the Green Climate Fund)	UK government	
Norway's International Forest Climate Initiative	Pledged US$3.48 billion (up to 2012) through bilateral and multilateral channels and civil society; Significant pledges made for REDD+ activities in Brazil, Indonesia, Tanzania, and Guyana	The Government of Norway	

Credit: Chris Chau, The University of Hong Kong

increase inclusion and accountability in multilateral fund governance. The following table summarizes the state of the UNFCCC financial mechanism (Table 4.2).

In Durban (COP16 in 2010), the UNFCCC established the Standing Committee on Finance to assist the COP in meeting objectives of its Financial Mechanism. The Committee is primarily responsible for the preparation of a biennial assessment of climate finance flows, with the first assessment published in late 2014 (COP20) and the second one in 2016 (COP22). Yet, a considerable volume of climate finance has flowed through institutions beyond the UNFCCC Financial Mechanism.

- Climate Investment Funds (CIFs): Established in 2008 with a pledge of US$ 8.14 billion, CIFs are administered by the World Bank, but operationally partner with regional development banks, including the African Development Bank (AfDB), the Asian Development Bank (ADB), the European Bank for Reconstruction & Development (EBRD), and the Inter-American Development Bank (IDB). In selected developing countries, CIFs finance intervention programs to explore the best models of

Table 4.2: Summary of the state of the UNFCCC financial mechanism.

Funding Source	Amount of Capital	Status/Purpose	Performance (as of 31 July 2017)
Global Environment Facility (GEF) (Established in 1991)	• 31 countries pledged over US$1 billion under the 4^{th} replenishment (GEF4) (2006–10) • 40 countries pledged US$ 1,350 million under GEF 5 (2011–14) • 30 countries pledged US$3.72 billion under GEF6 (2014–18)	• An operating entity of the financial mechanism of the UNFCCC • Funding allocated based on the environmental impact of proposed projects in developing countries	• GEF4: Most of the US$1 billion approved and disbursed to both mitigation and adaptation projects • GEF5: A total of US$909 million approved for 251 projects (of which US $494 million disbursed) • GEF6: US$1.1 billion to support climate change programs that targets multiple focal areas, including climate change, sustainable cities, and land use and forest
Least Developed Countries Fund (LCDF) Since 2001	• More than US$1 billion of voluntary contributions from donors • The Fund had given around US$1 billion for projects and programs in 49 countries by 2016 • Leveraging almost US $4 billion in financing from partners.	• Administered by the GEF under the guidance of the UNFCCC Conference of Parties (COP) • To support national adaptation plan development and their implementation (though largely through smaller scale projects with a country ceiling of US $20 million)	• At least US$981 million approved since its inception in 2002
Special Climate Change Fund (SCCF) Since 2001	• Have received more than US$351.77 million contribution from different countries • Total fund balance in 2017: US$130.58 million		• At least US$302 million approved since its inception in 2002 • Over 100 countries benefited from LCDF and SCCF
Adaptation Fund (AF) (Since 2009)	• Financed through (i) a 2% levy on the sale of emission credits from the Clean Development Mechanism of the Kyoto Protocol; and (ii) increasingly, contributions by developed countries • Over US$569 million raised	• Pioneered direct access to finance for developing countries through National Implementing Entities able to meet agreed fiduciary standards	• Over 54 projects financed for US$349 million

(*Continued*)

Table 4.2: (Continued)

Funding Source	Amount of Capital	Status/Purpose	Performance (as of 31 July 2017)
Green Climate Fund (GCF)	• Countries pledged to mobilize US$100 billion a year • An initial US$10.3 billion raised	• The primary channel for international public climate finance to flow over time • To fund the paradigm shift toward climate-resilient and low-carbon development in developing countries • Adopted a country-driven approach • Committed 50:50 allocation to financing adaptation and mitigation • Countries to access the GCF both through MDBs, international commercial banks and UN agencies, or directly through accredited national, regional and subnational implementing entities	• 33 Implementing Entities accredited by November 2016 • Over 54 projects for US $1,174 million approved

*As at 14 July 2017: https://www.greenclimate.fund/documents/20182/24868/Status_of_Pledges.pdf/eef538d3-2987-4659-8c7c-5566ed6afd19
Credit: Chris Chau, The University of Hong Kong

public finance deployment that supports transformation of development trajectories. CIFs consist of a Clean Technology Fund (US$ 5.47 billion), a Strategic Climate Fund, which comprises the Pilot Program for Climate Resilience (PPCR) with US$ 1.12 billion, the Forest Investment Program (FIP) with US$ 0.74 billion, and the Scaling-Up Renewable Energy Program for Low-Income Countries (SREP) with US$ 0.74 billion. CIFs will operate until 2019 by which the World Bank will review the sunset clause that hinges on the maturity of the global climate finance architecture.

- Multilateral Development Banks (MDBs): Climate change considerations have been incorporated into MDBs' core lending and operations. Some of them organize their own regional or thematic climate finance initiatives. For example, the World Bank's Carbon Markets and Innovation Unit has set up the Forest Carbon Partnership Facility (FCPF) to harness carbon market revenues for reducing emissions from deforestation and forest degradation, forest conservation, sustainable forest management, and the enhancement of forest carbon stocks (REDD+). Under the Carbon Markets and Innovation Unit is also the Partnership for Market Readiness, which helps developing countries establish market-based mechanisms to respond to climate change, as well as the Bio Carbon Fund, which is a public-private partnership that mobilizes finance for sequestration or conservation of carbon in the land-use sector.

Other examples are the Congo Basin Forest Fund and the Africa Climate Change Fund administered by the AfDB; and the EU Global Energy Efficiency & Renewable Energy Fund by the European Investment Bank. The International Fund for Agriculture & Development (IFAD) also manages the Adaptation for Smallholder Agriculture Programme (ASAP), which supports smallholder farmers in scaling up rural climate-change adaptation (Nakhooda et al., 2016). Figure 4.1 maps out the current global climate finance architecture.

4.3.3 Private Climate Finance

For the private sector, since the Climate Summit in New York in September 2014, climate action and voluntary commitments by the private financial sector have reached five inflection points; according to the report, "Trends in private sector climate finance," presented by Janos Pasztor, United Nations' Assistant Secretary-General on Climate Change in October 2015. The report demonstrates that financial institutions which had made announcements the previous year were delivering their commitments and had been joined by numerous other actors.

On one hand, a range of financial institutions worldwide had committed hundreds of billions of dollars in additional finance to support low-carbon and climate-resilient investments. On the other hand, a new green bond market had been created and was rapidly expanding; an increasing number of companies were adopting internal carbon prices; investors were expressing more concern about the activities of carbon-intensive assets and companies; and even the insurance sector was scaling up its efforts to respond to the climate impacts already observed. However, challenges still lie ahead for stretching such positive momentum toward non-OECD countries (United Nations, 2015). To enhance continuously the global climate finance architecture, better policy frameworks are needed to facilitate low-cost capital to flow toward these opportunities by reducing the costs and risks of climate investments, strengthening knowledge and technical capacity, and building the

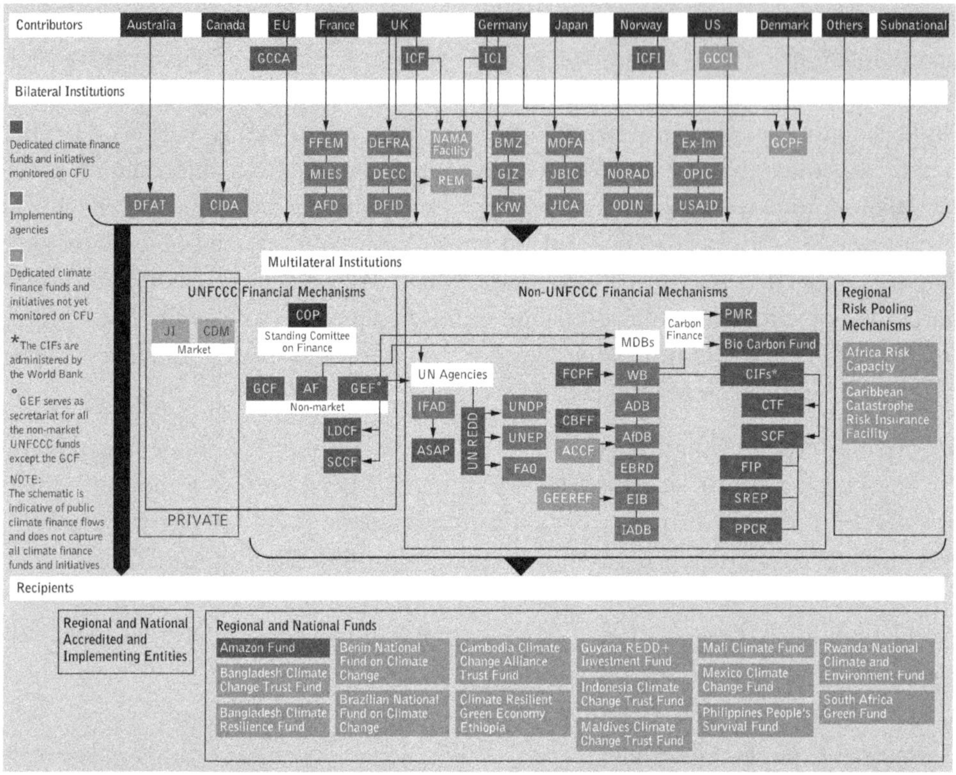

Implementing Agencies and Institutions	
AfDB	African Development Bank
AFD	French Development Agency
ADB	Asian Development Bank
BMZ	Federal Ministry of Economic Cooperation and Development
CIDA	Canadian International Development Agency
DECC	Department of Energy and Climate Change
DEFRA	Department for Environment, Food and Rural Affairs
DFAT	Department of Foreign Affairs and Trade (Australia)
DFID	Department for International Development
EBRD	European Bank for Reconstruction and Development
EIB	European Investment Bank
Ex-Im	Export-Import Bank of the United States
FAO	Food and Agriculture Organisation
FFEM	French Global Environment Facility
GIZ	German Technical Cooperation
IADB	Inter American Development Bank
IFAD	International Fund for Agricultural Development
JBIC	Japan Bank of International Cooperation
JICA	Japan International Cooperation Agency
KfW	German Development Bank
MIES	Inter-ministerial Taskforce on Climate Change
MOFA	Ministry of Foreign Affairs
NORAD	Norwegian Agency for Development Cooperation
ODIN	Ministry of Foreign Affairs
OPIC	Overseas Private Investment Corporation
UNDP	United Nations Development Programme
UNEP	United Nations Environment Programme
USAID	US Agency for International Development
WB	World Bank

Multilateral Funds and Initiatives	
AF	Adaptation Fund (GEF acts as secretariat and WB as trustee)
ACCF	Africa Climate Change Fund
ASAP	Adaptation for Smallholder Agriculture Programme
CBFF	Congo Basin Forest Fund (hosted by AfDB)
CDM	Clean Development Mechanism (implemented under the Kyoto Protocol)
CIF	Climate Investment Funds (implemented through WB, ADB, AfDB, EBRD, and IADB)
CTF	Clean Technology Fund (implemented through WB, ADB, AfDB, EBRD, and IADB)
FCPF	Forest Carbon Partnership Facility
FIP	Forest Investment Program (implemented through WB, ADB, AfDB, EBRD, and IADB)
GCCA	Global Climate Change Alliance
GCF	Green Climate Fund
GEF	Global Environment Facility
GEEREF	Global Energy Efficiency and Renewable Energy Fund (hosted by EIB)
JI	Joint Implementation (implemented under the Kyoto Protocol)
LDCF	Least Developed Countries Fund (hosted by the GEF)
PMR	Partnership for Market Readiness
PPCR	Pilot Program on Climate Resilience (implemented through World Bank, ADB, AfDB, EBRD, and IADB)
SCCF	Special Climate Change Fund (hosted by the GEF)
SCF	Strategic Climate Fund (implemented through WB, ADB, AfDB, EBRD, and IADB)
SREP	Scaling Up Renewable Energy Program (implemented through WB, ADB, AfDB, EBRD, and IADB)
UNREDD	United Nations Collaborative Programme on Reducing Emissions from Deforestation and Forest Degradation

Bilateral Funds and Initiatives	
GCCI	Global Climate Change Initiative (US)
GCPF	Global Climate Partnership Fund (Germany, UK and Denmark)
ICF	International Climate Fund (UK)
ICFI	International Climate Forest Initiative (Norway)
ICI	International Climate Initiative (Germany)
NAMA facility	Nationally Appropriate Mitigation Action facility (UK and Germany)
REM	REDD Early Movers (Germany and UK)

Figure 4.1

Global climate finance architecture. *Source: Overseas Development Institute & Heinrich Böll Foundation.*

track record needed to enhance confidence in such investments (which will be discussed further under "Blockchain – the trust machine" on p.[x]). Despite the efforts of the UNFCCC Standing Committee on Finance and a number of private sectors, there are still substantial gaps in data.

4.3.4 Civil Society Efforts

As the international community has reached a consensus that all countries have to accelerate the achievement of global climate finance goals in various climate agreements, the civil society has made substantial efforts to facilitate the release of capitals to climate projects worldwide. Below lists some recent nonstate-based international climate finance initiatives:

- **Mitigation Action Implementation Network (MAIN)** – The MAIN initiative, founded by the Center for Clean Air Policy, works to identify and highlight the most successful developing country mitigation policies and uses these lessons to assist other countries in refining, financing, and implementing their policy frameworks in order to achieve ambitious mitigation actions.
- **Non-State Action Zone for Climate Action (NAZCA)** – Established after COP20 in Lima, NAZCA captures the commitments to climate action by companies, cities, subnational, regions, investors, and civil society organizations. Today, it tracks over 12,549 mobilized actions that are helping countries achieve and exceed their national commitments to address climate change.
- **Low Carbon Investment (LCI) Registry** – Launched by the Global Investor Coalition on Climate Change before the UN Climate Summit in New York in September 2014, LCI is the first public, online database showing examples of global low-carbon investments made by institutional investors such as pension funds and asset managers (investors).
- **Transformative Actions Program (TAP)** – TAP is created to catalyse and improve capital flows to local and subnational governments by increasing their visibility and better positioning them for financing through targeted capacity building.
- **Global Innovation Lab for Climate Finance** – Created by the Climate Policy Initiative, the Lab is a global initiative that aims to drive billions in private investment into climate change mitigation and adaptation in developing countries.
- **Global Infrastructure Basel's SuRe Standard** (Standard for Sustainable and Resilience Infrastructure) – It is a global voluntary standard which integrates key criteria of sustainability and resilience into infrastructure development and upgrade, through 14 themes covering 65 criteria across governance, social, and environmental factors. It also provides a method for investors to choose sound investment opportunities in infrastructure.

Civil society efforts stretch to some very local initiatives too. In Cambodia, for instance, the nongovernmental **Waste-to-Energy Revolving Fund** has been created by Nexus and the Renewable Energy and Energy Efficiency Partnership. This Revolving Fund helps rice millers in Cambodia to access capital to purchase renewable energy generation technology. Rice milling requires huge amounts of electricity, which, in Cambodia and other developing countries, are expensive and rely on the burning of dirty fuels such as diesel. Through the unique Waste-to-Energy Revolving Fund, rice millers are accessing finance to purchase rice husk gasification technology, which converts currently discarded rice husks into clean renewable energy, benefiting both local communities and the environment.

4.4 Climate Finance Market Observations

Notwithstanding various climate finance estimates by different organizations, there are indeed many similarities in the general patterns observed. Most obvious are the facts that (a) most climate finance is mobilized and deployed domestically (e.g., through the Green Investment Bank in the UK); (b) most private climate finance is provided by corporations and households; and (c) most climate projects are financed and implemented by the same organization (Callaghan, 2015). These are the case for both developed and developing countries. In some developing countries from which information on domestic public climate finance is available, the data suggest that in these countries domestic public finance considerably exceeds the inflows of international public climate finance from bilateral and multilateral sources (UNFCCC, 2016).

Specifically, Callaghan (2015) has observed that climate investment concentrate in major economies and principally focused on "low-hanging fruit" in terms of technologies and financing instruments. A significant amount of investment has been made through corporate balance sheets and "mainstream" instruments such as bonds. But vital to smaller and poorer countries, some investments have been addressed with more complex, alternative instruments and structures, covering a range of activities from fully commercial through hybrid to concessional (i.e., different combinations of grants, loans and/or equity, etc.).

The general climate investment trends in both climate mitigation and adaptation could be learnt from the (Intended) Nationally Determined Contributions ((I)NDCs) submitted by state governments. An (I)NDC is like a "national action plan" by a State government to reduce the greenhouse gas (GHG) emissions that all signatory countries to the UNFCCC were asked to submit before the UN climate change conference (COP21) held in Paris in December 2015. Under the Paris Agreement, adopted in December 2015, an INDC becomes the first NDC when a country ratifies the agreement, unless it decides to submit a new NDC altogether.

(I)NDCs allow countries to "provide information to varying degrees based on their national circumstances." To this end, countries may choose to describe their INDCs in different levels of detail. Among other things, an (I)NDC may contain such information as a country's share of global GHG emissions; emissions per capita; start year for the contribution target; whether the plan is based on target reduction versus Base Year or "Business as Usual" projection; whether adaptation measures are included; and whether there are any relevant existing or planned climate change actions-related national strategies/ legislation, etc. To date, the UNFCCC has received 162 submissions representing all 190 parties to the Convention.

4.4.1 Climate Mitigation Investment Trends in Major Sectors

Generally, climate mitigation investment goes around the sectors of energy; industries and industrial processes; agriculture; land use, land use change and forestry (LULUCF); waste; and transport. As Callaghan (2015) suggests, energy and transport sectors dominate mitigation investment in Africa, Asia, and Europe as they are considered (a) "low-hanging fruit" in terms of achieving emissions reduction; and (b) sectors in which alternative technologies and solutions are relatively mature, whereas LULUCF also accounts for significant investment in Africa.

With regard to energy, the majority of the investments concentrate in solar and wind power generation and/or consumption, seeing the most advanced financing mechanisms. Governments, MDBs and some commercial banks also coinvest in large hydro plants. Many countries have set sector-wide renewable energy targets such as 80% by 2030 by Honduras or certain percentages by India and Thailand, though they would need international financial support and technology transfer. However, more funds are badly needed in the research and development of ocean energy and carbon capture and storage (CCS), in particular. Countries such as Myanmar invested significantly in energy efficiency to scale up clean cooking, off-grid domestic lighting, and insulation in the household sector (Callaghan, 2015). For energy efficiency and low-carbon technologies alone in the NDCs, at least US$16.5 trillion is required over the next 15 years to limit global temperature increase to 2°C (Climate Policy Initiative, 2017).

Next, the transport sector emits nearly a quarter of the world's energy-related CO_2. Climate finance opportunities involve investment in public transport (mass transits), improvement of the emission performance of private fleets, green infrastructure, and policy for traffic management. Studies have indicated that a green, low-carbon transport sector could reduce GHG emissions by 70% without major additional investment—a re-allocation of only 0.34% of global GDP. Thanks to powerful planning, incentivizing fiscal, and regulatory measures, transport is a sector where well-developed and understood technologies and business models are readily available. For instance, Ghana outlined a US$1.2 billion plan for expanding inter- and intra-city bus and rail links. Macedonia increases use of (electrified) railway, renewal of vehicle fleet and use of bicycles with complementary parking policy, though it needs international climate finance aid. The World Bank has contributed various green bond issuances for relevant investment projects in China. Foreseeably, international green bonds market should be expected to be the major source of finance for many public transport initiatives in the near future. Yet, noncommercial finance will also be essential for public education and institutional capacity building to change the economic perceptions over transport and; enable effective the transfers of technology, planning, and fiscal know-how built up in the developed world (Callaghan, 2015).

More than three quarters of African countries invest in LULUCF sector, including forestry, management of cropland and grazing; as well as prevention of desertification. Accounting for 29% of the global GHG emissions, much capital is deployed for REDD+ for its commercial potential in the carbon market, especially in Malawi where 78% of its emissions come from the forestry sector. Activities include protection and conservation of existing forests, afforestation, with unconditional commitments for sequestering carbon and capacity building support. There are highly innovative financing mechanisms available through Althelia Climate Fund, for example, that enables individual investors to commit €15m by means of "Nature Conservation Notes" to create carbon offset credits through protection of forests, and generate revenues from sustainably certified products such as cocoa, coffee and wood.

Conversely, the industrial processes, which mainly rely on businesses to replace outdated equipment, are so far the least invested sector. It is probably attributed to the sector (a) being of less importance as an emission source in smaller economies; and (b) where companies rather than governments would like to be seen as the key investors with their own balance sheets and access to finance. Currently, with some governments setting energy performance standards, such investment is supported by commercial banks or manufacturers of equipment, export credit guarantee, or other trade finance providers (Callaghan, 2015).

Mitigation-focused finance represented more than 70% of the public finance in developing countries reported during 2011–14, according to the UNFCCC. Adaptation finance provided to developing countries by climate funds and bilateral concessional channels accounted for only a quarter of the total finance. And more than 80% of MDB investments concentrated in mitigation, whereas less than 20% on adaptation (UNFCCC, 2016).

4.4.2 Adaptation Investment Trends in INDCs

Despite the inclusion of adaptation actions being optional in the INDC process, 84% of countries included adaptation plans in their submissions, and those which did not were principally the most developed countries. In developing countries which submitted national adaptation plans (NAPs), adaptation measures are often instrumental to supporting their national economic development such as through the introduction of climate-resilient agricultural practices, which could absorb extra billions of dollars of revenue or investment a year. Undeniably, well-targeted early investment to increase climate resilience, no matter in infrastructure development, technology advancement, capacity building, shifts in systems and behaviours, or risk transfer measures, is probably more cost effective for the international community than complex post-event disaster relief efforts.

In fact, adaptation priorities vary from region to region. African countries prioritize energy over other sectors more often than countries in other regions. In Asia, countries considered oceans and coastal, plus assessment and monitoring as priority sectors. Similarly, countries in Latin America indicated oceans and coastal, as well as terrestrial ecosystems as priority sectors. Generally, adaptation investment falls into two categories: (1) physical interventions; and (2) capacity building.

Physical interventions refer to the undertaking of adaptive actions in infrastructure, construction, production or development program. Example sectors include agriculture/fisheries, where farmland could be made more arable, fisheries or land management more sustainable, irrigation systems more efficient with drying trends in mind, and better crop storage or food processing facilities designed with warming trends in mind. Other interventions are found in coastal/marine, housing/urban, tourism, health, and insurance sectors. In the insurance sector, product innovations include parametric-based crop and weather insurance, health and disaster insurance for poorer communities, etc.

The World Bank classifies adaptation measures and options in the following ways (World Bank Group, 2017):

- **Incremental, transitional, and fundamental actions:** Based on the depth of the change, measures can be incremental when they imply simply increasing the frequency or quantity of existing interventions that may or may not consider climate change. Changes are said to be transitional if they deliberately take into consideration climate change and expected health outcomes due to climate change. Fundamental is change that is classified as transformational and permanent.
- **Short- versus long-term measures**
- **Proactive versus reactive measures:** Proactive measures are taken to prevent events that have not happened yet but for which there exists a risk; reactive measures address events that have already happened and are likely to recur with greater or lesser intensity.
- **"No regrets," "low regrets," and win-win solutions:** In terms of cost-benefit, "no-regrets" adaptation measures are those whose socioeconomic benefits exceed their costs, regardless of what happens to the climate. Measures are considered "low-regrets"

when the associated costs are somewhat low and the benefits are expected to be large if the projected climate change occurs. Win-win options are those that minimize social risk and/or exploit potential opportunities and also have other socioeconomic or environmental benefits.

- **Local and general actions:** In terms of geographical scope, measures can be local (such as vector control in an area), or general/systemic.
- **Sector-specific or broader:** Adaptation measures may be taken either in the health sector or other sectors.

A basis for public sector reforms, **capacity building** involves technology transfer, risk management, government, and civil society. As with other training efforts under foreign aid programs, these are educative actions to improve the adaptation capacity of government or business sectors by:

a. securing technology and related knowledge transfers;
b. advancing research and development of relevant technologies by and for the business sectors;
c. improving risk management;
d. creating early warning systems;
e. addressing vulnerabilities based on gender, age groups and geographical areas; and
f. increasing public awareness of climate change (Callaghan, 2015; UNFCCC, 2017).

Among others, it is the government capabilities in planning and legislation which counts the most for determining the level of resilience of a country to climate change.

4.5 Barriers for Unlocking Climate Finance

For public climate finance, two factors—"austerity" and "national politics"—can explain most of the difficulties encountered at national level. At subnational level, the transfers of climate finance are often hindered by the following factors:

a. *Lack of communication between financial institutions and local governments* – For example, in Indonesia, a resort would not have been repetitively destroyed by wildfires if the investors had been informed by the community and local government about the needs to adapt to fire risk at that location.
b. *Uncertainty over regulatory and tax policies* – For example, in the South African government, there is only one unit dedicated to public-private partnerships with the sole authority to initiate projects, which means very few can move forward.
c. *Lack of capacity and expertise* – Developing world tend not to be able to effectively implement their vision and demonstrate their creditworthiness (CityTalk, 2016).

Nevertheless, for private climate finance, which will account for a significant proportion of global climate finance in the future, the story is much longer and more complicated. Briefly speaking, Callaghan (2015) considers the barriers to mobilizing private climate finance being three-pronged:

a. **"Viability gap"** – The "viability gap" in the financing of many climate investments exists between costs and revenues. From investors' perspectives, the costs of capital for such investments as renewable energy projects may be so high that the profit margins would be uncertain given the revenue streams being perceived as not sustainable. This gap should be bridged by concessional finance from governments in the form of tax incentives, feed-in tariffs, concessional loans, or what CPI calls in its 2014 report "public framework expenditures" that improve the investment climate.
b. **Risk aversion** – (Foreign) investors tend to be very sensitive to risks that would impact the returns for their investment. Their appetite for risk associated with climate projects in developing countries, which are characterized by various uncertainties, is relatively low. These risks include the familiar **adverse currency movements** which are ever-present for external investors and may be mitigated with short-term commercial hedging products. Another example is **political risk** which can be reduced by political risk insurance or export credit guarantees from governments or development agencies such as Overseas Private Investment Corporation in the US and the World Bank's Multilateral Investment Guarantee Agency.
c. **Visibility of investment opportunities** – The major constraint for private climate finance is the low visibility of climate investment opportunities, leading to mismatch between assets and liabilities. For instance, in many (I)NDCs, micro-insurance products that cover crop risks have never been mentioned despite rapid growing demand.

An example of a sector where there has been a significant climate finance shortfall is in the clean cooking sector. In this sector, finance is needed for improved biomass and clean-fuel supply chains, working capital for improved cookstove producers and distributors, support for market transformation programs and enabling infrastructure. According to the World Bank, estimates of the total funding gap vary, but they suggest that the sector is significantly underfunded with roughly US$70 million of investments globally in clean cooking interventions by donors (Putti, Tsan, Mehta, & Kammila, 2015). This contrasts with IEA estimates that annual investments of US$4.7 billion are needed globally to ensure universal application of clean cooking fuels through 2030. To address the significant shortfalls, the World Bank calls for a significant increase in clean cooking solutions, potentially via smart and targeted subsidies, and continued investment in improved cookstoves.

4.6 Current Climate Finance Instruments to Bridge the "Gap"

Great coordination and intermediation are what is required to facilitate climate investment flows and hence, scale up global climate finance in order to align the interests of stakeholders, including governments, regulators, affected communities, and investors taking risks of different kinds and levels. The table below summarizes five major categories of useful climate finance instruments: policy incentives, risk management instruments, grants, low-cost project debt, and capital instruments (Table 4.3).

Table 4.3: Major categories of climate finance instruments.

Instrument	Definition/Coverage	Examples
Policy incentives	Income-enhancing mechanisms, such as feed-in tariffs, tradable certificates (e.g., emission trading schemes), tax incentives and, clean energy subsidies	• **Feed-in tariffs** are policy mechanisms designed to accelerate investment in renewable energy technologies by offering long-term contracts to renewable energy producers, often based on the cost of generation by each technology. • Tradable certificates – **Renewable Energy Certificates** (RECs) are tradable, nontangible energy commodities in the US that represent proof that 1 MWH of electricity was generated from an eligible renewable energy resource—solar photovoltaic, thermal, wind, geothermal, hydropower, and biomass. – **Carbon pricing** such as the European Union Emission Trading Scheme (ETS) is a government-mandated, market-based approach to controlling pollution by providing economic incentives for achieving reductions in GHG emissions through the trading of limited number of carbon permits among major emitters. Carbon pricing systems are policy tools designed to devolve governance over the atmosphere to supra-national and nonstate actors and to the market. Pricing carbon can motivate changes in investment behaviors by providing a financial incentive for investors to allocate resources to lower carbon sources. As of 2017, over 40 national and 25 subnational jurisdictions are putting a price on carbon, covering around 8 gigatons of carbon equivalent or 15% of global GHG emissions.

(Continued)

Table 4.3: (Continued)

Instrument	Definition/Coverage	Examples
Risk management instruments	Guarantees, insurance policies, contract-based instruments	**Chile Geothermal Risk Mitigation Program** Funded by the Climate Investment Fund's Clean Technology Fund (CTF), this program was designed to stimulate investments in geothermal power with risk transfer mechanisms to reduce exploration and development costs and risks, and mobilize private capital to ensure sustainable, long-term growth. The risk mitigation instruments used include insurance, guarantees, mezzanine financing, or capital contributions to projects.
Grants	Covering incremental costs associated with low-carbon and climate-resilient activities (including cash transfers or the provision of in-kind support for which recipients incur no legal debt)	**Egypt Wind Energy Initiative** The CTF approved a US$1 million preparation grant to support the technical feasibility study for Egypt's first wind public-private partnership project. The feasibility study includes wind resource assessment, which typically requires at least one year of wind or meteorological data measurement on the site.
Low-cost project debt	Concessional debt (i.e., financing provided at terms preferable to those prevailing on the market such as longer loan tenor, grace periods, or lower interest rates)	**Morocco Wind Energy Plan** The CTF financed ONE's (Morocco's National Utility) 2,000 MW wind energy plan, including support for transmission infrastructure and water pumping and storage facilities in combination with wind energy for pumping by 2015. Concessional terms from CTF funds helped lower the cost of capital and thus enable ONE to pass on the savings directly to the price per kWh.
Capital instruments	Transformation of capital contributions into shareholder ownership (equity), creditor claim (debt, loans, bonds, etc.) and hybrid capital instruments (used with investment expecting market rate returns), including: – **Project-level market rate debt** – **Project-level equity** – **Balance sheet financing**	**World Bank Green Bonds** Green bonds raise funds from fixed income investors to support World Bank lending for eligible "green" projects. One World Bank Green Bond was designed in partnership with the Skandinaviska Enskilda Banken (SEB) to respond to specific investor demand for a triple-A rated fixed income product supporting climate projects. The expanding green bond market is anticipated to reach US$120 billion in 2017. Since the World Bank's first green bond launched in 2008, the World Bank has issued 127 green bonds in 18 currencies, totalling approximately US$9.6 billion equivalent. Green bonds have also been issued by a number of corporations in Asia (especially in China), but their proceeds are confined to financing green projects.

In general, mitigation projects tend to be financed with a mix of equity and loan instruments (both concessional and nonconcessional) supported by various types of policy incentives. Besides, investments in climate resilience tend to be supported with grants and low-cost loans due to the generally higher incremental cost component.

Instruments not only help channel investments to low-carbon and climate-resilient projects, but also serve an important function for public climate finance actors in mobilizing climate finance and achieving green investment at scale. There are various examples of instruments that can address investor-specific needs, align public and private interests and enable scaled-up low-carbon and climate-resilient investments, the magnitude of critical mass needed to achieve transformational shifts toward climate-friendly investment decisions. However, given the complexities in coordination and intermediation, stakeholders (in developing countries) face hurdles which have to be overcome with technical assistance and capacity building alongside financial support to unleash the full potential of the above instruments.

4.7 Problems with Climate Finance Tracking

Although billions of dollars of climate finance have gone to many countries for reducing their GHG emissions (mitigation finance) and helping them cope with rising climate impacts (adaptation finance), the unstated purpose of climate finance is to building "**mutual trust**" needed to continue the international climate negotiations. But activists, academics, and recipient governments have long been disputing the amounts of climate finance contributing countries have disbursed. How much is being given? Where exactly is it going and what impact is it having? These create the problem of clarity and transparency. The ability to understand and measure the impact of climate finance on sustainable development is a fundamental condition for increasing finance in a way that is commensurate with the task.

Huge "information gaps" still exist in the global climate finance landscape. The task of tracking, including monitoring, reporting and verification (MRV), of climate finance is exceptionally challenging. Information gaps limits what can be tracked and captured annually by international efforts. Climate finance MRV, despite some improvement seen recently, remains far from perfect. The application of different definitions associated with climate finance can yield vastly different calculations of climate finance flows and its uses. On one hand, tracking private finance ultimately depends on voluntary reporting of data such as through press releases, financial statements and presentations, etc. On the other, tracking public finance relies on the existence of public support mechanisms that require disclosure of project details such as those under the Clean Development Mechanism.

Many important research reports in the past 5 years have tried to put a number on international climate finance flows, but have decried the lack of solid data and transparency in international climate adaptation funding owing to fragmentation, lack of regulation and an adequate system for defining, categorizing, tracking, and evaluating climate finance to guarantee the data reported is true and accurate. In fact, the availability of robust data varies significantly across countries. National-level indicators may also conceal major subnational differences. For example, in an assessment of the design of emissions trading systems in China, the analysts highlight issues with data quality and quantity. They explain that this is because, in the case of China, there are, "small penalties for obstruction of inspection and falsification of data, as well as vague and ineffective rules to fight "statistical corruption." In addition, "beyond national level emissions data, regulators at local levels have not collected data at the facility level for very long" (Munnings, 2014).

4.7.1 Adaptation Finance Transparency

A 2015 report on adaptation finance transparency by the Adaptation Watch have unearthed the severity of these problems. Its main findings include:

- Under the OECD Rio marker system two-thirds of the projects (especially those by the EU institutions) categorized as having adaptation as principal or significant objective were only tenuously linked to climate change adaptation (e.g., a multiyear funding package to enhance trade and development by supporting the World Trade Organization Global Trust Fund was coded by the donor country as having a climate change adaptation objective);
- Developed countries' Biennial Reports demonstrate wide variation in transparency in contributor-country reporting due to lenient UNFCCC guidelines—Other than Germany, Canada, and Ireland, countries do not report sufficient data at project level while the US counts some funding to the OPIC and Export-Import Bank in its reports of climate finance;
- The planning process within the Least Developed Countries Fund (LCDF) and National Adaptation Programmes of Action (NAPA) was hampered by the complexity of the systems involved. The process of funding the urgent adaptation priorities of NAPAs through the LDCF has been characterized by lengthy delays, a chronic lack of funds and a lack of transparency in reporting the relationship between NAPA priorities and LCDF projects. Almost half of countries, particularly Latin American countries, prioritized at least one project under their NAPA that had no climate adaptation or vulnerability justification at all (Adaptation Watch, 2015).

Echoing the first two findings above, more recent research reveals that donor countries as a group are becoming less transparent in reporting their public climate finance investments.

Despite the increased attention paid to the transparency gap at COP21 in 2015, the distance between donor countries' pledged climate adaptation finance and trackable reality show that the "finance gap" has widened. The research found that donor countries' average level of compliance toward UNFCCC climate finance transparency requirements declined from a poor average of 58% for reports filed in 2014 to an even worse mean of 52% in 2016 (Brown University, 2015). If the mechanism for accountability on climate action–upon which the success of the Paris Agreement hinges, in the stead of any binding legal agreement—is to have any hope for success, transparency in the reporting of climate finance is absolutely crucial.

The third finding above, indeed, reflects some structural problems with the current intermediaries-centred transaction mechanism. Many of the smallest and most at-risk countries, and especially subnational entities, now claim that they do not have the means to access the international climate funds directly. The inability of poor countries to promptly satisfy many funds' stringent accreditation requirements means that they cannot develop smarter renewables grids, credible cap-and-trade schemes, and other essential climate investment projects in time. Such "accreditation culture" in international climate finance management may stem from the traditional, centralized transaction mechanism in which all exchanges of financial resources between donors/investors and recipients must go through financial "intermediaries." In global climate finance these intermediaries are typically international (climate) funds that manage public investments; and financial institutions that handle private investments.

Going forward, new data and information flows are needed to address the huge information gaps that exist in the global climate finance landscape. To be effective in addressing these information gaps, a new approach is needed to yield data and information that is: relevant, analytically sound; and timely. The importance of this type of data is described further below.

To be relevant, climate finance data need to provide a representative picture of environmental conditions, pressures on the environment or a society's responses. To be relevant, the data need to be simple, able to show trends over time; be responsive to change, provide a basis for international comparisons; and have a threshold or baseline value against which to compare it, so that users can assess the significance of the values associated with it.

To have analytical soundness, climate finance data should be theoretically well founded in technical and scientific terms; be based on international standards and international consensus about its validity.

An important criterion affecting the usefulness of information about climate finance is the timeliness of the data. The timing between the period to which the data refer and their release date should be as short as is practicable. As discussed previously in this chapter, the current timeliness of climate finance data often remains insufficient for evaluation purposes.

4.8 Blockchain—The Trust Machine

Once called a "trust machine" in an *Economist* article in October 2015, Blockchain can address the above institutional problems by enabling trustworthy and transparent "peer-to-peer" transactions that allows for the generation and communication of data that are relevant, analytically sound, and timely. With a Blockchain transaction, every participant in a network can transact directly with every other network participant without involving a third-party intermediary. Transactions are no longer stored in a central database but encrypted and distributed to all participating computers, which store the data locally. Where a provider and a customer agree to enter into a transaction, they determine the variables of this transaction by specifying the recipient, the sender and the size of the transaction, among other things. All information relating to an individual transaction is then combined with the details of other transactions made during the same period to create a new block of data. This is comparable to sending emails, which are also split into separate data blocks. Blockchains are different in that this process relates to a single standardized transaction.

The data stored in a block are verified using algorithms, which attach a unique hash to each block. Each such hash is a series of numbers and letters created on the basis of the information stored in the relevant data block. If any piece of information relating to any transaction is subsequently changed as a result of tampering or due to transmission errors, e.g., the exact amount of the transaction, the algorithm run on the changed block will no longer produce the correct hash and will therefore report an error.

All number/letter combinations are continuously checked for correctness and the individual data blocks are combined to form a chain of individual data blocks—the Blockchain. Owing to the interlinking of these number/letter combinations, the information stored on the Blockchain cannot be tampered with (at least this would require a great deal of effort). The verification process ensures that all members can add to the Blockchain but no subsequent revisions are possible. In other words, information on new transactions is integrated in a precise, time-stamped, interlinked manner, which secures integrity and makes fraud easy to detect. Therefore, Blockchain enables direct, peer-to-peer transactions between persons or organizations accustomed to the services of an intermediary in order for their transactions to be legitimately recorded. For example, although a bank is currently needed as an intermediary to effect a financial transaction between two parties, the same transaction can be executed and documented directly between the two parties if a Blockchain is used (PwC, 2016). (*See further description in Sebastien Meunier's chapter on the workings of Blockchain.*)

The first high-profile Blockchain application was Bitcoin. Bitcoin has subsequently spawned other Blockchain applications, most of which are being developed in the banking and financial services industry. Currently, the five major developments in Blockchain use cases are smart contracts, smart assets, clearing and settlement, payments, and digital identity, which will diminish the importance of traditional financial intermediaries.

For instance, SWIFT is developing projects around Blockchain for payments transactions. Its lead player, R3, backed by 42 banks, already started experiments with Microsoft's Azure-based Blockchain-as-a-Service to develop Ethereum as a bank-to-bank global transaction system in spring 2016. This new system will not only facilitate global bank-to-bank transfers, but also any exchange of value over the internet. The emerging value exchange models between players in Africa are likely to result in digital currencies that avoid the high exchange costs of Western Union or other remittance players (Skinner, 2016). A Santander's study suggests that Blockchain technology can save banks US$20 billion a year by 2022 by eliminating central authorities and bypassing slow, expensive payment networks (Bello Perez, 2015). Against this backdrop, if Blockchain is deployed for green and sustainable finance, the implications on filling the current climate finance gap are likely to be tremendous.

Climate finance and green investment provide the best ground on which to apply Blockchain as a "fintech" to combine technology and finance. For example, the capitals for a renewable energy project currently come from banks only, although the latest information technology provides a bonding between the industry and finance. As a loop, the energy sector provides valuable production data very useful for the research and development of financial products, whereas new financial products benefit the energy sector reciprocally. Among the fintechs including big data, cloud computing, machine learning, distributed computing technology, Blockchain is the most impactful and revolutionary to the bottom-level (green) finance architecture, especially in lowering regulatory costs and expanding regulatory boundaries.

The Paris Agreement alone represents a US$23 trillion green investment market between now and 2030 (Stoilijkovic, 2017). The emerging demand for green investment requires significant investment in social capital—the networks of "trusting" relationships among investors and enterprises that facilitate finance flows to climate/green projects. Despite green finance being a priority in many governments' agendas, it has not been developed at a pace expected because: (a) the cost of developing green finance products (e.g., green bonds) is higher than traditional ones; (b) information asymmetry is greater; and (c) third-party certification and regulatory systems have been inadequate. These are exactly the pain points to which Blockchain technology can provide innovative, far-reaching solutions. Its key features—distributed data, cross-sector process cooperation, and low-cost third-party certification platform—could turbo-boost global climate finance and green investment.

Sections 2, 3 and 4 of this book will introduce to readers some pioneer Blockchain use cases or ideas for a smarter renewable energy sector; smoother international climate finance transfers; and fraud-free emissions management. Section 5 will present authors' insights regarding new regulatory frameworks for Blockchain-based climate finance architecture.

References

Adaptation Watch. *Toward mutual accountability: The 2015 adaptation finance transparency gap report.* (2015) Policy Briefing: <http://www.AdaptationWatch.org>

Bello Perez, Y. *Santander: Blockchain tech can save banks $20 billion a year.* Santander. (2015). <http://www.coindesk.com/santander-blockchain-tech-can-save-banks-20-billion-a-year/>

Brown University. *Climate finance 'less transparent' since Paris.* Institute at Brown for Environment & Society. (2015). <https://www.brown.edu/academics/institute-environment-society/news/story/climate-finance-%E2%80%98less-transparent-paris%E2%80%99>

Buchner, B., Trabacchi, C., Mazza, F., Abramskiehn, D., & Wang, D. (2015). *Global landscape of climate finance 2015.* Venice: Climate Policy Initiative.

Callaghan, I. (2015). *Climate finance after COP21 – Pathways to the effective financing of commitments and needs.* London: Investor Watch, Ian Callaghan Associates Limited.

CityTalk. *Is communication the greatest barrier to unlocking climate finance?* CityTalk: A blog by ICLEI, (2016). <http://talkofthecities.iclei.org/is-communication-the-greatest-barrier-in-unlocking-climate-finance/>

Climate Policy Initiative. *Climate finance landscape.* CPI. (2017). <http://www.climatefinancelandscape.org/>

DARA. (2012). *Climate vulnerability monitor: A guide to the cold calculus of a hot planet (2nd ed.).* Madrid: DARA Group & Climate Vulnerable Forum.

DCDC. (2006). *The DCDC strategic trends program 2007–2036.* Shrivenham: Development, Concept & Doctrine Centre.

Dyer, G. (2011). *Climate wars: The fitht for survival as the world overheats.* London: Oneworld Publications.

Economist. (2016). *The green light.* Economist.

Gilding, P. (2011). *The great disruption.* London: Bloomsbury Press.

Green Climate Fund. *Resource mobilization.* Green Climate Fund, (2017). <http://www.greenclimate.fund/how-we-work/resource-mobilization>

IEA. (2009). *World energy outlook.* Paris: OECD/IEA.

Jowit, J., & Wintour, P. (2008). *Cost of tackling global climate change has doubled, warns Stern.* The Guardian.

Lo, J. (2016). *Why can't poor countries access the climate finance they were promised?* The Guardian.

Munnings, C. *Assessing the design of three pilot programs for carbon trading in China. Discussion Paper 14–36.* Resources for the Future, (2014). <http://www.rff.org/files/sharepoint/WorkImages/Download/RFF-DP-14-36.pdf>

Nakhooda, S., Watson, C., & Schalatek, L. *The global climate finance architecture.* Overseas Development Institute, (2016). <https://www.odi.org/sites/odi.org.uk/files/resource documents/11021.pdf>

Putti, V., Tsan, M., Mehta, S., & Kammila, S. (2015). *The state of the bloal clean and improved cooking sector. ESMAP Technical Paper No. 007/15.* Washington DC: World Bank.

PwC. (2016). *Blockchain—an opportunity for energy producers and consumers?* London: PwC.

Skinner, C. *The five major use cases for financial blockchains.* FinYear, (2016). <http://www.finyear.com/The-five-major-use-cases-for-financial-blockchains_a35655.html.>

Stern, N. H. (2007). *The economics of climate change: the Stern Review.* Cambridge: Cambridge University Press.

Stoilijkovic, N. *The Paris Agreement is a $23 trillion investment opportunity. How can we unlock it?* World Economic Forum, (2017). <https://www.weforum.org/agenda/2017/01/unlocking-23-trillion-of-climate-investment-opportunities-is-mission-possible/>

UNFCCC. (2016). *Summary and recommendations by the Standing Committee on Finance on the 2016 biennial assessment and overview of climate finance flows.* Bonn: UNFCCC.

UNFCCC. *INDCs as communicated by Parties.* UNFCCC, (2017). <http://www4.unfccc.int/submissions/indc/Submission%20Pages/submissions.aspx>

United Nations. (2015). *Trends in private sector climate finance.* New York: United Nations.

World Bank Group. (2017). *Geographic hotspots for World Bank action on climate change and health: Investing in climate change and health series.* Washington DC: International Bank for Reconstruction and Development/World Bank.

World Resources Institute. *Climate finance.* World Resources Institute, (2017). <http://www.wri.org/our-work/project/climate-finance/climate-finance-and-private-sector.>

INTERLUDE II

SECTION 2

Blockchain for Smarter Renewable Energy Deployment

> To be sure, reaching the goal of 100 percent renewable and truly clean electricity within 10 years will require us to overcome many obstacles… Today, our grids are antiquated, fragile, and vulnerable to cascading failure. Power outages and defects in the current grid system cost US businesses more than $120 billion dollars a year. It has to be upgraded anyway.
>
> ***Al Gore, speech at Constitution Hall in Washington, DC, July 17, 2008.***

The dawn of this much-needed upgrade may have happened in Brooklyn, New York, in April 2016, finally. A resident who owns a roof-top solar panel sold the world's first few kilowatt-hours of locally generated surplus solar energy to a neighbor through an Ethereum Blockchain smart contract, rather than fed it in the national grid to sell to a utility company. This is the first-ever, Blockchain-supported peer-to-peer energy transactions system. Behind this trailblazing milestone is a start-up company—LO3 Energy—which developed what it calls the "Brooklyn Microgrid." But what does this mean to the energy sector? It is exactly what this section is to examine.

Today's electricity grid was not designed to support millions of producers and consumers with bidirectional power flows or manage oft-inconsistent demand in millions of localities. It is a centralized model that transmits power only from power plants to end users. Blockchain is set to decentralize the energy sector and create a more distributed model in favor of "prosumers" (a portmanteau of producer and consumer). It will revolutionize how energy producers and consumers interact with each other in the market.

Thanks to the defining feature of Blockchain that eliminates the need of traditional intermediaries in transactions, on the supply side, it will overcome the market entry barriers for smaller generators. Blockchain technology can efficiently codify energy transactions between two parties on a platform which is more open, transparent and hence, verifiable and trustworthy than ever. Smart contracts are pre-programmable that can automatically trigger transactions and reduce transaction costs partly because Blockchain eliminates the need of archaic and costly meter readings.

In Chapter 5, How Blockchain Can Democratize Global Energy Supply, Adam Woodhall will discuss how Blockchain can "democratize" global energy supply considering the experience of three case studies, including LO3's Brooklyn Microgrid, The Sun Exchange, and SolarCoin. In Chapter 6, How Blockchains Will Industrialize a Renewable Grid, Paul R. Brody will share his insights on the various ways in which Blockchain makes an impeccable enabler of a distributed energy grid. In Chapter 7, Efficient Power Markets: Reimagining the Global Market With Ethereum, Adam Richard will introduce Blockchain-enabled energy management systems developed by Volt Markets to support the transition of the existing energy market, which will become decentralized, highly efficient, and transparent. He will explore new paradigms for the renewable energy certificates market.

On the demand side, Blockchain technology can provide consumers with a more streamlined and accurate billing or accounts experience. For prosumers, smart contracts allow them to export excess energy into the grid through Blockchain-enabled meters, while energy buyers and sellers are speed-matched by algorithms locally in real time. This set-up help balance the geographical mismatch of supply and demand. In Chapter 8, Flexibility Trading Platform—Using Blockchain to Create the Most Efficient Demand-Side Response Trading Market, Joanna Hubbard and Paul Ellis will introduce Electron's Blockchain trading platform to facilitate a more liquid, transparent, and efficient market for demand-side response actions.

Integrating Blockchain in the energy sector means more accurate energy generation and consumption data will be available. It is because energy sources can be assessed more effectively and all stakeholders benefit from fairer prices than they are under a feed-in tariff system. A plausible use case in this respect is NRGcoin that transforms the way in which both local production and consumption of renewable energy is rewarded. Dr. Mihail Mihaylov, Dr. Iván Razo-Zapata, and Prof. Ann Nowé will introduce this concept in details in Chapter 9, NRGcoin—A Blockchain-Based Reward Mechanism for Both Production and Consumption of Renewable Energy.

CHAPTER 5

How Blockchain can Democratize Global Energy Supply

Adam Woodhall

Advisor, Activator, Author: Inspiring Sustainability[1], London, United Kingdom

5.1 Introduction

> **Democracy:**
>
> *"Favouring or characterized by social equality; egalitarian"*
>
> *The Oxford English Dictionary*

This chapter investigates how Blockchain-enabled enterprises can democratize energy supply by making it:

- More accessible in all markets globally;
- Distributing benefits more equally; and
- Decentralizing decision-making.

It will also consider the lesson learned from the historical arc of the Internet; that a technology which appears to promise decentralization can paradoxically lead to more centralization, and by extension, less democratic outcomes.

The aim is for the chapter to be accessible to a nontechnical audience, but it does assume the reader either has an existing basic understanding of the Blockchain (e.g., that part of its benefit is that data are held across a distributed network, rather than one server), or has read the introduction to this book.

[1] www.inspiring-sustainability.com

Transforming Climate Finance and Green Investment with Blockchains.
DOI: https://doi.org/10.1016/B978-0-12-814447-3.00005-7

5.1.1 Three T's

Blockchain technology has three fundamental features which enable it to be a transformative tool, which are summarized as the three T's:

- Transparency between users;
- Transaction costs to be negligible; and
- Tracking of end to end transactions accurate, easy, and permeant.

The first section of this chapter lays out the opportunities and challenges faced by Blockchain technology as it relates to energy supply, and explores the four assertions, and verdict, detailed below. The second section highlights case studies that demonstrate how innovative companies are utilizing these key features of Blockchain to democratize energy supply.

5.1.2 Assertion One: Centralized Power Equals Magnified Inequalities

Based on this assertion, decentralization therefore equals more equality, which increases democracy (as defined by the dictionary definition above). Energy supply is currently overwhelmingly centralized and so struggles to be democratic.

5.1.3 Assertion Two: The Paradox of the Internet

By decentralizing participation, the Internet (via the "giants" such as Apple, Google, etc.) centralizes authority. The "giants" do this by; making the Internet user-friendly, controlling data, owning transactions, and vertical integration of a markets supply chain, which eliminates middlemen.

5.1.4 Assertion Three: Ensure both Decentralized Participation and Authority

If we are to fulfil the democratic promise of Blockchain technology, we must ensure that we create decentralized participation and decentralized authority. To enable this, the key principles of an effective use of Blockchain technology are: generating trust (or reducing the burden of trust), reducing costs (low barriers), and giving the control of data back to the participants.

5.1.5 Assertion Four: Energy Supply is Ripe for Democratization by Blockchains

Numerous factors lead Blockchain technology to be ideally suited to democratizing global energy supply. Examples are: rapidly shifting energy systems; current grid architecture being incompatible with the dynamic needs of the future; opportunity for Global South due to plentiful renewable resources.

5.1.6 Verdict: Who Decides the Rules?

Based on these four assertions, the key verdict when considering the current situation is: who decides the rules will dictate if Blockchain is to successfully democratize global energy supply.

5.2 Section 1: Blockchain for Energy's Opportunities and Challenges

Utilizing Blockchain technology on renewable projects creates a considerable opportunity to democratize global energy supply. However, we need to remember these lessons from the growth of the Internet:

1. That it grew exponentially when made easily understandable and accessible to the mass population.
2. A technology that initially predicts increased decentralization and therefore democratization, can create increased centralization, and therefore become undemocratic overall.

Blockchain for energy is in a comparable situation to where the Internet was in the early-to-mid 1990s; for most people, an obscure and difficult to understand technology. Just like the Internet, before it, there is considerable optimism regarding the possibilities of Blockchain as both technologies had two promises inherent in them. First, in how they can transform our world, and second lead to a more democratic and equal society, due to their decentralized nature.

The Internet has amply fulfiled its first promise: our world is a very different place to where it was at the dawn of the Internet age over 20 years ago; few would have predicted applications such as Airbnb, Uber, and Facebook. Likewise, while it is impossible to know what imaginative minds will create with Blockchain-based applications over the next decade or two, we can be confident they are likely to change the world. The Internet went through a rapid adoption of innovation cycle, and this has consequently changed the world. Figure 5.1 illustrates this; with only "Innovators" accessing it in the early 1990s but even the most resistant "Laggards" online by the 2010s. Blockchain is currently on the far left of the diagram: with only a small percentage of even the innovators engaging with the technology.

5.2.1 Centralized Power

Making information free and easily accessible is one of the major changes the Internet has delivered. A few clicks of a button and Google offers access to more information than any library can hold. This is democratic, as historically only associating with an institution such as a University could give access to such information. However, Google being the go-to place for information, and *de-facto* monopoly, is undemocratic as it gives them huge centralized power and control.

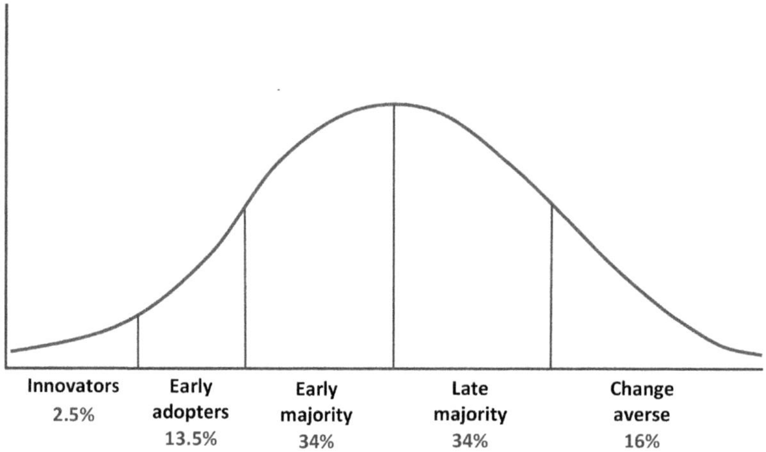

Figure 5.1
Adoption of innovation curve.

The founding Blockchain platform, Bitcoin, is beginning to experience the impacts of centralization. The huge amount of processing power required to profitably mine Bitcoins require significant capital investments and creates a barrier to entry. Consequently, mining power becomes centralized. If we are to realize Blockchains promise to democratize global energy supply, we must design platforms to avoid centralization by aligning incentives which promote decentralized authority and democratic principles. The case studies in Section 2 examine various approaches toward this goal.

5.2.2 Equals Magnified Inequalities

The Internet has failed though, in its second promise of a more democratic and equal society. The reality is the Internet has magnified inequalities by centralizing control over data. For example, Amazon and Airbnb have decentralized participation by creating a rich and wide marketplace, but centralized power by controlling and monetizing the data of its users. The issue is that one company is the middleman, concentrating power rather than distributing it.

The giants of the Internet also have two other aspects that make them so successful: they both track all the interactions and "own" the transactions conducted on them. This gives them a near monopoly in their fields, centralizing control, and decision-making and therefore making them intrinsically undemocratic. It is also possible for them to utilize democratic technology to further their aims; for example, Google's Android system is based on the unpatented, free, and open-source software: Linux.

There is considerable concern about centralization from many well-respected sources such as the MIT Technology Review (Technology and Inequality), CNN Money (Stephen Hawking), and the World Bank. A report in 2016 from the latter comments on the risks of the Internet age such as: "the growing concentration of the [digital] industry, increasing inequality as some types of jobs get automated and disappear, and the threat that the Internet will be used to control information instead of sharing it" (World Bank report).

5.2.3 The Paradox of the Internet

How the Internet has centralized power is a cautionary tale for Blockchain. This is because, while it could enable democratic relationships between users (e.g., energy consumers) and creators (e.g., prosumers: people who are producers and consumers of energy), it could recreate a different type of centralization. Theoretically, we have a level playing field as we enter the Blockchain era. This is because Blockchain technology is low cost and relatively easy to employ; even small organizations without much capital muscle, can access it. However, because certain players, such as large global corporations, have vast R&D budgets, they could work out ways of hording many of the benefits for themselves.

The organizations that have gained the greatest traction by using the Internet, such as Airbnb, Uber, Amazon, Google, and Facebook, are also the ones that have been best at making it easy to use and accessible to everybody. It is unlikely the general populace will ever understand the detail of how the Internet, or Blockchain, works. For example, how many people know what "http" stands for, much less what it does? But this lack of knowledge doesn't mean they don't use it. Part of the reason for the phenomenal success of the Internet giants is they've worked out how to make the internet easy, familiar, and trusted, which makes it far less mysterious to the average person.

5.2.4 Ensure both Decentralized Participation and Authority

When trading together, people typically want two things: to trust each other and the cheapest and easiest platform to do business on. These desires are the foundation of what makes the Blockchain such an exciting technology as a transaction platform: building trust by creating a reliable record of undeletable transaction, and massively reducing costs of micro trading. This could lead to an exponential growth of the Blockchain, just like the current rise of solar PV (as shown in Figure 5.2) and the historic growth of radio, TV, and Internet.

In the rapidly changing globalized world we live in, people are struggling to know who to trust, with trust in the political system, media, and many other institutions falling. The giants of the Internet know this too, and they have made massive investments in working out how to build trust so that, for example, we now allow strangers to sleep in our spare

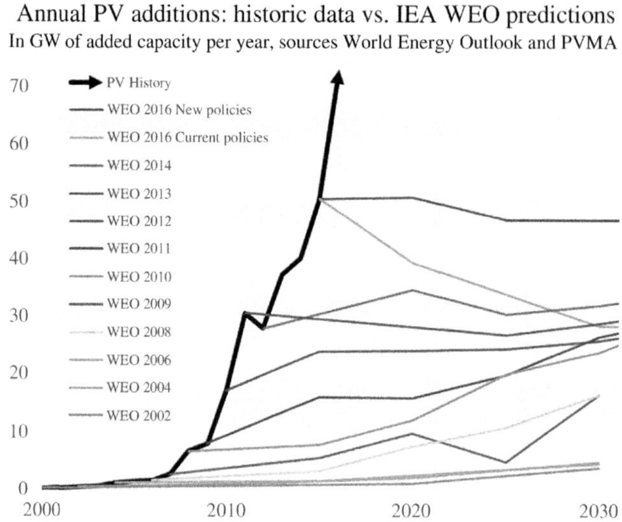

Figure 5.2
Printed with permission of Auke Hoekstra IEA WEO referred to above is: "International Energy Authority World Energy Outlook."

rooms and drive us home. Furthermore, due to their scale, the giants have reduced costs; e.g., the mammoth size of the Amazon operation means many small businesses can successfully trade globally on it.

Blockchain technology is in its essence the opposite of this; the ability to both own your data and track it transparently means that theoretically anybody, anywhere, can enter the energy services market. However, I'd argue when the Internet first started to develop, it was in a similar place; but, then clever individuals and corporations worked out how to own the data and relationships, and have grown exponentially. This isn't intrinsically wrong; however, it does have consequences, and the growing income inequality (The World Economic Forum's "Global Risks Report 2017") we see on average across the globe is one of those.

Making information less available, and therefore increasing equality, is one of the counter-intuitive opportunities of the Blockchain. This is because it can enable everybody to retain, then sell, their own "data," and therefore create value from it. An example offered by Molly Webb, founder of Energy Unlocked, will be how the Internet, via iTunes, Spotify, and similar platforms, disrupted the music industry by making digital music accessible *en masse*. This provided massive revenues to these music platforms, but tiny returns to most artists.

To remedy this inequality, Blockchain technology could create direct peer-to-peer financial relationships between artists and listeners: this would turn the music into "data," traded on the Blockchain between the producers and consumers. A clear analogy is in the energy market with the enabling of individual prosumers who own their own rooftop solar PV sell their "data," in this case, solar power, direct to other individuals.

5.2.5 Energy Supply is Ripe for Democratization by Blockchains

Energy systems worldwide are undergoing a rapid shift from fossil fuel to renewables, outstripping predictions. Figure 5.2 demonstrates this, showing the exponential growth of solar PV installations; both in the existing grids in the Global North (also known as developed countries) and the new power additions which are growing in the Global South (developing countries). Energy security and the need to respond to climate change are drivers for this, but the key enablers are plummeting costs for solar, and comparatively easy access to the relevant technologies.

The prevailing historic mindset has been energy supply needs the centralized solution of "big generation," i.e., a nationwide or regional utility "big" grid fed by "big" power, such as nuclear or coal power stations. Many localized, decentralized clean energy systems are appearing around the world to challenge this orthodoxy. The biggest opportunity is in the Global South where the "big generation" systems haven't become fully entrenched, and there are plentiful renewable resources, especially solar.

We are seeing clear signs of the wish to democratize energy supply. As Paul Ellis, CEO of the Blockchain start-up featured elsewhere in this book, Electron, comments: "We also see a shift in attitude towards energy: increasing numbers of companies and households are able and want to participate in the new energy market [e.g., via rooftop PV] and are demanding their share of its value" ("Interview with Paul Ellis CEO Electron"). Blockchain will work best to deliver on this desire by connecting renewables and decentralized energy generation with other fast-growing solutions such as smart devices, smart grids, and energy storage.

5.2.6 Who Decides the Rules?

It is critical therefore that we direct the democratic essence of Blockchain technology toward giving more equal benefit across the value chain; from individuals in rural Africa to large corporations in the USA, and everything in between. Those who decide the rules, regulations and gateways that guide its use will be the ones with power and control over the outcomes. Will that be policy makers and regulators, the multitude of start-up enterprises who use it, or a small number of global corporations? The race is on to see whose interests will prevail.

It is certain that there will a need of cooperation between regulators, start-ups, and corporations if we are to scale the opportunity Blockchain offers to democratize global energy supply. An example of this is the setting up of the nonprofit Energy Web

Foundation (EWF) in May 2017 to accelerate Blockchain technology across the energy sector. This initiative has the backing of some of the largest energy companies in the world. The EWF's opportunity, and challenge, is to balance the demands of the incumbents with those of the start-ups and the desire for democratized energy.

Co-Founders of Electron, Paul Ellis, and Jo-Jo Hubbard, encapsulated the opportunity of Blockchain well when they stated on the World Economic Forum website:

"The transition to a democratized energy system will require;

a. transitioning to smart grids capable of local optimization,
b. better enabling and incentivizing desired consumption behaviour, and
c. re-distributing the benefits of those actions in a fairer, more transparent way" (Paul Ellis & Jo-Jo Hubbard "The Democratisation of Energy").

Dozens of initiatives are being launched to exploit the opportunities of Blockchain for energy. The following case studies about The Sun Exchange[2], Brooklyn Microgrid[3], and SolarCoin[4] are proofs of concepts that will summarize Ellis and Hubbard's three requirements, and demonstrate how it is possible to democratize energy supply using the Blockchain.

5.3 Section 2: Case Studies

5.3.1 Brooklyn Microgrid

In today's competitive marketplace, brands are increasingly competing for consumer's cash by appealing to their ethics; for example, promoting their environmental sustainability or connection to the local community. Until recently, when it came to purchasing energy, it was possible to buy environmentally (e.g., green tariffs from large utility companies), but not locally. This is because those people who are purchasing a "green tariff" are paying their energy company to invest in renewables. However, the actual electrons fed into their property will come from the nearest power source, which is often fossil fuel based.

5.3.2 Incentivizing Prosumers

A microgrid is an energy system consisting of energy generation, storage, and users operating in parallel with, or independently from, the main power grid. In New York, USA, the Brooklyn Microgrid[2] (BMG) has used the Blockchain to create a virtual microgrid to enable residents to trade in a marketplace and buy solar power produced on their neighbor's

---

[2] https://thesunexchange.com/
[3] https://www.brooklyn.energy/
[4] https://solarcoin.org/en/node/6

rooftop. Plans for a physical microgrid are also progressing, enabling delivery of electricity directly to houses or businesses and therefore decentralizing power distribution. In addition to localizing energy distribution, it also gives incentives for residents who are consumers to become prosumers and invest in solar and battery storage. This is because the BMG offers a market to sell excess power in a way that is more compelling than selling back to the regional utility company.

Although there are many benefits to setting up a physical microgrid, security of supply and energy resilience were the direct inspiration for the launch of the BMG. In 2012 Superstorm Sandy led to complete power loss for thousands of homes across the USA and Canada, including many in Brooklyn. The founders of BMG, LO3 Energy, recognized that creating a microgrid was an effective response to the loss of supply they had experienced, and have collaborated with the global conglomerate, Siemens on this project. This is because a microgrid is a miniature utility grid that continues to deliver power to users, even when power outages compromise the main power grid due to issues such as a terrorist incident or extreme weather. Furthermore, the main power grid suffers transmission loss from sending the energy over long distances, with the global average loss currently around 8% (The World Bank). A significant added advantage of the microgrid is that it considerably diminishes this loss.

Lawrence Orsini, founder of LO3 Energy, recognizes that BMG is a proof of concept: "In the world of finance, Blockchain technology is rapidly advancing across many sectors, but in the energy market, things are comparatively different. With our microgrid solution in Brooklyn, we'll demonstrate just the beginning of what Blockchain can do in the transactive energy world."

5.3.3 Reflecting Desires of Communities

BMG is creating a new market for local energy that reflects the desires of individual communities to localize decision making and therefore democratize energy supply. This though is an evolution, not a revolution, as they aren't intending to disconnect from the grid or work outside existing regulatory structures. The US regulations currently limit prosumers to "net metering," which means counting electricity sent to the grid from rooftop PV as a credit against the prosumers' own consumption. However, there are proposals that could enable US prosumers to sell production within a peer-to-peer energy market or store the energy in a battery located on or off-site.

At the centre of this process is a TransActive Grid element (TAG-e). This device measures both: energy production and consumption, and shares and acts upon this information with other TAG-e in the network over Wi-Fi. The data shared over this network then feed into a virtual market. The LO3 Blockchain engineers have set this market up to be transparent, auditable, nonrepudiable, peer-to-peer, and a cryptographically secure smart contract platform. This enables real-time purchase and sale of kilowatt hours and trading on an open, decentralized energy exchange.

Part of the reason why Blockchain is key to enabling peer-to-peer trading over a microgrid is because every community has different requirements and brings a different mix of energy assets to the equation. San Francisco, or even Manhattan, for example, will offer a very distinct challenge compared to Brooklyn. The existing grid gets around this challenge by using utility scale big generation; with excess capacity for times when needed. Blockchain can help overcome the challenges of a decentralized microgrid's variety of assets, by storing and tracking all the data cheaply, securely, and transparently.

5.3.4 Building Trust, Making Trading Easy and Engaging

By utilizing the Blockchain to enable peer-to-peer transactions, BMG is using many of the tools that the Internet giants have used to build trust and make trading energy easy. This is both in the practicalities of their offering and the way they promote themselves. Practically speaking, they make it accessible via a smartphone app and by offering a low cost to entry for investing in a community solar project at $25.

When promoting themselves across their website, social media, and videos, they talk in the language of their target market; about becoming an investor in energy assets, revitalizing neighborhood, and pollution-free power. BMG have numerous engaging videos, focusing on personal stories of the microgrid users. However, the videos rarely mention Blockchain technology, even though it is key to making BMG work.

The app has a user-friendly interface, which makes it possible for consumers to set the preferred daily maximum cost, and they can either "set it and forget it" or change every day depending their financial needs. The prosumers can use it to see what times they pull electricity from the grid or export green electricity back onto the grid. Furthermore, the app puts consumers and prosumers in league tables and encourages reporting of energy-efficiency efforts, therefore using modern "gamification" techniques to make functionality fun and engaging.

With an eye on the growth of the microgrid, the app enables residents to cast votes on suitable sites for new PV installations. After reaching a threshold, BMG reaches out to the site owner, then pulls together local investment, and finally works to build the new solar generation site.

The Brooklyn Microgrid[2] model will truly utilize the Blockchain to democratize energy supply when consumers have an experience of buying energy from BMG that is at least as cheap and user friendly (e.g., straightforward sign up, paying by direct debit) as the traditional big utilities provide.

The initial steps detailed above are very encouraging, and has led to investment in LO3 by global players, because as Lawrence Orsini says: "We don't get rid of fossil fuels in one fell swoop, it takes 100 incremental steps, and one step is to harness the buying power of ethically conscious consumers to tell people who are thinking about solar power 'Hey if you put it in, I'll help you'."

5.4 The Sun Exchange

> *"South Africa has the greatest potential in Sub-Saharan Africa to make energy universally accessible. Realising this potential will only be possible however, if the public sector creates the enabling environment, entrepreneurs enter the market with innovative models, and communities are prepared to adapt to alternative solutions."*
>
> Tanner Methvin and Felix Phillip, "Energy at the Base of the Pyramid Report" (Tanner Methvin and Felix Phillip of Impact Amplifier)

South Africa is one of the most consistently unequal countries in the world (Income inequality in South Africa), but a social enterprise based there is using Blockchain technology to drive solar adoption in one of the sunniest places on earth.

Most of South Africa receives double the sunshine hours compared to the UK and Germany. However, there are comparatively few solar installations, in part because solar panels have a high up-front cost. Many rural areas rely on expensive and poor-quality fuels such as kerosene and candles. This is typical across sub-Saharan Africa where about two-thirds of the population, over 600 million people, lives without electricity.

South Africa is the world's seventh largest producer of coal and coal generates three quarters of the country's electricity. Consequently, there is a comparative lack of political motivation to promote small-scale off-grid solar projects (Energy alleviating poverty). In addition to problems caused by greenhouse gas emissions of coal, the political inertia of traditional models increases social disparity.

Recognizing an opportunity, Abraham Cambridge moved from the UK to South Africa to found The Sun Exchange[1]. The Sun Exchange[1] is a marketplace with Blockchain at its heart which connects solar energy projects with investors around the world. The platform enables people anywhere to crowd-fund solar cells and lease them back to organizations. Recent projects funded through this model include an elephant park, tyre recycling factory, school in Cape Town, and village in Lesotho.

5.4.1 Solar Nano-leasing Made Viable

Into the gap created by the comparative lack of government interest stepped The Sun Exchange[1], using the Blockchain to facilitate the funding. They have automated the payments of their solar plants with smart contracts on the Ethereum Blockchain. Once the solar plant is in operation, the owners of the solar cells, wherever they are in the world, earn a healthy income of 10% per annum from the rental payments of the solar energy user (e.g., school or village) over a 20-year period.

In addition to using Blockchain based smart contracts, The Sun Exchange[1] also can receive and issue payments in Bitcoin. The radically reduced costs and processing time of Blockchain makes this process of solar nano-leasing viable compared to traditional payment methods, enabling The Sun Exchange[1] to take their fee's transparently from the projects and still provide above market rate returns.

They are also not the intermediary in the transactions, meaning the person who has bought the solar panel in Africa doesn't have to rely on any third parties. Once the initial crowd funding process completes, it doesn't matter if The Sun Exchange[1] itself stops functioning, the Blockchain ensures the payments continue. The combination of this, along with The Sun Exchange[1] performing due diligence on all the projects they choose for their platform, is particularly important because it mitigates concerns about corruption that many African countries experience. As an added bonus, buying solar panels through The Sun Exchange[1] entitles owners to receive an allocation of the digital token, SolarCoin[3], detailed in the next case study.

5.4.2 Benefiting from the Renewables Revolution

Delivering access to electricity has proven to be one of the best tools for reducing poverty (Wikipedia), as it supports study, commercial activity, and improves safety. Thus, this is an innately democratic use of the Blockchain. Furthermore, installing solar is neither physically nor economically viable for many people in the Global North. However, with The Sun Exchange[1] everybody is now able to directly participate in, and benefit economically from the global renewables revolution.

The Sun Exchange[1] drives solar adoption by employing a decentralized finance system to support decentralized energy systems. As Abraham Cambridge comments: "Crowdfunding is enabling the democratisation of infrastructure development. We're not relying on political will-power to go solar; it comes down to the inclination of people. It offers the added benefit of knowing exactly where the money is going."

Also as Cambridge states: "Funders start seeing returns the moment the projects are turned on, because due to the lack of transaction fees it means we can send payments out per kilowatt hour of electricity consumed."

In the Global North, hard currency and big generation founded the energy systems. The Sun Exchange[1] demonstrates that in the Global South, particularly in Africa, whole regions and even countries can go decentralized, using mobile payment systems with digital cryptocurrency and powering themselves with Blockchain enabled energy systems. This gives them the opportunity to leapfrog the technological progression the Global North has been through.

5.5 SolarCoin

Air miles have incentivized frequent flyers to travel more since the early 1980s, and now a Blockchain-based initiative is stimulating solar production to create a virtuous loop of use and investment. Air miles are a type of limited virtual currency, as they give points to exchange for goods, with no direct exchange back into money. These frequent flyer programs have led to around 20 trillion points accumulated since then, equivalent in value to approximately USD$1 trillion.

5.5.1 Rewarding Solar Generation

The SolarCoin[3] Foundation (SCF) has set up a similar free reward programe, but as a cryptocurrency, to incentivize global solar electricity generation. For every 1MWh of energy verified as produced using solar panels, individuals or organizations can claim one SolarCoin[3]. This currency circulation mechanism is similar to how Bitcoin rewards miners to commit computer processing power to the network by rewarding them with Bitcoins. The difference is each SolarCoin[3] rewards the solar energy production equivalent to that required to power a home for one month.

Launched in 2014, the aim is to support the generation of 97,500 TWh of solar power over 40 years. Anybody can claim SolarCoins; from prosumers with panels on the roof to large solar farms, with 240,000 of them going to solar power producers in 39 countries since launch. In contrast to air miles, SolarCoin's are a full cryptocurrency as they can buy services and products and exchange directly into traditional currency.

Unlike Bitcoin, SolarCoin[3] is also more environmentally friendly. Bitcoin, the original cryptocurrency, uses considerable energy to maintain its network system, with some estimates suggesting it requires the same amount of power as Croatia, 15 TWh per year (Bitcoin energy usage). As of the writing of this document, estimates suggest it costs $1.6 billion per year maintain the Bitcoin network of roughly 10 million users. To service a similar network, SolarCoin[3] requires a much smaller resource, estimated at $100,000–200,000 annually. The SolarCoin[3] Foundation acts an auditor, certification, and verification agency for the issuance of circulating SolarCoin[3]. It intends to fully distribute these roles to their community in the future.

The Blockchain is essential for SolarCoin[3] to work, and it is a good example of democratization. Historically, only large organizations, such as national or regional governments, or big airlines, had the scale to set up national or transnational reward mechanisms such as Air Miles, feed-in tariffs, and Renewable Energy Certificates. A small group of volunteers from the SCF, who live across three continents, has created a reward mechanism with the theoretical potential to incentivize as much solar installation as a large country achieves with feed in-tariffs. While it is important for governments to continue supporting renewables, initiatives such as SolarCoin[3] mean that in addition to the plummeting price of solar PV there are other market driven activities to motivate adoption and use.

5.5.2 Aiming to Support 30% of Global Power Production

By 2018, the SolarCoin[3] Foundation expects to have 2,500 participating facilities and issued two million SolarCoins (which uses the code "SLR" on the crypto-exchanges). This is just the proof of concept stage; for example, they have a goal, with partners, to build the world's largest solar monitoring platform, and by 2020 expect to be reaching up to a million solar-producing households. SolarCoin[3] aims to be incentivizing the predicted three terawatts of solar PV installed by 2050; with this delivering 30% of global power production.

Currency works like a language: the more people who speak the language, the more valuable it becomes, and the more people believe in it, creating a virtuous loop of use and belief. An example would be the protocol, or language, that connects computers across the Internet; TCP/IP. This language has massive value because it enables all the computers to speak to each other. A contributing factor to the US dollar being the most valuable currency for a long time was because so many people across the globe believe in its value and are willing to trade in it.

Like its big cousin Bitcoin has already done, SolarCoin[3] will come into its own as a currency once it reaches a large enough scale. As the number of SolarCoins increases as people claim them, and then people use it to trade, they then increase in value due to increased spendability, further incentivizing the generation of solar energy, creating a virtuous cycle.

Over the year previous to writing this case study, the price for the SolarCoin[3] had oscillated between $0.03 and $0.30, with it being in the mid-point between those two at time of writing. The eventual aim is for it to grow to $30 for a SolarCoin[3]. As the average cost of solar currently ranges from $30–200 per MWh, an extra $30 per MWh can be a significant incentive to either increase the number of panels or go ahead with a project. With an intention to freely distribute 98 billion coins over the next 30 years, this would lead to

SolarCoin[3] being worth $2.94 trillion US dollars. This will happen when it generates a large and robust ecosystem and there are ways of becoming a true alternative currency for both buyers and sellers of goods.

While the financial value of the SolarCoin[3] grows, offering an option for investors, its main value is in leveraging the social added value of incentivizing solar energy production. There are several established voices endorsing this value, such as The International Renewable Energy Agency and Solar Power Europe, and the SCF has ongoing projects with the United Nations Development Programme and United Nations Framework Convention on Climate Change.

5.5.3 Spinning the Solar Wheel

Speaking to members of the European Parliament, François Sonnet, cofounder of the SolarCoin[3] Foundation stated: "The technology is there; the will and the money isn't yet. We're at the very beginning of the value creation of SolarCoin[3]. It takes time, obviously, but it's like a spinning wheel: once the cogs are in place, it starts spinning by itself."

Frequent flyer rewards from a single airline are useful, but they really come into their own when they can accrue points from other carriers, through airline alliances and partnerships with other retail and service outlets, such as hotels, car-hire, and supermarkets. Similarly, while SolarCoin[3] is a great stand-alone solution, and there are users who are choosing to claim them directly, it will only truly gain traction when integrated into and interacting with many other platforms. Think about how many outlets globally accept the US dollar, and how it is now possible to buy a coffee with a Bitcoin. Detailed below are enterprises that are at the beginning of the innovation cycle in supporting SolarCoin[3] on its Blockchain journey of democratizing energy.

5.5.4 Lykke Global Marketplace

The first step of any currency is to be transferrable into other widely accepted currencies. The Lykke Exchange is first platform enabling trading of SolarCoin[3]. The goal of the Swiss company behind this is to build a global marketplace for the free exchange of financial assets. The company uses Blockchain technology which eliminates market inefficiencies and promotes free access from anywhere in the world, including on Android, iOS, and laptops.

5.5.5 Smappee Energy Monitor

The main selling point of the Smappee energy monitor, or datalogger in technical parlance, is that it identifies electrical appliances, reporting exactly how much they consume; it also registers how much solar energy those users with PV panels are receiving. The latter attribute means it can integrate SolarCoin[3] reporting into its application programming interface (API) (i.e., its set of commands, functions, and protocols), making Smappee the first in the world to do this.

There are two significant advantages to this. The first is that it makes it considerably easier for prosumers to claim the coin: people buy a Smappee device because of its other functionality, and the SolarCoin[3] is just an added benefit which they don't need to even think about; they just receive the rewards. Prior to this, they had to go through a process obtaining a specialist wallet or datalogger, now the Smappee takes care of it all. The second is that previously SolarCoin[3] users registered their solar energy production only every six months. The integration with Smappee allows the SolarCoin[3] Foundation to monitor real-time solar energy production and gather data to track climate change and air pollution levels.

5.5.6 EkWateur Exchanging SolarCoin for Energy

EkWateur is a French 100% green energy supplier launched in 2017, and it is the world's first retailer to accept SolarCoins in exchange for energy. This is key for supporting the SolarCoin's growth as a currency, as it will only become truly valuable when used to directly buy goods. Keeping the SolarCoin[3] loop going, this ground-breaking supplier had early funding via renewable energy crowdfunding platform Lumo, who also offer SolarCoins to their investors.

By making it easier to use, access, and gain value from SolarCoin[3], all these platforms help make it easier to access SolarCoin[3], and therefore help democratize Blockchain for energy.

5.6 Conclusion

Ellis and Hubbard of Electron observed in the introduction that the transition to a democratized energy system requires smart grids capable of local optimization, better incentivizing of desired behavior and redistributing the benefits in a fairer and more transparent way. The preceding case studies illustrate how it is possible for Blockchain to democratize energy supply. They demonstrate that Blockchain technology enables building of trust by creating and tracking a reliable record of undeletable transactions and massively reduced costs of micro transactions.

The Blockchain has enabled entrepreneurs to enter the market to create proofs of concept, enabling anybody, anywhere to participate in and make money from the surge in solar generation globally. Therefore, rather than concentrating the economic benefits for a small number of centralized large organizations, the Blockchain enables wide redistribution of these benefits.

Both the Brooklyn Microgrid[2] and The Sun Exchange[1] have used it to create direct relationships between producers and purchasers, as neither enterprise is the intermediary in this relationship; the Blockchain holds all information, not on the businesses servers. These two innovative enterprises show how the Blockchain can store and share secure information about a wide variety of assets in an exceptionally inexpensive manner, and the Brooklyn Microgrid[2] is beginning to illustrate how smart grids can be locally optimized. SolarCoin's are an example of how Blockchain technology can incentivize behavior, created by a small group of volunteers using a process historically only open to governments and large corporations.

All three case studies exploited the Blockchain's three T's of transparency between users, negligible transaction costs, and easy end-to-end tracking. They also show how to build virtuous loops of use and investment in the energy-supply market using Blockchain technology. Finally, all three made access to decentralized energy supply easier.

The lessons of the Internet's tendency to centralize power and control is a cautionary tale when considering the future of Blockchain for energy. The three enterprises highlighted, along with many other similar initiatives, have considerable potential to democratize energy supply, but we could lose this globally; especially, if the rules governing Blockchain for energy become antidemocratic.

It is crucial that key stakeholders support the opportunity presented by Blockchain technology to enable more open and equal access to the energy supply market. This support will create many more true stories of how Blockchain is democratizing global energy supply.

References

Bitcoin energy usage <http://digiconomist.net/bitcoin-energy-consumption>. (accessed 2017).

Energy alleviating poverty: *International institute for applied systems analysis – "Global Energy Assessment", 2012* <http://www.iiasa.ac.at/web/home/research/Flagship-Projects/Global-Energy-Assessment/GEA-Summary-web.pdf>. (accessed 2017).

Income inequality in South Africa, 2015 <http://theconversation.com/factcheck-is-south-africa-the-most-unequal-society-in-the-world-48334>. (accessed 2017).

Interview with Paul Ellis CEO Electron, 2017 <http://asia.blog.terrapinn.com/emerging-tech/2017/07/19/interview-with-paul-ellis-ceo-electron/>. (accessed 2017).

Paul Ellis & Jo-Jo Hubbard "*The Democratisation-of Energy*", 2017 <https://www.weforum.org/agenda/2017/06/the-democratisation-of-energy-four-key-technologies-for-follow-it-through/>. (accessed 2017).

Stephen Hawking: *Technology is making inequality worse, 2015* <http://money.cnn.com/2015/10/12/news/economy/stephen-hawking-technology-inequality/index.html>. (accessed 2017).

Tanner Methvin and Felix Phillip of Impact Amplifier: *Energy at the base of the pyramid report, 2017* <http://www.impactamplifier.co.za/our-insights/reports/energy-at-the-base-of-the-pyramid-report/>. (accessed 2017).

David Rotman, *Technology and Inequality*, 2014 <https://www.technologyreview.com/s/531726/technology-and-inequality/>. (accessed 2017).

The World Bank, *Electric power transmission and distribution losses (% of output), 2014* <https://data.worldbank.org/indicator/EG.ELC.LOSS.ZS>. (accessed 2017).

The World Economic Forum's 'Global Risks Report 2017': *Rates rising income and wealth disparity as the number one trend that will determine the shape of the world in the next decade, 2017* <https://www.weforum.org/reports/the-global-risks-report-2017>. (accessed 2017).

Wikipedia: *Solar power in South Africa - Adoption and government conflict of interest* <https://en.wikipedia.org/wiki/Solar_power_in_South_Africa>. (accessed 2017).

World Bank report *Digital dividends, 2016* <http://www.worldbank.org/en/news/feature/2016/01/13/digital-revolution-needs-offline-help-to-realize-its-potential>. (accessed 2017).

CHAPTER 6

How Blockchains Will Industrialize a Renewable Grid

Paul R. Brody

Principal & Global Innovation Leader, Blockchain Technology, Ernst & Young[1]*, San Francisco, CA, United States*

In 1830, the world's first intercity rail line opened: the Liverpool and Manchester Railway. Using a steam-powered engine on metal rails, the rail line ushered in a new era of high-volume, high-speed intercity transportation. Investors took note and by 1835, there were 54 intercity railway projects underway, and by 1870, Britain had just over 13,000 miles of track with more than 300 million annual passenger journeys. That worked out to 14 rail journeys per person.[2] Water may have powered the earliest stages of the industrial revolution, but it was the coal-powered steam engine that catapulted the British economy ahead of the rest of the world. **In the coming years, the world will undergo another enormous energy transformation; this one from coal and other fossil fuels to renewables**. Blockchain technology will play a key role in scaling up and funding this massive transformation. And although this may seem like a big leap, in fact, major energy and industrial revolutions have always depended on information technology, however crude, to scale up. Renewables may be new, but the pattern for growth and investment is not.

Applying coal power and steam engines to transportation set off a powerful virtuous cycle. Transportation cost more than mining and the efficiencies brought by rail lines cut the cost of coal delivered to users, both household and industrial customers, in half between 1820 and 1840. The price drop set off, in turn, a boom in both manufacturing and transportation (Clark and Jacks, 2007). Coal output, which had been growing steadily since the invention of the steam engine was running at around 30 million tons a year in 1830, at the start of the railway era. By 1900, the virtuous cycle of lower transport costs and higher demand had taken that up to 200 million tons a year, almost all of which was moved by rail ("History of Coal," 2017).

[1] http://www.ey.com/gl/en/industries/technology/ey-blockchain-reaction-tech-plans-for-critical-mass.

[2] https://en.wikipedia.org/wiki/File:GBR_rail_passengers_by_year_1830-2015.png.

Transforming Climate Finance and Green Investment with Blockchains.
DOI: https://doi.org/10.1016/B978-0-12-814447-3.00006-9

Up until this time in history, no society or economy had ever seen this kind of explosive growth in economic or productive output. Even when innovations were groundbreaking, the ability to scale up enterprises and accumulate capital for investment was limited. Early railroads were small affairs, moving freight and passengers on limited point-to-point trips. The speed and scale of this transformation was something new. And it was driven, in no small part, by the simultaneous industrialization of information technology: starting with the **telegraph**.

As passenger and freight volumes soared, railroads struggled to manage track access across long distances. Trains moved too quickly, over distances far beyond the line of sight from origin to destination, and at speeds that made quick stops impossible. Controlling access to sections of track became a critical gating factor in sustaining the growth of the industry. With the telegraph, however, information could move even faster than the trains.

In 1838, the world's first commercial telegraph was installed by the Great Western Railway in London to manage track signals on the line between Paddington Station and the town of West Drayton, 13 miles to the west ("Electrical Telegraph," 2017). Lengths of rail line were sectioned off into blocks, with access limited to one train at a time. Telegraph messages were used to announce the movement of trains in and out of blocks and control signals along the line.

The result was an astonishingly successful innovation that made train travel in the UK efficient and safe. Between 1830 and 1850, the British rail industry grew from 1 million annual passenger trips to 57 million passenger trips, with only one fatality related to trains colliding where better signaling could have kept them apart ("List of Rail Accidents," 2017). Increasing track congestion diminished that enviable record, particularly after 1850, but the consequence of that was to combine further improvements in signaling methodology with expansion of the telegraph network and the first stirrings of government regulations on safety and operating standards. By 1900, there were more than 15,000 telegraph stations in the UK, controlling track access and signals serving over 1.2 billion passenger journeys annually ("Electrical Telegraph," 2017).

Tracking shipments, bills, and payments became an equally large challenge. In the 1890s, the New York Central Railroad was handling around four million freight waybills a year, all by hand (Fierheller, 2014). To get a handle on this enormous task, in 1895 the Railroad became the second commercial customer for an entirely new information technology: the **Hollerith Punch Card Tabulating machine**. This same machine was employed by the US Census in 1890, and the company that made it would, in time, come to be known as IBM.

Punch cards evolved into **digital computing systems**. One of the first general purpose transaction processing systems was the **Public Utility Customer Information & Control System**, known as PU-CICS, and later simply as CICS. Developed in 1968 for customers in the oil and public utility industries, it quickly became IBM's single most successful product, generating over $60 billion in sales for the company ("CICS," 2017).

From the telegraph to the Hollerith machine to mainframe computing, **information technology has been the key to managing and financing vast industrial transformations**. All of them built not on a few major customers but on thousands and eventually millions of customers and investors.

6.1 From One to Many to Millions

Despite the cost of renewables having plunged in recent years, the speedy adoption of renewables is not guaranteed. However low the cost threshold, renewables break the current model on which our US$2 trillion dollar electric grid was built (Rhodes, 2017). The electricity grid as we know it today is a largely one-way affair with energy flowing from centralized generators to industrial and residential consumers with all grids built on predictable models of demand. A shift to renewables changes this long-established model and creates a number of problems that, if unsolved, could slow or halt the adoption of renewables.

First and foremost, the future of the grid is one with decentralized production, storage, and consumption. Where integrated utilities once controlled the entire system, a large marketplace for production and consumption is now emerging. Consumers and enterprises that used to only consume power will now both produce and consume (or become so-called "prosumers"), depending on their own usage and the output of solar panels or batteries in their possession. Electricity deregulation in many mature markets shifts the model from a single company to a marketplace of tens of different companies.

Though a single company typically retains control over the distribution infrastructure, multiple generation companies can compete in a market for power. A market with a limited number of players looks easy to manage compared to the next iteration for the grid where millions of rooftop solar installations and car batteries are linked into a digital market.

In addition to keeping the power grid stable, network operators must make sure that all participants in the market pay and are paid according to their participation. Financial incentives are keys to building a shared, distributed infrastructure that provides resiliency at a reasonable cost.

The second major challenge of a rapid shift to renewables is a significant mismatch between peaks of supply and demand and their predictability. Peak power consumption is typically in the early evening, after the sun sets and long after peak solar cell output. If we all start charging our cars at night, the current mismatch may become dramatically worse. At the same time, the sun and wind are not as predictable as a gas turbine.

While grid-scale storage and new infrastructure that supports daytime car charging at offices and workplaces will help, batteries are a very costly form of power "generation." We must consequently learn to become much more responsive to rapid shifts in localized supply and demand. Today, utilities have centralized demand-shedding processes in place with both household (air conditioners and electric vehicles) and industrial customers. As with the market itself, today's programs are limited in scope and participation. In order to scale, they have to go from dozens of centrally managed elements to potentially millions of small responses.

The management infrastructure that underpins today's electric grid simply was not designed to support millions of suppliers as well as consumers to have power consumption flow in both directions or to shape demand in millions of small, localized decisions. The transaction processing infrastructure that grew up in the 1960s is unlikely to scale well into this new era: it is centralized technology in an increasingly decentralized marketplace that was designed to record usage in a predictable environment rather than actively shape and clear a dynamic marketplace.

6.2 Distributed Grid Meets Distributed Computing

Blockchain technology, which grew out of the innovations in Bitcoin, turns out to be a remarkably good fit with the evolving requirements of an electric grid powered by renewables. To understand why it is such a good fit, it is useful to first dive into what Blockchains do and how the technology works.

Blockchains are, first and foremost, a transaction processing technology. As with Bitcoin, any banking system or the original CICS, Blockchains are able to track transactions across a network and update accounts and payments. What Blockchains can do is unremarkable. How they do it explains why they are such a good fit for this new decentralized environment.

There are three key characteristics that differentiate Blockchains from prior transaction processing technologies. The first is the **distributed ledger, which keeps track of the digital coins (tokens) that each users has in their possession**. Although most systems keep a single, centralized copy of transactions and accounts, Blockchains spread those across all the key points in the network. The result is that every location in the network has all the information it needs to function autonomously and tampering with transactional data is nigh impossible because of the massive redundancy of good copies.

As transactions in the distributed ledger can originate anywhere and propagate across all the points in the network, Blockchains are an excellent fit with a decentralized model of production and consumption. Every power meter, battery and solar panel can have an identity and transaction history, all recorded in a tamper proof flow of transactions.

The second key Blockchain characteristic is the concept of **smart contracts**, also known as the "programmable ledger." Whereas traditional systems record numbers in ledgers, Blockchains allow participants to exchange digital agreements alongside economic value. For example, a utility and a customer can agree upon a discounted price for power if the customer agrees to shut-off their water heater during peak consumption periods. This agreement is then distributed across all the network nodes and enforced automatically. The result is an elegant integration of actions in the real world with the exchange of value and payments online.

As with the distributed ledger, smart contracts can be between any participants in the network. IBM's early autonomous decentralized peer-to-peer telemetry (ADEPT) Blockchain prototype showed how solar panels, batteries, and home appliances could be linked together in a network of smart contracts that worked to adjust electricity usage in the homes based on output from the solar panels and to minimize the cost of drawing power from the public grid. Trials using Blockchain technology are underway in Germany around tracking electric vehicle charging ("Germany's Energy Giant," 2017) and in Brooklyn to track sharing of rooftop solar power across a community (Cardwell, 2017).

Finally, Blockchains approve and record transactions through a process known as a **consensus algorithm**. Blocks (groups) of transactions are batched together and distributed for approval along to all nodes in the network which confirm them. Where there is conflict, the version of the truth backed by the majority of the network wins. The result is a system that is remarkably resilient to attack and one that gets stronger as the network grows larger. Any attacker must compromise the majority of the network rather than only a single point of failure in the middle.

Until recently, even if the concepts of Blockchains had been well understood, they would have been impossible to implement. At the dawn of the computing age, when CICS was created to manage meter readings, computing power was expensive. An architecture that took every transaction and copied it to thousands of other locations for extensive verification would cost more than the electric grid decades ago.

Today, computing isn't just cheap, but for all practical purposes it is nearly free. Most of the world's computers are, in fact, idle most of the time. With Blockchains we can "spend" the nearly free computing power of millions of grid-connected smart devices to buy something extremely valuable: **reliable, trustworthy, business transactions**. Distributed computing and distributed grids are an elegant fit.

6.3 Financing the Future

Beyond tracking and processing the work done in a distributed energy production and consumption environment, there will be many other applications for Blockchain technology in scaling our transition to renewables. Blockchains make it possible to take anything that can be digitally represented and turn it into a tradable, saleable asset and do so simply, securely, and at extremely low cost.

A recent study found that consumers who lease their solar panels pay up to 50% more for electricity than those who have access to low-cost financing in order to buy them outright (Lee, 2015). Traditional approaches to financing capital purchases are costly, especially when it is done on a one-on-one basis to individual buyers.

The biggest innovations in business and industry have always been about the ability to achieve scale through the aggregation of many small transactions with very high efficiency. Joint stock companies helped businesses become larger than the capital stock on any one entrepreneur and; the explosive growth of the fossil-fuels based energy infrastructure came one railway journey or electric meter at a time. Nowadays, millions of customers and investors are indeed financing a huge collective investment.

Historically, though the customers may have been in the millions, the investments were highly centralized. Money has flowed into aggregators who built massive power stations. To some extent, that pattern will continue in the era of renewable energy. Some of the biggest investments to-date have been in utility-scale solar farms and batteries. As the cost of renewables has come down, large utilities have responded by treating solar panels and batteries in the same way they used to buy gas-fired power stations.

With Blockchains, however, there may be a new kind of opportunity—to inexpensively yet effectively finance a widely distributed network infrastructure. Take the cases of electric cars with large batteries, as most people return home in the evenings, the opportunity exists to discharge car batteries into the grid at peak evening times. Today, though this scenario is feasible, payment to the car owner for discharging the battery is done at the time when it "generates" power. Discharging only part of the battery each evening could earn the car owner US$300—500 over a 5-year period, depending on the "feed-in" tariff rate[3].

If it happens only on a monthly basis, it is not appealing enough to change a consumer's decision to buy an electric car. If it is bundled altogether as a US$500 rebate on a new electric car, however, it is more meaningful. With Blockchain technology, it is possible to create digital tokens that represent asset ownership rights and income streams and then sell those into an investor market. On a blockchain technology platform, coin converts a real-world asset into a tokenized ownership. More precisely, then a token is a crypto or digital asset which actually represents a share of a user in a specific asset or property. Therefore, tokens can serve as a proxy of ownership rights in an asset or a property. Millions of investors can now finance and manage millions of assets efficiently. The future of energy finance lies in millions of producers, millions of consumers, and millions of investors working together.

6.4 Some Assembly Required

Blockchains are new and trendy. But frankly, as a technology, it is still immature. The vision laid out in this chapter assumes a future in which significant progress is made in both

[3] The earnings projection assumes US$0.05 feed-in tariff per kwh and 5 kwh/day, 200 days/year.

Blockchain technology and related areas. Specifically, there are four key challenges that must be addressed for Blockchains to become the technology that scales our renewable future.

The first challenge is in transaction processing speed and scale. Blockchains, in their present form, copy all transactions to all nodes in the network. This makes records in these networks immutable—tamper-proof—because there are so many redundant copies. Given the potential billions of transactions to come in a distributed energy grid future, this is probably technically infeasible. Future Blockchain solutions are likely to preserve immutability by copying all transactions to many locations, but not every location. How many copies to make as well as how to manage the consensus-based decision-making are the subjects of intense debates and numerous research projects?

The second critical challenge for Blockchain technology is around privacy and secrecy. We think of privacy and security as two things that go hand in hand. Blockchains have typically involved an explicit trade-off: less privacy for more security. The visibility of account balances is a key tool that enables the consensus algorithm: each network node can inspect transactions for validity against existing contracts and account balances. If you cannot read those account balances or digital contracts, you cannot verify a transaction.

As with transaction processing speed, there are numerous different solutions being developed to these questions. Some involve restricting the readability to contracts and data to a smaller number of trusted parties although others involve special types of encryption that allow queries to be answered without decrypting the content.

The third challenge for Blockchain solutions is that the success and growth of Blockchains would also be dependent on two other important technologies: *machine learning* and *mobile networks*. Machine Learning, or *Artificial Intelligence* as it is often referred to, is likely to be critical for enabling effective decision-making in large-scale networks. Not only will Blockchain-based energy networks have millions of devices, those systems will need to make billions of decisions each day.

Traditional approaches to computer-based decision making are rule-based and structured on an "if-then-else" conditional construct; if (x) happens, then do (y). However, as transaction systems become more complex, it becomes harder and harder to think through all scenarios and special cases that may occur, never mind develop the programing required to execute those rules. Machine Learning systems are not programed; however, they are trained. And, once trained, Machine Learning algorithms can respond reasonably even to entirely novel situations.

Self-driving cars are one of the most visible applications of machine learning technology. Humans on the road behave unpredictably, but with sufficient training, self-driving cars are able to handle nearly all routine driving scenarios. Machine Learning algorithms can make simple work of all kinds of complex energy consumption decisions that are too small to warrant human attention. For example, a dip in solar panel output in a house can be handled by reducing demand or drawing more power from the grid. Pausing the laundry cycle for 2 min or choosing to switch off the hot water heater for the same time period would just do the trick and save a miniscule amount of money—probably, less than one penny.

But the decision is complex: what time of the day is it? Has everyone already left the house after taking a shower? In which case, there is no rush to re-heat the water. Is it Friday evenings when people are waiting on a favorite shirt to be washed and dried before going out? It would be intolerable for a person to make all these decisions and impossible to program the details of every scenario. With Machine Learning, it will be possible for smart homes and appliances to build their own decision criteria (within limits) from observations. Where Blockchains are key to tracking transactions and moving around value, machine learning is critical to managing billions of small decisions.

The last challenge is that Blockchains needs to become the postmodern equivalent of the telegraph. Wireless networks today are optimized for humans carrying smartphones watching adorable movies. There are small numbers of users consuming huge amounts of data. A distributed, renewable energy grid will be a network of billions of devices, all of which handle many small transactions, and need to be connected and distributed across the network. Mobile networks specific to the Internet of Things (IoT) are now emerging and optimized for these use cases. Narrow-band LTE devices will have a range of up to 10 km and data speeds of up to 1 MB, even battery life of up to five years (Nokia, 2017). Though these devices are still relatively costly, forecasts suggest their price level going down to just US$3–4 per unit by 2020.

With so many unsolved problems, it may seem outlandish to envision a future renewable energy industry completely enabled by Blockchain technology tomorrow, but the history of technology is the story of such massive improvements. A lot of problems had to be solved before the British got to 15,000 telegraph stations across 13,000 miles of railway track, but the journey started with just one telegraph line across 13 miles of track. In other words, betting against the ability of new technology to mature and scale has, historically, been a losing proposition.

References

Cardwell, D. (2017). *Solar experiment lets neighbors trade energy among themselves. New York Times.* <https://www.nytimes.com/2017/03/13/business/energy-environment/brooklyn-solar-grid-energy-trading.html?_r = 0>. Accessed on 10.10.17.

CICS. (2017). <https://en.wikipedia.org/wiki/CICS>. Accessed on 10.10.17.

Clark, G., & Jacks, D. (2007). Coal and the industrial revolution, 1700–1869. *European Review of Economic History*, *11*(1), 39–72. Available from https://doi.org/10.1017/s1361491606001870.

Electrical telegraph. (2017). <https://en.wikipedia.org/wiki/Electrical_telegraph>. Accessed on 10.10.17.

Fierheller, G. A. (2014). *Do not fold, spindle or mutilate: the "hole" story of punched cards* (2nd ed). Markham, Ont.: Stewart Pub.

Germany's Energy Giant Launches 100s of Ethereum Based Electric Cars Charging Stations. (2017). <http://www.trustnodes.com/2017/04/29/germanys-energy-giant-launches-100s-ethereum-based-electric-cars-charging-stations>. Accessed on 10.10.17.

History of coal mining. (2017). <https://en.wikipedia.org/wiki/History_of_coal_mining#cite_note-21>. Accessed on 10.10.17.

Rhodes, J. (2017). *The old, dirty, creaky US electric grid would cost $5 trillion to replace. Where should infrastructure spending go?* <http://theconversation.com/the-old-dirty-creaky-us-electric-grid-would-cost-5-trillion-to-replace-where-should-infrastructure-spending-go-68290>. Accessed on 10.10.17.

Lee, M. (2015). *Rooftop solar loans beat out leases. San Diego Union Tribune.* <http://www.sandiegouniontribune.com/sdut-sun-shines-on-loans-2015feb12-story.html>. Accessed on 10.10.17.

List of rail accidents in the United Kingdom. (2017). <https://en.wikipedia.org/wiki/List_of_rail_accidents_in_the_United_Kingdom#1830.E2.80.931922:_Pre-grouping>. Accessed on 10.10.17.

Nokia. (2017). *Nokia white paper: LTE evolution for IoT connectivity*. Espoo: Nokia.

CHAPTER 7

Efficient Power Markets: Reimagining the Global Market with Ethereum

Adam Richard and Rob Hitchens

CTO & Co-founder, Volt Markets[1], Houston, Texas, United States

7.1 What's Wrong with the Power Market Today—The Birth of Volt Markets

The release of the Blockchain paper by Satoshi Nakamoto has put in motion a wave of intellectual exploration to identify new ways for distributed Blockchain technology to transition systems toward censorship resistance and hence, improving transactions of every kind. The journey thus far from simple cryptocurrencies to reimagining how assets transact in the physical world through complex processes has shown that Blockchain technology will reach far across industries, from macro to micro levels where trust is required to exchange value. In the energy market, the pervasive nature of electricity generation, transmission, and the oversight to maintain reliability makes it a daunting undertaking to engineer a realistic solution that rewards the participants with empowerment, disintermediation, scalability, integrity, and immutability.

Current smart grid, Internet of Things (IoT) and utility ecosystems rely on centralized, brokered communication models. These models are slow, encumbered by manual trading instead of autonomous agents, and do not accurately reflect real-time transmission, resulting in unnecessary costs and inefficiencies. Existing energy IoT solutions are expensive and lack a single, interoperable platform that connects all devices. Multitiered power supply chains create massive electricity loss in long distance transmission, whereas congestion on any particular branch of the network thwarts economic dispatch.

There is a patchwork of standards that vary across regulatory jurisdictions. This creates a systemic break in data governance and an inability to trade assets at a global scale. Complex onboarding processes for even the smallest participants are prosumer-prohibitive. This

[1] https://voltmarkets.com/

Transforming Climate Finance and Green Investment with Blockchains.
DOI: https://doi.org/10.1016/B978-0-12-814447-3.00007-0

impacts their ability to participate and unlock economic incentives, and renewable energy development needs more accessible incentives to accelerate investment and adoption.

The renewable energy credit market is complex, discouraging small and mid-size companies from participating in and benefiting from the marketing opportunities available to business that wish to participate and back office costs are exorbitantly high for current users. Auditing the origination and chain of custody is arguably impossible for ordinary consumers and difficult even for those who transact power and energy attributes on a large-scale every day. These challenges lead to double accounting and which undermines the credibility of Certificates that represent value.

Today, reconciliation for power transactions is a slow, costly process that results in lost time and operating capital. Blockchain technology can resolve these inefficiencies with automatic reconciliation.

Industry experts are optimistic that Blockchains can provide a promising solution by embedding the asset registry and settlement process into the movement of power from generators to consumers. This idea has systemic implications which move away from a dual system process, where the physical and digital systems run in parallel, into a single system, where everything is digital in nature. The result is utopian, but the effort is arduous. To begin an undertaking of this nature, Volt Markets was established to conduct original research and engineer the foundational infrastructure that allows stakeholders to alter their business models and processes to transact with counterparties through newly aligned interests.

As with the finance sector which has been experimenting, collaborating, and trialing Blockchain solutions for letters of credit, securities, and derivatives over the past few years, revolutionizing a traditional giant business like the electricity market with Blockchain would imply voluntary commitment from some organizations or a mandate from a major market participant to changing the way communications and transactions are conducted between counterparties.

7.2 Public Blockchain for Global Power Trends

Although scalability and privacy have been driving the research and development of private Blockchains, Volt Markets see that the industry-wide efforts put forth for unrolling Blockchain solutions in power markets have surprisingly not been so biased, with many ventures in the ecosystem opting for public Blockchains, which could facilitate sector-scale, aggregated transformation with minimum level of resistance. For Volt Markets, the decision to implement public Blockchain infrastructure rests on the timeline expectations for the Ethereum developer community to accomplish scalability and privacy functionalities in a coinciding time for industry adoption.

The flourishing development of public Blockchains that offer scalability and privacy is fast approaching validation in production environments. zk-SNARKs, which allows one party (the prover) to prove to another (the verifier) that a statement is true, without revealing any information beyond the validity of the statement itself, offers privacy for Ethereum and economic clusters, snapshotting, and permnodes offers scalability for IOTA. Ethereum Metropolis gives Volt Markets users the ability to utilize zk-SNARKs for addressing counterparty identity and transaction privacy. State channels (i.e., Raiden Network (Hees et al., 2017)) and Plasma (Poon and Buterin, 2017) will provide the technology to scale interaction with the system and the transactions per second to near-real-time while simultaneously decreasing costs the longer counterparties transact off-chain.

Anticipating future development of public Blockchains, Volt Markets decided to use the public Ethereum Blockchain to solve the problems identified. With Blockchain as the underlying infrastructure, Volt Markets focuses our efforts on developing a more efficient, universal power trading regime that may be executed by autonomous hubs, reflecting real-time transmission, minimizing costs, increasing interoperability, and minimizing electricity transmission dissipation.

Indeed, Volt Markets' innovative work builds around the concept of the EEI Master Contract, which is a power-purchase agreement created by the Edison Electric Institute. According to an EEI Master Contract, invoices are "due and payable in accordance with each Party's invoice instructions on or before the later of the twentieth (20th) day of each month, or tenth (10th) day after receipt of the invoice or, if such day is not a Business Day, then on the next Business Day. Each Party will make payments by electronic funds transfer, or by other mutually agreeable method(s), to the account designated by the other Party" (Edison Electric Institute and National Energy Marketers Association "EEI Master Contract.", 2000). The Volt Markets solution allows for real-time settlement with its single system for asset transfer and settlement, as well as considerations for addressing letters of credit between counterparties. This ensures large network participants' processes are not disrupted too drastically upon adoption.

7.3 The Vision: Trading Renewable Energy Certificates on a Global Blockchain Platform

Volt Markets' vision is to disintermediate traditional energy markets with a public, decentralized, Blockchain platform on which to monitor, manage, and trade energy and energy attributes. Smart contracts in the platform will (i) ensure jurisdictional regulatory compliance; (ii) open markets to new asset classes; (iii) incentivize renewable energy generation, trading, and usage; and (iv) provide end-use customers the ability to trade the energy they produce and consume in real time at nearly every scale.

A Blockchain-supported trading platform empowers the people and companies with ability to control their energy generation and consumption, greater flexibility for transmission and distribution grids, enabling self-supply with renewables, batteries, control system implementations, and securing demand via the stimulation of new consumption.

Volt Markets' smart contracts communicate with points of power generation to issue tokens on the Blockchain. The tokens are unique to their provenance and represent energy assets on the Ethereum Blockchain. Governance smart contracts give users the ability to comply with jurisdictional regulatory compliance and authenticate participants' equipment. The system can originate power (MWh) and energy attributes (Renewable Energy Certificates (RECs), Renewable Identification Number (RINs), carbon credits), certify assets, and track the chain of custody from generation to consumption. Volt Markets facilitates peer-to-peer transaction settlement via the user's choice of decentralized exchange that is secure, auditable, scalable, and capable of demand response load balancing by leveraging IoT devices that strive to source affordable and renewable energy from the shortest distance determinable.

The foundation of Volt Markets consists of an ecosystem of Ethereum smart contracts which provide certainty of what can and cannot happen next with incorruptible business logic and records. The smart contracts contain code functions and can interact with other contracts, make decisions, store data, and send value to others. The network participants are governed by approved parties to certify power generators and their equipment to originate Ethereum ERC20 tokenized assets. Those assets are then sold in a decentralized market where VOLT tokens are payable to interact with the system. A client-side decentralized app can transact directly with smart contracts, the market, and the assets.

The aim is to transition and substitute existing systems with a decentralized and highly efficient market to be characterized by substantially reduced onboarding friction, auditable code, direct line-of-sight from buyers to origination documents and elimination of market inefficiencies arising from counterparty risk, opaque information siloes, accidental and intentional misrepresentation, and double accounting.

7.4 Design Philosophy for Energy Assets in Peer-to-peer Trading Systems

The Volt Markets team spans across many disciplines, and together have reimagined commodity markets without abstracting too far away from what the industry is so used to. The Ethereum Blockchain and the smart contracts that give it utility can take many forms, and its adoption may have deep societal-level impacts. The Volt Markets' Blockchain infrastructure for power markets is offered as a completely transparent, highly efficient, accessible, and global public resource that will assist humankind as renewables flourish, the grid edge evolves, and companies enjoy greater profits.

Energy grids are no more than a complex web of machinery, infrastructure, people, laws, commercial interests, government organizations, and processes that has evolved to its present state over the past century. The processes themselves are valuable. Even as Volt Markets proposes a more efficient, more inclusive and more transparent system, there is considerable effort spent to incorporate the best processes the existing system has to offer. It is ultimately Blockchain that makes this possible.

Blockchain technology is unlike any computing platform that came before it. **Blockchains are incorruptible.** Records that meet the strict requirements of Smart Contract logic are immutable. Data stored in a Blockchain has met all necessary requirements, as specified in publically viewable code. Results simply must be correct, because there is no other economically feasible possibility.

Blockchains link value with data. As compared to earlier generations of databases, data in a Blockchain database are not about the asset, data are the assets, as set forth by Ian Grigg's Ricardian Contracts and Nick Szabo's concept of turning digital registered assets to digital bearer assets. The strength of incorruptible records is aptly demonstrated by the emergence of cryptocurrencies such as Bitcoin. Currency is by no means the only asset class open to digital representation on Blockchains. It is merely the simplest to demonstrate and appreciate. **Virtually all commerce can benefit from absolute and immediate certainty** about such matters as identity (who am I dealing with?), origin (where did this come from?), and ownership (who owns this?). This is a sound foundation for efficient settlement of trade across all asset classes that can be digitally represented, including energy assets.

Smart Contracts provide certainty about what will happen next. Smart Contracts add incorruptible logic to incorruptible records. The word "contract" should not be taken to construe an equivalence between legal contracts and computer software. This description, namely "Smart Contract" has taken hold because it conveys the unique properties of this entirely new and novel form of software.

These programs themselves consist of immutable data and therefore are permanent members of the Blockchain. They provide certainty about what will happen next. Thus, if Alice agrees to sell electricity to Bob, and Bob agrees to send money to Alice, both parties can be certain that their counterparty is incapable of noncompliance. The contract acts as the escrow hub and executor, and because it is incapable of diverging from the preset terms, both parties can proceed with confidence. This is unprecedented. It means that actual asset exchanges, and not merely claims, can be linked to trigger events using unstoppable rules that are plain for all to see.

7.5 The Rise of New Assets and Market Efficiency

Financial contracts for commodities are directly representing the thorough details of the underlying commodity particulars. Futures never do, and forwards do sometimes, but not all the times. However, in a Blockchain system, the value linked directly with the data makes transparency inherent to the system. The results have beneficial outcomes for commodities such as RECs and carbon credits which are traded every day in global markets with little or no transparency.

Volt Markets' approach to long-term adoption falls across three major foci. The first focus is on the Blockchain infrastructure, the second one will be to prove compatibility and utility for IoT devices, and the third one testing and integrating for autonomous IoT powered by machine learning or artificial intelligence (AI). These three emerging digital technologies, when combined to act in concert, will hopefully achieve unparalleled market efficiency.

The IoT component of this approach will offer unique financial structures for project development and energy management, which will result in less stress on our energy infrastructure while accelerating and easing the world's energy transition. Frankly, our vision to create a utopian energy trading platform is just the beginning of this energy revolution, and requires a lot more hard work to be done.

References

Edison Electric Institute and National Energy Marketers Association "EEI Master Contract." (2000). <http://www.eei.org/resourcesandmedia/mastercontract/Documents/contract0004.pdf> Accessed 08.08.17.

Hees, H., Soleimani, A., & Hack, A. (2017). Payment-channel network for Ethereum. <http://raiden.network> Accessed 13.08.17.

Poon, J., & Buterin, V. (2017). Plasma: Scalable autonomous smart contracts. <http://plasma.io/plasma.pdf> Accessed 13.08.17.

CHAPTER 8

Flexibility Trading Platform—Using Blockchain to Create the Most Efficient Demand-side Response Trading Market

Paul Ellis and Joanna Hubbard

Electron, London, United Kingdom

8.1 Introduction

Electron[1] is a London-based start-up that designs and builds Blockchain systems with and for the energy sector. One of our key developments is a trading platform for demand-side response (DSR).

DSR is a tool for balancing the grid when demand and supply are out of sync. Most balancing tools require generators to increase or decrease their output in response to signals from the System Operator (SO). These tools are most effective for fossil fuel generators that can ramp up or down with ease, newer and cleaner technologies such as wind and solar cannot change their output so easily. To cut ties with fossil fuel generators permanently, newer and better solutions must be developed to balance the grid.

DSR is an emerging solution that uses flexible consumption from the demand side to balance the grid. Users increase, decrease, or shift their energy usage to balance the grid and are financially compensated for doing so. When DSR is used, dirty generators are not asked to increase their output, and clean generators are not curtailed (turned down). In the future, it could also mean that locational loads can be managed to prevent expensive assets becoming overstressed and shortening their working lives such as the case of transformers. The potential DSR market is far larger than the current version, as there are many players who do not have access to trade or who face significant barriers to participation.

Our DSR trading platform leverages Blockchain technology to create the incentives for all players in the market to trade in a single venue, without establishing a new monopoly or conflict

[1] www.electron.org.uk

Transforming Climate Finance and Green Investment with Blockchains.
DOI: https://doi.org/10.1016/B978-0-12-814447-3.00008-2

of interest between the market operator and participants. Bringing all liquidity into a coowned, cooperated trading platform enables greater coordination and the ability to trade collaboratively.

As multiple parties can benefit simultaneously from a single DSR action, enabling collaborative trading is the key to maximizing the value and liquidity of this market. Moreover, a large, liquid DSR market is also critical to the integration of a high proportion of intermittent renewable sources into our energy system and ensuring security of supply in the most cost-efficient way. It will also unlock new revenue streams for industrial and commercial players, and potentially even for aggregated domestic loads in years to come.

In competitive energy markets, multiple service providers are delivering a fungible commodity over a shared infrastructure. In this way it is much like financial markets. As in finance, some of this infrastructure is physical (ATMs and cash *versus* pipes and wires), though much of it is virtual, such as trading platforms and protocols. Blockchain technology presents a way for market participants to cooperate over that shared virtual infrastructure in a more efficient and innovation-friendly way.

In this chapter we will explore how a Blockchain platform could be used to enable a better coordinated, collaborative traded marketplace for DSR. The first half will explain the DSR market and its increasing importance in our electricity system. Those already familiar with this market may wish to skip this section. The second half will examine the peculiarities of this market, the nonrival value components of a single DSR action, and how Blockchain technology presents the opportunity for multiple parties to coordinate and share the cost and value of those actions. We refer to this as "collaborative trading."

Nonrival Value Component

A nonrival good is one that can be used by multiple parties without prohibiting others' use of it. A classic example is television: anybody can watch a show without impacting the ability of another user to also watch that show.

In this specific example, the same action is valued by different parties for different reasons. These reasons are the "value components," and include location, speed and length of reaction, etc. These "value components" can be nonrival, i.e. the same action can simultaneously deliver benefits to multiple parties.

Examples and statistics in this chapter will focus on the UK: a deregulated, competitive electricity market in which 34% of installed electricity generation capacity is renewable—accounting for 25% of all generation output (National Grid, 2017). This is forecast to rise to account for as much as 60% capacity and 67% generation by 2050 (National Grid, 2017). Parallels can be drawn with other similarly high penetration of renewable generation and deregulated markets, such as the US state of Texas and Germany.

8.2 What is Blockchain?

Blockchain is a protocol, or set of rules, that allows network participants (peers) to reach consensus on the state of a system without resorting to an intermediary. This enables value or ownership to be digitally transferred, not copied, without relying on a trusted central intermediary such as a bank. This was the problem that bitcoin, the first and most famous instance of Blockchain, set out to solve.

But Blockchain is not just about the digital transfer of value. It also presents a new way to build transaction (or interaction) platforms without a central operator in a privileged position. Rather, platforms can be coowned and cooperated by the users of that platform themselves. In this way, Blockchain is not just a new technology, but also a completely new way of conducting business transactions.

8.2.1 What Does Blockchain Fundamentally Change?

Firstly, the role of a central trusted intermediary is no longer required. This has the potential to remove a conflict of interest, a cost base, and an innovation barrier or information silo.

Secondly, Blockchains can create new levels of trust and transparency. Rather than simply expecting platform operators to behave honorably, desired properties can be built inherently into a system such that it functions with the guarantees we expect. Moreover, all transactions have immutable auditable trails and can be inspected.

Thirdly, platform users can extend functionality themselves by deploying their own smart contracts, or pieces of code, to automate business logic. This automation functionality, coupled with the trust and transparency guarantees, explains why Blockchain is commonly reported as a key enabler of the machine-to-machine economy.

Finally, Blockchain presents new models for data ownership and monetization. Because there is no central platform owner such as Bloomberg or Amazon to take control of all data, users retain control of their own data. The platform design can enable data producers to choose from keeping their data private or donating, lending, selling, or renting it, which eventually monetizes the data through direct transfer of value on the platform.

8.3 What is DSR?

In the electricity system, supply and demand must be balanced on a moment-by-moment basis to maintain the frequency of the system and, quite literally, keep the lights on. The frequency of the system falls if demand is greater than supply, and vice versa. In the UK,

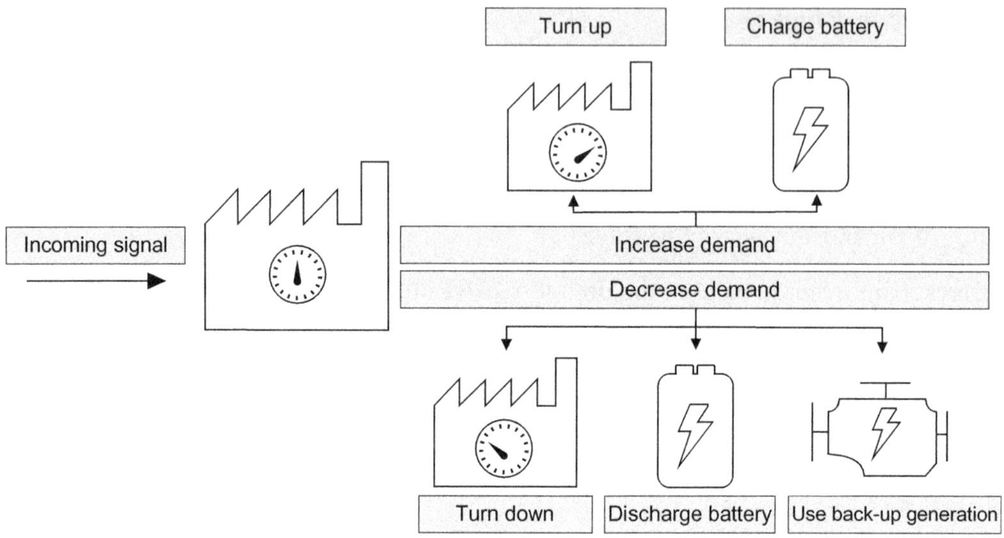

Figure 8.1
Infographic of DSR.

the National Grid performs the role of the Systems Operator (SO) and is responsible for retaining this balance and a system frequency of 50.00 Hz ± 0.2 Hz.[2]

Balancing, or flexibility, services can be provided either by generators adjusting output on the supply side or by end-consumers adjusting demand. DSR is when end-consumers are financially incentivized to turn up, down, or shift their electricity usage in response to explicit signals.

DSR is typically achieved in four main ways—see Fig. 8.1 (Charles River Associates, 2017):

- Turn-down DSR: reducing demand during peak times or shifting demand away from peaks
- Turn-up DSR: increasing demand during periods of excess supply
- Battery-led DSR: temporarily shifting demand to on-site energy storage to reduce consumption from the grid
- Stand-by DSR: temporarily shifting demand to on-site generation to reduce consumption from the grid.

In this chapter we use the term "DSR action" to cover any of the above, an example of which could be a "turn-down action" that lessens the load on a stressed wire and decreases consumption to help balance the grid.

[2] On the national grid, the operational range is 49.8–50.2 Hz whereas the statutory range is 49.5–50.5 Hz.

With regard to turn-down and turn-up DSR, there are typically two types of flexible load with energy intensive processes that are not time-sensitive, and cross-sectional technologies such as heating, cooling, motors, pumps, and compressors.

8.4 The Need for a Liquid, Efficient Flexibility Market

Historically turn-down, turn-up, and battery-led DSR have only accounted for around 5% of balancing actions by value in the UK market (Charles River Associates, 2017). The vast majority of balancing services are still sourced from the supply side, typically from gas and coal plants capable of ramping generation up or down quickly.

However, the impetus to decarbonize is rapidly reducing the flexibility of the country's installed generation base. Generation technologies such as wind and solar cannot increase output on demand. They can only be turned-down, which is referred to as "**curtailment**." Curtailment is a suboptimal grid balancing tool as not only is it costly, but it also decreases the percentage of electricity generated from renewable sources.

Systems with a high proportion of renewable generation such as solar and wind suffer from increased intermittency, and some (those without moving or spinning parts, particularly solar) lack inertia. Inertia acts as a "shock absorber" which helps to lower the system's rate-of-change-of-frequency (RoCoF). If RoCoF is too high, some large energy generators and users will automatically disconnect from the system to prevent internal damage. This causes further dramatic changes to the system and can render imbalance situations exponentially harder to rectify.

The result is an increased need for demand-side flexibility and rapid response technologies to enable a higher penetration of renewables while maintaining security of energy supply at a reasonable cost. DSR, thus, touches all the three points of the *reliable, affordable, and clean energy trilemma.*

It is estimated that a large, liquid DSR market could unlock £1 billion savings today; a net systems benefit of £1.4–2.4 billion per year by 2030, and an overall saving of £17–40 billion by 2050[3] (Carbon Trust, 2016). These cost savings are achieved through:

- Reduced capital expenditure on generation;
- Lower operating costs in maintaining flexible demand-side capacity vs. conventional generation;
- Reduced network reinforcement and management costs[4]; and
- Lower-cost security of supply.

[3] The saving is projected on the assumption of the electricity carbon emissions intensity target being 100g/kWh in 2030.

[4] DSR can lower loads on stressed assets, thereby, extending their useful lifespan.

These savings could reduce energy bills for customers, assuming savings be passed along.

Carbon savings are also realized through greater uptake of DSR. These savings are achieved through:

- Decreased reliance on peaking plants (often fossil fuel powered);
- Reduced requirement for installing new generation or network assets (e.g., transformers, wires); and
- The ability to deploy renewables at fully installed capacity with less curtailment.

Case Study

A study performed in a cement plant showed that the site has potential to achieve a 4.2% decrease in electricity cost and 4% decrease in electricity-driven emissions by "load shifting"—changing the time of energy usage to low-carbon periods (Applied Energy, 2016). Importantly, these savings were achieved without any loss of productivity. Scaled to values from a typical plant, this would amount to savings of £350,000 and 2000 tonnes of CO_2 annually.

The SO has published targets for DSR to provide 30–50% of capacity in the electricity balancing market by 2020, and more than 50% by 2030 (Association for Decentralised Energy, 2016). Achieving these targets is estimated to require between 10.3 and 12.7 GW in DSR capacity by 2030 compared to just over 2 GW of actively engaged capacity today (Poyry & Imperial College London, 2017; Charles River Associates, 2017).

These targets are ambitious but potentially achievable within estimates of 14.7 GW of commercial and industrial capacity being engageable by 2020 (Charles River Associates, 2017).

8.5 Flexibilty Market Redesign

The DSR market has evolved rapidly over the last 10 years to keep pace with faster-than-expected renewable deployment. This rapid evolution has unintentionally resulted in a highly complex system, which is skewed toward large-scale national (vs. local) actions and an unattractive innovation environment.

Today, the SO contracts for, and actions, balance grid services in three broad categories: frequency; voltage; and RoCoF control. These services have evolved into over 20 separate submarkets for DSR or flexibility products, often created to cater to specific technologies, and each procured in a different manner (e.g., monthly, bi-weekly or annual auctions, tenders, and bilateral agreements). Each submarket has very specific requirements for participation. As a result, there is very little diversity of participants within them, which again restrains competition and restricts liquidity and innovation.

Furthermore, the complexity has created knowledge barriers to entry; discouraged the participation of new technologies with new performance profiles; and frustrated price discovery and transparency (while auctions results are published, many bilateral agreements are not). Such a market will struggle to scale and the SO has already commenced work to simplify this product offering and improve market signals.

However, there is another significant change underway for which this system redesign should cater. Though the SO remains all but the sole purchaser of flexibility today, other players are beginning to introduce new routes to market. Several distribution network operators (DNOs) are developing pilot projects to capture the local network benefits of DSR actions and some suppliers are already contracting directly with customers to use their flexibility to balance their trading positions.

Some key considerations in this respect are as follows:

- The current system has one main buyer who engages with multiple sellers via multiple competing platforms—this compromises efficiency; hampers market liquidity; and does not allow for price discovery. If the market shifts to become multilateral, then liquidity will be further constrained as new entrants will introduce yet more trading venues and both buyers and sellers of DSR will have limited visibility of the market.
- Moreover, new market entrants will assign different values to the various value components of a DSR action (including action size, lead time, length, cost, reliability, location and associated supplier, etc.), effect on system inertia and carbon intensity. This could risk creating yet more products and further complexity, fracturing the market further.
- Finally, a greater degree of coordination will be required to ensure that actions by DNOs and/or suppliers/traders do not conflict with those of the SO, threatening security of supply and increasing the cost of balancing the grid.

Table 8.1 summarizes key characteristics of the current DSR market and options for how this might evolve.

The current direction of travel toward multiple competing trading platforms has some advantages in terms of competition and impetus to innovate and adapt. In equity markets, there are a number of competing trading platforms, driving down the costs of transacting and allowing traders to select the best venue for their specific instrument and trade size, etc. Equity traders can use trading aggregation tools that collate all trading liquidity from individual platforms so they can see the complete market. When they trade, they need to use order routing systems which ensure that trades are sent to the right venue and that they have a clear sight of their overall position.

This structure works well for financial instruments as they are concerned with rival goods, meaning that the instruments can only be possessed or consumed by a single party and their value cannot be shared. However, this aggregation of trading interest across multiple platforms will not address the liquidity challenge of the DSR market in the same way.

Table 8.1: Market structures.

	Current DSR Market	Multiple Platforms, Single or Multiple Buyers	Single Centralized Platform, Multiple Buyers	Single Blockchain Platform, Multiple Buyers
Liquidity	Low-single buyer	Fractured	High	High
Levers of power	SO	Multiple parties	Single-platform operator	Multiple parties & platform users
Operating cost	Unknown	Competitive	Monopolistic	Competitive
Innovation speed	Reliant on SO	Competitive	Reliant on platform operator	Fastest innovator
Transparency and auditability	Reliant on SO	Challenging to compare	Reliant on platform operator	Transparent and auditable
Coordinate & trade collaboratively	No	No	Yes	Yes

DSR actions possess multiple, nonrival value components, such that different types of industry players can benefit simultaneously from the same DSR action. In order to maximize the value and liquidity of a market for a nonrival good, collaborative trades need to be assembled in a manner which enables multiple purchasers to contract for the same action (see next section).

Collaboration across multiple bilateral platforms would be extremely difficult in practice and impossible to implement in a transaction-safe manner. Thus, in order to enable collaborative trading, a single trading venue needs to be established.

8.6 Collaborative Trading

When multiple parties simultaneously benefit from a single action, the value of that action is maximized when parties are able to share its cost. Conversely, there may be instances in which the same action adversely affects another party. Allowing all affected parties to express their preferences as a value will enable the optimal action set to be identified.

In the DSR market, an action ordered by the SO and undertaken by a consumer will also impact the trading position of the associated supplier and the assets of the local DNO. If the SO had the opportunity to pool trading interest with other affected parties, opportunities could arise for taking actions that benefit all involved.

Let us consider a situation in which total demand on the grid exceeds generation. The SO would like to send a turn-down signal to an energy consumer in order to bring the system into balance and there are a number of potential consumers who might be instructed to take this action. Assuming that all of these consumers were able to respond reliably, today the principal selection criteria would be the cost of a turn-down action. A bilateral trade would then take place. However, what has in fact been selected here is the consumer with the lowest cost to deploy an action and not necessarily the action with the most overall value to the grid.

But flexibility actions will impact the local distribution networks as well as the overall system imbalance position. This bilateral trade has not taken this into account. It might be that the lowest priced action was beneficial nationally but detrimental locally. In this case, then the cost of that action might be much higher than its face-value.

On the other hand, higher priced actions may also offer more overall value to the system. Consider that a local distribution network operator was in particular need of a turn-down action in a certain location to manage a stressed distribution asset. That distribution network operator would be willing to contribute to, and thus subsidize, a suitably located response in order to prolong the life of that asset. Likewise, a supplier who has underestimated the amount of electricity they require during a certain period will be facing an imbalance charge and willing to pay for a turn-down action (up to the price of the associated penalty) in order to correct it.

Enabling collaborative trading in the DSR marketplace will:

- Unlock trades that were not viable on bilateral trading basis;
- Release cost savings for flexibility purchasers—see Fig. 8.2; and
- Result in more trades overall, more market liquidity, more end users: a virtuous circle.

In order to assemble collaborative trades, DSR actions are disaggregated from their various value components, such as locational delivery components, associated supplier, inertia and carbon intensity, etc. Purchasers of flexibility are then able to ascribe value to these components individually. These components are then reassembled by a trading engine to propose optimal coalitions according to a predefined utility function.

Let us consider how this might work for locational value components of a DSR action as an example. Transmission and distribution network operators would be given the option to price at which they would sponsor a DSR action that affected their transmission or distribution assets.

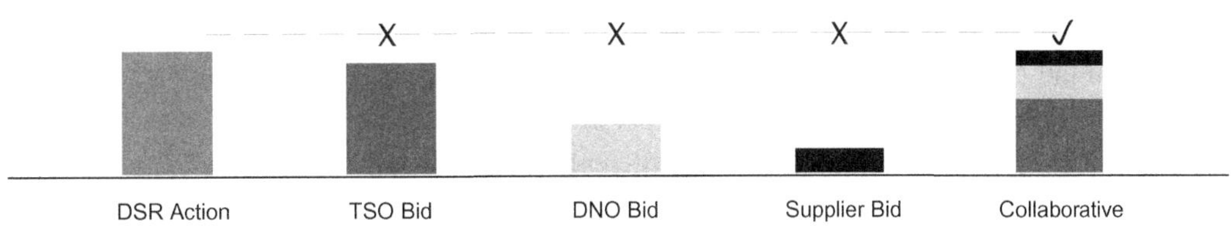

Figure 8.2
Example of a collaborative bid.

Network operators would have the option to price: regionally, providing a blanket price for a wider area (e.g., at a time of anticipated stress); or nodally, submitting a price for a particular asset (e.g., for a transformer associated with new electric vehicle charging points in order to study new load stress). Ultimately, we expect the system to evolve toward nodal pricing as assets become capable of self-pricing. Accordingly, we have integrated machine learning into our platform design.

Combining this trading interest will essentially establish a negative cost to deliver a DSR action in that particular location and, thus, make it more likely that the action will happen in the location where delivers the highest collaborative value.

Once a sufficiently liquid DSR market has been established, the opportunity for DNOs to submit locational pricing interest should help to resolve potential conflicts with actions dispatched by the SO. Actions in areas that DNOs are willing to support are more likely to be selected and the reverse is true of actions in areas where those DNOs are keen to discourage.[5]

8.7 Centralized Monopoly vs. Decentralized Blockchain Platform

Given that collaborative trades cannot be facilitated between parties on different platforms, maximum liquidity and market value will likely require all market participants to be trading on a single platform. Historically, this would have required a trusted intermediary or a monopoly to run the platform.

Single-platform ecosystems can create a number of ownership issues such as extracting monopoly rents and other costs from market participants, and freeriding—lack of incentives for transparency or innovation.

Blockchain presents an elegant alternative by allowing for a single-platform system that does not necessitate a monopoly or central party to run it. A Blockchain platform could:

- Allow for all DSR products to be routed through a single venue to ensure the best possible liquidity and combination of trades is achieved;
- Establish trust among collaborators by proving protocols are being followed and the value of flexibility is being allocated fairly;
- Ensure the lowest cost of execution, through competitive operation and lack of a centralized cost structure;
- Maintain the capacity and incentive for platform users to innovate, enabling them to extend functionality as required—a key for markets in transition with high growth forecast; and
- Enable data traders to maintain control of their own data, and retain ownership without compromising system auditability.

[5] This same approach might be applied to trading exported energy; particularly in microgrids, in order to establish a delivery price.

8.8 Transformation Potential of Blockchain in DSR Trading

The market for DSR requires participants to collaborate in order to deliver optimum value. This collaboration will unlock cost savings for purchasers of flexibility and drive wider market liquidity by enabling the execution of trades that would not have been possible on a bilateral basis.

Blockchain technology enables the market to trade collaboratively without handing the levers of power to any single entity. Moreover, the platform is cost-competitive and demonstrably fair with its open protocols creating a transparent execution and governance structure. All parties can see the rules that they are signing up to in advance and verify correct value allocation for themselves.

Furthermore, a collaborative DSR market which allows for location pricing also has the potential to proactively establish new value propositions for local energy markets.

Finally, as our electricity system becomes increasingly renewable and distributed, centralizing market decisions and control could not only be suboptimal, but it might also become infeasible.

8.9 Closing Remark

Given rapid and global advances in both Blockchain technology and flexibility markets, we have taken to date tagging our contributions. This chapter was written by Joanna Hubbard in July 2017. Electron[1] is not responsible for the content of other chapters.

References

Association for Decentralised Energy. (2016). *Flexibility on demand: Giving customers control to secure our electricity system.* <https://www.theade.co.uk/resources/publications/flexibility-on-demand-giving-customers-control-to-secure-our-electricity-su> Accessed on 14.08.17.

Carbon Trust, Imperial College London. (2016). *An analysis of electricity system flexibility for Great Britain.* <https://www.gov.uk/government/uploads/system/uploads/attachment_data/file/568982/An_analysis_of_electricity_flexibility_for_Great_Britain.pdf> Accessed on 08.09.17.

Charles River Associates. (2017). *An assessment of the economic value of demand-side participation in the Balancing Mechanisms and an evaluation of options to improve access.* <https://www.ofgem.gov.uk/system/files/docs/2017/07/an_assessment_of_the_economic_value_of_demand-side_participation_in_the_balancing_mechanism_and_an_evaluation_of_options_to_improve_access.pdf> Accessed on 14.08.17.

National Grid. (2017). *Future energy scenarios 2017.* <http://fes.nationalgrid.com/fes-document/> Accessed on 14.08.17.

Poyry & Imperial College London. (2017). *Roadmap for flexibility services to 2030.* <https://www.theccc.org.uk/wp-content/uploads/2017/06/Roadmap-for-flexibility-services-to-2030-Poyry-and-Imperial-College-London.pdf> Accessed on 14.08.17.

CHAPTER 9

NRGcoin—A Blockchain-based Reward Mechanism for Both Production and Consumption of Renewable Energy

Mihail Mihaylov[1,3], Iván Razo-Zapata[2] and Ann Nowé[1]

[1]Department of Computer Science, Vrije Universiteit Brussel, Brussels, Belgium [2]Luxembourg Institute of Science & Technology, Esch-sur-Alzette, Luxembourg [3]Enervalis, Houthalen-Helchteren, Belgium

To mitigate our dependence on fossil fuels, new schemes to reward the production and consumption of renewable (green) energy are required. Green energy needs to effectively offset the consumption of grey energy, i.e., energy from mixed sources. The NRGcoin concept supports the integration of renewable energy by incentivizing its local production and consumption within the low-voltage grid. It creates not only a more secure and transparent energy provisioning mechanism, but also promotes a local ecosystem in which key stakeholders (prosumers, consumers, and utilities and grid operators) can mutually benefit from each other. After discussing the main issues with traditional support policies, this chapter presents the NRGcoin concept in details. It describes the problem being solved while emphasizing the main benefits and potential challenges for NRGcoin's adoption. Finally, we also contribute four important questions, which help set a reasoning framework to clarify and justify the use of Blockchain in addressing various challenges.

9.1 Introduction

The decentralization of renewable[1] energy production comes with numerous challenges for the traditional electricity grid. These challenges threaten to undermine investments in the renewable energy economy, prompting the search for alternatives. Many believe that Blockchain technology can address those challenges and, thus, its attractiveness has gained popularity in the energy sector. This revolutionary technology has spawned a multitude of ideas and projects each of which could be a promising use case. Peer-to-peer energy trading appears to be the most popular use case in this sector, with numerous projects and companies competing to be the first to commercialize its potential. Meanwhile, other

[1] Throughout this chapter, we use the terms "renewable" and "green" interchangeably.

Transforming Climate Finance and Green Investment with Blockchains.
DOI: https://doi.org/10.1016/B978-0-12-814447-3.00009-4

projects are exploring complementary use cases away from the "mainstream hype." NRGcoin is one of these projects.

NRGcoin is a residential support policy for renewable energy exchange. It is a decentralized mechanism based on smart contracts that rewards prosumers for their injected green energy and makes green energy more economically attractive to consumers. In doing so, NRGcoin aims to offset the consumption of grey energy, i.e., energy from mixed sources, and helps increase the share of renewable energy sources (RES) in the overall energy mix (European Commission, 2011). NRGcoin offers a more scalable alternative to traditional support policies, such as net metering and feed-in tariff, as well as to other existing reward programs. As such, NRGcoin is more than a peer-to-peer energy market, but an alternative that involves not only the buyers and sellers of energy, but also other key stakeholders in the low-voltage grid.

The NRGcoin concept has been mainly developed at the AI Lab of the Vrije Universiteit Brussel[2] and is currently in the phase of real-world deployment by the industrial partner, Enervalis.[3] Up-to-date information on the progress of this project is available on its website.[4]

Besides NRGcoin, there are several other projects and companies that exist at the intersection where Blockchain meets the energy sector. Each of those initiatives at its core aims to exploit the capacities of Blockchain for transforming the smart grid and global climate finance. With ever-growing interest in this multidisciplinary domain, it becomes difficult to discern between the value propositions of various projects of similar nature, i.e., the differentiator or added value of a given project compared to other similar projects in this field, and how they relate to each other. To answer this question, any given project should be able to provide answers to four basic questions. We list these questions here and, later, in this chapter we will demonstrate how NRGcoin answers them.

A. **Which problems are you solving?**

   You are developing a product or a service that has a particular aim or direction. You aim to achieve something or improve a given situation. You take design decisions with a certain goal in mind. It is important to keep those goals clear and state them explicitly. However, the scope of the problems needs to correspond to the size of the project—they should neither be too small to be trivial, nor too big to be unreachable. Is your end goal to tackle climate change or poverty? These problems are too broad and it is unlikely that one single project can adequately address such problems.

[2] AI Lab of the Vrije Universiteit Brussel - http://ai.vub.ac.be

[3] Enervalis - http://www.enervalis.com

[4] NRGcoin Project website - http://www.nrgcoin.org

B. **Why those problems need to be solved?**

Whereas some issues do not need much explanation, other problems are less evident as to why they need to be solved. You are solving the problem that peer-to-peer energy trading did not exist before. A little motivation would help one understand why this solution is needed in the first place. No matter the problems in scope, the motivation for solving them should be adequately expressed. Motivating the need for solving the given problem will also ensure that the problem itself is not contrived or inessential, and that devoting resources to address it is justified.

C. **How exactly are you solving them?**

Now that the problems are properly sized and the need to solve them is clear, one should describe how the proposed project aims to address those problems. This is where one can observe the added value of a given project and the particular approach that is applied.

D. **Why do you need Blockchain to solve those problems?**

Blockchain has a number of characteristics that distinguish it from other technologies. Blockchain participants append information to a decentralized immutable database, where those participants are either not known or not trusted, and where the system does not rely on central third parties. A project that claims to apply Blockchain to solve a given problem needs to demonstrate that the above characteristics are *essential* to the proposed solution. If any of these characteristics are not present (e.g., there is a need for trusted third parties), the use of Blockchain is not justified due to the computation and communication overhead, which is needed to achieve decentralized consensus. If, on the other hand, all of the above characteristics are present, the project inherits a variety of benefits from the Blockchain technology, such as decentralization, censorship resistance, absence of the need for trust, permissionless participation, etc.

Note that these questions are applicable not only to the Blockchain-energy sector, but also to all other cross-disciplinary fields that involve Blockchain technology. Within the energy sector, moreover, there should be some other questions to be addressed. For instance, what kind of financial schemes should be in place to support the integration of renewables? How can we make sure that the financial schemes create value for all stakeholders? As an attempt to answer such questions, governments worldwide have adopted an array of renewable energy support policies (Butler & Neuhoff, 2008). Although these policies have contributed to an incipient boost in renewable generation, they soon need to be replaced as they are not able to cope with future scenarios in which renewable energy sources (RES) would be massively deployed not only at the high-voltage grid, but also at the medium-voltage and low-voltage grids (International Energy Agency, 2015a).

9.2 Traditional Support Policies

Committed to some emission reduction targets, policy makers design and implement a series of support policies for renewables generation (Butler & Neuhoff, 2008), which motivate residential prosumers to feed their energy into the grid. Prosumers—a portmanteau of "producer" and "consumer"—are homes that not only consume energy, but also produce renewable energy with, for examples, photovoltaic panels or micro-windmills to be fed into the grids. We outline here two of the most commonly applied support policies, namely net metering (NM) and Feed-in tariff (FiT). FiT is by far the most widely adopted support instrument (REN21, 2016), whereas NM is a policy still prevailing in numerous countries such as Belgium, Denmark, Italy, and the Netherlands. Their incentives have contributed to increasing amount of installed residential RES capacity and to growing number of prosumers (International Energy Agency, 2015a; International Energy Agency, 2016a; REN21, 2016). There have been, however, discussions on the need to phase out these traditional subsidy schemes in favor of mechanisms that scale better with growing decentralized generation (Cerd & del Ro, 2015; Couture, Cory, Kreycik, & Williams, 2010; European Commission, 2013; Greer et al., 2014). Here, we identify various issues with NM and FiT and analyze on the possible mid- to long-term implications for different stakeholders.

9.2.1 Net Metering

On an annual basis, Electricity Provider (EP) inspects the energy consumption readings of households. The invoice is based on these readers and a fixed rate per kWh of energy consumption, whether it is a peak/off-peak tariff. When household producers inject energy into the grid their energy meter readings fall. In other words, the meter counts forward when energy is "imported" from the grid, and it counts backwards at the same rate when energy is "exported" to the grid. Remunerating prosumers at retail price allows them to use the grid as a virtual storage. However, in order to prevent households from becoming net producers, the meter readings of households at the end of the year cannot be lower than those at the start of the year; and usually reimburses each prosumer only for injected energy that does not exceed her annual consumption.

9.2.2 Feed-in Tariff

FiT is a standard support policy through which different types of RES are guaranteed a fixed sale price for certain contractual periods (Couture et al., 2010; Jacobs et al., 2013). Unlike NM, FiT rewards all injected energy regardless of prosumer's own annual consumption, but it does so at prices lower than the retail price of electricity. For this reason, the inflow and outflow of electricity need to be measured separately (e.g., by a separate meter), which cannot be achieved with the traditional single bidirectional "spinning" meter. The actual FiT must be decided by policy makers, such that it is high enough to be attractive for RES investments, but not too high that prosumers are over-compensated. This support scheme allows policy makers to adjust the tariff at any time in order to control the speed of RES integration. Nevertheless, great caution ought to be exercised in these adjustments, as too large, too frequent or retroactive tariff changes undermine investor confidence, which is detrimental to the adoption of renewables (De Boeck, Van Asch, De Bruecker, & Audenaert, 2016; Rathmann et al., 2011; REN21, 2016).

9.2.3 Drawbacks of NM and FiT

NM and FiT are "blind" to *when* and *where* energy generation takes place (International Energy Agency, 2016b). The *when* refers to the temporal space in which generation occurs, whereas the *where* the location/physical space within which energy is generated. These two main drawbacks give rise to a number of issues.

For instance, as NM and FiT have no connection to temporal information (e.g., real-time market price signals), they reward prosumers at a rate disproportionate to the actual demand for energy at a particular point of time, which may lead to over-consumption or over-production. Prosumers can be either underpaid or overpaid, the unnecessary costs of which are eventually passed onto end consumers.

Moreover, these mechanisms do not incentivize the consumption of green energy injected in the local grid, but only its production. Also, neither NM nor FiT considers grid stability and scalability for future scenarios.

9.2.3.1 Over-Consumption

NM caps the amount of green energy for which a given prosumer is paid subject to one's own consumption. This policy encourages prosumers, who often pursue value for money, to withdraw from the grid at least as much energy as they inject on an annual basis, because the excess green energy that prosumers inject in summer will discount their winter consumption of grey energy. A more environmentally-efficient policy should not encourage prosumers to use the grid as a virtual buffer, storing "free energy" for the months when low production or high consumption is expected, due to the seasonal stress it exerts on the grid infrastructure.

FiT, on the other hand, rewards prosumers at a rate lower than the retail value of electricity. This policy encourages consumption of own-produced energy and feeding only the surplus energy into the grid.

9.2.3.2 Under-Payment

Although FiT rewards all injected energy, NM caps these rewards to a prosumer's own annual consumption. Consider a district with a majority of consumers and only few prosumers who have large production capacity (present scenario), despite the large demand created by district-wide consumers, those few prosumers will not be paid for all the energy they have injected if their own annual consumption levels remains lower than their own production levels. In other words, NM does not reward excess annual supply per home even if it can actually be consumed locally in the district. NM, therefore, does not sufficiently encourage the offset of grey energy in favor of green.

9.2.3.3 Over-Payment

NM and FiT reward injected energy without considering the actual energy demand. Consider again a district where most homes are prosumers with sufficient production power (future scenario), the total consumption from the grid during daylight hours will be negligible compared to the amount of injected energy. All prosumers will be rewarded for the injected energy, even though there are not enough consumers to withdraw it. As the number of prosumers continues to increase (International Energy Agency, 2016a; REN21, 2016), both NM and FiT will become costly to the energy providers. These costs will naturally be passed onto end consumers and, hence, the overall cost of electricity is very likely to rise.

9.2.3.4 Grid Stability

The above described drawbacks have implications on grid stability. Because NM rewards injected energy at retail price, prosumers are indifferent between feeding their energy or self-consuming it. Using the grid as virtual storage exerts stress on the grid infrastructure comprising cables and transformers, etc., means it requires more frequent maintenance and reinforcements to prevent blackouts.

Although FiT does incentivize self-consumption by offering lower-than-retail price for fed-in green energy, it rewards all injected energy even when it is in surplus, which again strains the grid.

Both traditional policies, therefore, incentivize green energy without considering its impact on the grid infrastructure.

9.2.3.5 The Need for an Alternative

Despite their drawbacks, NM and FiT have contributed to the significant growth in residential RES capacity. These policies have been the first step toward RES integration. They perform reasonably well when the number of producers is far lower than that of consumers. Although this assumption of producers outnumbering consumers has been true until recently, the number of prosumers is now on the rise sharply (Rickerson et al., 2014; REN21, 2016). The growing decentralization of renewable production and number of prosumers will exacerbate the problems described above and impact all stakeholders on the low-voltage grid. Experts advocate replacing these traditional support policies with market-based incentive mechanisms in order to improve the incentives for renewable energy generation and its consumption (Cerd & del Ro, 2015; Couture et al., 2010; European Commission, 2013; Greer et al., 2014).

Although NM and FiT represent examples of top-down incentives that aim to increase the share of renewables to offset grey energy consumption and reduce greenhouse gas emissions, bottom-up or local actions are also required to much better adapt to the impact of climate change across different regions (International Energy Agency, 2017). The NRGcoin concept represents one of those alternatives. Unlike international climate funds (e.g., Green Climate Fund) that require strong commitment of national governments, NRGcoin offers a mechanism that requires the commitment of mostly local stakeholders. It incentivizes not only consumers, producers, and energy utilities, but also grid operators and governments because they do not need to make huge investments to support the penetration of RES in the energy market.

9.3 NRGcoin Concept

Similar to NM and FiT, NRGcoin is not a peer-to-peer energy trading mechanism and it does not rely on an energy market with locally produced renewable energy fed into the grid, and withdrawn by consumers (Mihaylov et al., 2014a; Mihaylov, Jurado, Van Moffaert, Avellana, & Nowe, 2014b). NRGcoin, however, provides **incentives distributed near real-time and based on the total supply and demand in the local district on an interval of 15 min**, rather than based on the individual's annual supply and demand in the way under NM. In case of energy, prosumers are paid proportionally to the total demand in the district at that point of time, unlike the FiT regime which pays for all injected energy indiscriminately. In NRGcoin system, any over-produced energy that exceeds the local demand is not paid. By so doing, prosumers are always rewarded for the maximum amount of renewable energy that is withdrawn by consumers at the time it is injected and thus they can never be overpaid or underpaid. In contrast to NM and FiT, therefore, under the NRGcoin mechanism, it matters *when* the energy is injected and what the supply-demand balance is at the time of injection (Mihaylov et al., 2014a; Mihaylov et al., 2014b). **This model of rewarding reflects the real temporal value of renewable energy and offers a much better granularity to help balance local production and consumption**.

9.3.1 Mechanism

The NRGcoin mechanism is implemented as a network of hardware gateway devices (e.g., RaspberryPi or any other board) that interacts with the energy installations of residential homes, one device per home. These gateway devices have current sensors to measure the energy "imported" from the grid as well as the energy "exported" to the grid. They send these measurements at regular intervals to the NRGcoin smart contract. A smart contract is a decentralized software running on a Blockchain (e.g., Ethereum or any other Blockchain that supports smart contracts). The NRGcoin smart contract in particular can be running either on an existing (public) smart contract-capable Blockchain, or on a Blockchain-supported network of gateway devices used for current flow measurements. The smart contract implements the protocols of the NRGcoin mechanism to be described in this section.

As shown in Figure 9.1, the NRGcoin mechanism works as follows:

1. Each prosumer injects one's green energy surplus in the low-voltage grid of the distribution system operator (DSO). Likewise, each consumer withdraws energy from

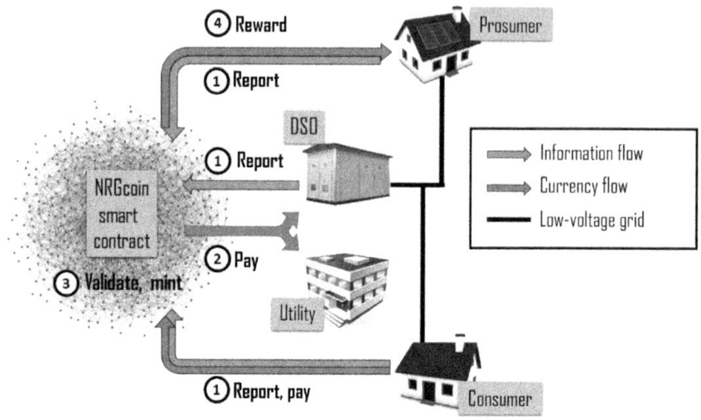

Figure 9.1
NRGcoin mechanism.

the grid. Here, the low-voltage grid encompasses the network of cables supplying residential homes with electricity. At the end of every time slot (15 min), a number of steps below take place.

The gateway device at each home reports the measurements of injected and consumed energy for that time slot to (trigger) the "NRGcoin smart contract." In parallel, each substation of the DSO reports the observed measurements in the local district to the smart contract in order to deter users from tampering with their measurements. At this point, the smart contract knows the amount of green energy injected in the local district within the last 15 min. Note that the role of the DSO in reporting measurements is neither central nor crucial to the operation of the mechanism because it can be replaced with machine learning algorithms for tamper detection. Along with the reported measurements, the gateway device of each consumer pays certain units of NRGcoins equivalent to one's green energy consumption, where one kWh of green energy always costs one NRGcoin. The payment process is automated, i.e., NRGcoins are sent from the digital wallet of the consumer to the smart contract as a Blockchain transaction. The NRGcoins are not destroyed upon payment, but remain in circulation. Any consumption exceeding the locally available green energy, which the smart contract computes based on all reports, is paid for in the traditional way—directly to the utilities as per the conventional legal contracts between them and the consumers. Although the NRGcoin mechanism primarily supports green energy, utilities are free to decide whether they accept NRGcoin for mixed energy usage and at what rate. However, utilities cannot mint NRGcoins or change the smart contract in any way.

2. Once all consumed green energy is paid to the smart contract, the smart contract then pays all necessary grid costs and fees to the DSO, plus a small percentage to the utility, which has balancing responsibilities. These payments are automated and sent

in NRGcoin as a Blockchain transaction. In case the DSO and/or Utility choose not to accept the NRGcoin currency, a third-party payment processor can convert the NRGcoins to fiat currency by trading them on a currency exchange market (not shown in Figure 9.1). In other words, the NRGcoin mechanism pays the utility for balancing or buffering the renewable energy grid and the DSO for using its low-voltage grid infrastructure.

3. After consumers have paid for their green energy consumption with a portion of revenue paid to the utilities, the remaining coins are reserved for prosumers, but not sent yet. The smart contract first cross-checks the reports of individual gateways with the report of the DSO to determine whether any user has tampered with the measurements. The validation protocol consists of comparing whether the sum of individually reported consumption and injection measurements matches with the aggregate report of the DSO. Injection measurements are also checked against local weather data. If tampering is detected, this information is passed on to the utility and to the DSO for follow-up and dispute resolution. For each prosumer for whom no tampering is detected, the smart contract mints new NRGcoins according to: (i) the amount of injected energy that was actually consumed in the last time slot (computed in step 1) and; (ii) the current NRGcoin minting rate (see below). The newly generated coins serve to bring new currency in circulation without a centralized issuer. Any injected energy that is in excess of the local demand does not produce new NRGcoins and, therefore, is not paid for.
4. The freshly minted coins together with the existing NRGcoins paid by consumers (see step 3) are rewarded to the prosumers. These amounts are sent automatically from the smart contract as a Blockchain transaction to the NRGcoin wallets of the individual prosumers.

9.3.1.1 Minting Rate

As stated above, the smart contract mints new NRGcoins when injected green energy is consumed locally and those coins will be paid to prosumers. The currency is minted by the smart contract and, thus, it is decentralized and not owned by any central institution or organization. The amount of newly minted coins is such that the final income to prosumers (step 4 above) is less than one NRGcoin per injected kWh. Thus, although consumers always pay one NRGcoin per kWh of green energy, prosumers receive a lower amount in order to incentivize self-consumption; and ultimately avoid wastage caused by over-production.

The minting rate for new NRGcoins decreases over time for three main reasons:

- Firstly, decreasing issuance limits inflation. As this cryptocurrency is never destroyed, the ever-increasing number of prosumers would lead to hyper-inflation if the issuance rate remains the same over time.

- Secondly, diminishing returns for new prosumers drive consumers to invest earlier in renewable generators and therefore contribute to a faster shift toward renewables in order to limit the environmental implications of grey energy production.
- Lastly, after sufficient number of years, the decreased issuance rate will result in lower incentives to become a prosumer. Although the target is to achieve 100% renewable energy consumption, this does not necessarily require 100% prosumers. When there are already enough prosumers that provide 100% green energy, new prosumers would only contribute to excessive production peaks, causing more stress on the grid and ultimately higher costs for the end consumers.

The NRGcoin smart contract allows the minting rate to be programed to follow a particular pattern (e.g., decreasing logarithmically over time). Once this is set at launch, owing to the unique attributes of a Blockchain network, it cannot be changed by any single party unless there is a (super)majority consensus for this parameter to be altered.

9.3.1.2 NRGcoin Value

Independent from its function of supporting renewable energy transaction, NRGcoins can be traded on open currency exchange markets for their monetary equivalent (e.g., Euro, Dollar, Pound, etc.). These markets can be either traditional centralized Foreign Currency Exchange (FOREX) markets, or decentralized currency markets, such as Bisq.[5] Buy and sell orders for NRGcoins are matched using an "orderbook." Because consumers do not own renewable generators, they can only obtain NRGcoins from the currency market, which is where prosumers can sell their coins.

Although the fiat value of NRGcoin is determined based on free market rules and on the principle of supply and demand, it is unlikely that the value diverges from the retail cost of energy. One NRGcoin always buys 1 kWh of green energy as stipulated in the smart contract. As 1 kWh of energy has particular fiat cost on the retail market, it is expected that arbitrage drive the NRGcoin price toward the energy price. For example, if the price of energy is 25 Euro cents/kWh, a rational consumer will not purchase NRGcoin for a higher price, as one always has the alternative of paying her bill in euro directly to the utility for that price. Naturally, any user can still decide to buy/sell NRGcoin for a price higher than the retail value of electricity if one considers additional aspects that contribute to its value, such as the importance of green energy, the need for decentralization of the energy market or the price of carbon credit (when emission trading schemes cover households), etc. Arbitrage would also prevent the value of NRGcoin from dropping too low, although the minting rate ought to be carefully determined to prevent over-supply of the cryptocurrency. Nevertheless, the retail price of energy will always remain a reference point for the value of

[5] Bisq (https://www.bisq.io) is an open-source desktop application that allows you to buy and sell Bitcoins in exchange for national currencies, or alternative cryptocurrencies.

NRGcoin, unlike other cryptocurrencies whose values are mostly subject to speculative activities.

Because the retail cost of electricity differs among countries and regions, it is reasonable that each country/region has its own NRGcoin regime which is not necessarily interconnected with others. For example, it is unlikely that one NRGcoin bought on the German exchange market can be spent to buy green energy for a house in Spain. It is not the aim of the NRGcoin concept, nor there is a need, to create one global currency for green energy as it will not reflect *where* the energy was injected. Rather, **NRGcoin aims to boost the *local* renewable energy production and consumption**. Therefore, the smart contract in each region will accept only the NRGcoins which the smart contract itself mints. Nevertheless, cross-border NRGcoin trading will still be possible using bilateral exchange rates that reflect the difference between electricity prices in the respective countries in order to keep the local renewable energy sectors competitive.

9.3.2 Benefits and Drawbacks

NRGcoin aims to be a more scalable alternative to the traditional residential support policies, including NM and FiT. The mechanism considers not only prosumers, but also takes into account other key stakeholders in the low-voltage grid, such as consumers, utilities, and grid operators. NRGcoin offers numerous benefits to these stakeholders, but it also comes with a number of drawbacks.

9.3.2.1 Benefits for Prosumers

NRGcoin greatly reduces the risk of government policy changes for prosumers. Remuneration for green energy is governed by an immutable smart contract that cannot be changed by market actors. As opposed to trust-based promise, therefore, prosumers have Blockchain-level guarantee that the support policy will execute exactly as coded in the smart contract. In other words, because the Blockchain technology guarantees that no single party can modify or tamper with the smart contract, prosumers can be certain that the policy encoded in the smart contract will not change either. Moreover, **prosumers will always be paid for the maximum amount of their injected energy** that is usable for other consumers in their districts at the time it is produced. Their rewards do not depend on their own annual consumption, but on the real-time consumption of all their neighbors.

9.3.2.2 Benefits for Consumers

Given the general trend of rising electricity prices (U.S. Energy Information Administration, 2017; PwC, 2017), **NRGcoin can lead to lower green energy costs for consumers**. One unit of NRGcoin is always equal to 1 kW green energy, regardless of the electricity price in fiat currency. When the electricity price rises, one NRGcoin bought at a lower price will still buy 1 kWh of green energy. In addition, NRGcoin gives consumers a financial tool that allows them to directly express their valuation for green energy. Using NRGcoin supports the renewable energy economy, but above all **it supports their local district's shift toward renewables**.

9.3.2.3 Benefits for Utilities

NRGcoin offers faster cash flow for utilities. Instead of receiving payment for energy once a month, the Blockchain-enabled home gateways can automatically pay for consumption every 15 min at virtually zero overhead costs. Moreover, NRGcoin lowers the costs of utilities, because utilities do not need to pay prosumers for their energy as per the current support policies such as FiT. Instead, the smart contract mints the currency and pays prosumers.

9.3.2.4 Benefits for Governments

NRGcoin reduces the financial burden for governments which implement renewables support schemes. The Smart Contract generates new currency and rewards prosumers for their green energy. Thus, **NRGcoin can save a huge amount of public money without lowering the incentives for renewables**.

9.3.2.5 Benefits for Grid Operators

As NRGcoin can lower the costs of green energy for consumers, they are incentivized to shift their consumption to periods when green energy is produced. This shift results in lower peak demand which helps stabilize the grid and **minimize the stress on the grid infrastructure**. Also, NRGcoin has built-in incentives for self-consumption. Prosumers receive less than one NRGcoin/kWh and hence they will maximize the value of their renewable generators by storing or self-consuming their produced energy as much as possible before feeding the excess electricity into the grid.

9.3.2.6 Drawbacks

Although NRGcoin offers numerous benefits to different stakeholders, there are a number of shortcomings too. The concept relies on Blockchain technology and smart contracts, which are themselves very new and rapidly changing technologies. They are quite different from established technologies and relatively complex to grasp. Besides, energy market regulations which have not been reformed to suit the forthcoming "Blockchain era" explain why the uptake rate is slow. These factors render the NRGcoin concept too early for mass adoption, though the prospect of Blockchain technology and smart contracts in the energy sector is very positive. For the NRGcoin concept itself, there are still some aspects of the mechanism that need to be properly worked out. Some of these are related to the economic model, the security of transactions, data privacy and tamper prevention, among others.

9.3.3 Answers to the Four Key Questions

At the beginning of this chapter, there are four key questions about the value addition to which an innovative idea needs provide answers. Here are the answers about the NRGcoin.

9.3.3.1 Which Problems Does NRGcoin Solve?

The NRGcoin concept is mainly motivated by the following issues in the energy sector:

1. Current renewables support policies are inadequate and do not scale to a large number of prosumers.
2. Prosumers-to-be face the risk of retroactive policy change.
3. There is a general trend of rising electricity prices for the end consumers.
4. The existing initiatives to offset the consumption of non-renewable energy are insufficient.

9.3.3.2 Why These Problems Need to Be Solved?

The above four problems are motivated as follows:

1. Inadequate subsidies for renewables. As outlined in Section 9.2, two of the most widely applied support policies—NM and FiT—are inadequate in supporting renewable generation. A subsidy that provides *too little* support will lead to slower integration of renewable energy, as fewer consumers will have sufficient incentives to install renewables. *Too much* support, on the other hand, will incentivize many consumers to become prosumers possibly resulting in excess production, and hence, causing stress on the grid infrastructure. High stress on the infrastructure may lead to blackouts and higher costs for the end consumers. Therefore, subsidies ought to scale well with the number of prosumers and offer *just enough* support to promote a gradual integration of renewables that matches the demand for energy.
2. Risk for policy change. Retroactive changes in support policy undermine the confidence of investors and lead to slower adoption of renewables. Such changes have already occurred in a few regions and have a clear impact on investment decisions (International Energy Agency, 2016c). A good support policy should provide strong guarantees, i.e., guarantees not based on trust that the existing reward mechanisms will not be changed. It is therefore necessary that the policy scales well in future scenarios (see Q1) in order to avoid such changes.
3. Rising electricity prices. Although numerous factors contribute to the rising electricity costs, a key factor is the rising number of prosumers. The grid infrastructure has not been designed for decentralized generation. Injecting renewable energy at the low-voltage grid destabilizes the supply-demand balance calls for additional balancing services, which are costly. The growing imbalance also causes more supply and demand peaks, which threaten to damage equipment and cause blackouts. To handle peak

production and consumption, the grid infrastructure needs to be reinforced, which introduces further costs to the end users. An adequate policy, therefore, will not contribute to such imbalances, e.g., by "blindly" promoting renewables, but will effectively provide the right incentives to balance supply and demand.

4. Insufficient incentives to consume renewable energy. Merely subsidizing production is not sufficient to mitigate our dependence on fossil fuels. Green energy needs to effectively offset the consumption of grey energy (Bertoldi, Rezessy, & Oikonomou, 2013; Phatak, Robucci, & Banerjee, 2014; Razo-Zapata, Mihaylov, & Nowe, 2016), i.e., energy from mixed sources. Although NM, FiT as well as other programs motivate the injection of clean energy, they do not make the injected energy more economically attractive than grey energy from consumer's points of view. An adequate support policy should not look only at prosumers in isolation and merely incentivize self-consumption, but it should also provide sufficient incentives to consumers to offset the consumption of grey energy in favor of green.

9.3.3.3 How Does NRGcoin Solve the Stated Problems?

Each feature of NRGcoin addresses one or more of the above issues, therefore the explanation below is not organized by problem.

Prosumers are paid in NRGcoin for every 1 kWh of injected green energy that was consumed locally in the last 15 min. Any excess energy that does not match demand is not paid. Thus, NRGcoin provides economic incentives for consumers to install renewable generators and become prosumers for as long as there is sufficient demand for energy. Once there are sufficient producers, i.e., the locally produced green energy completely covers local demand, there will no longer be incentives for becoming a prosumer. In this way NRGcoin automatically scales to future scenarios by adjusting the level of subsidies proportional to the local balance between supply and demand. This feature also helps minimize the stress on the grid infrastructure and, therefore, reduces the need for costly maintenance.

The NRGcoin mechanism is built in an immutable smart contract, which cannot be changed by the utility or any other individual entity. In other words, the subsidies that prosumers receive are set in stone and no single party can alter them. Moreover, smart contracts provide reliability that the code will always execute exactly as written, so prosumers need not place their trust in any central entity.

Consumers always pay one NRGcoin for every 1 kWh of green energy they consume, regardless of current electricity price in fiat currency. So, regardless when a consumer bought a coin, it is always worth 1 kWh of green energy, which can be redeemed whenever such energy is available locally. Unlike the vast majority of cryptocurrencies, NRGcoin is not designed as a speculative asset, but it is tied to energy. Therefore, it is expected that as a result of arbitrage the exchange rate between NRGcoin and fiat currency will remain stable or close to the fiat value of energy on the retail market. Although NRGcoin cannot directly prevent energy price rise, it can offer a stable alternative, in which each minted NRGcoin has the value of exactly 1 kWh.

NRGcoin provides incentives to both producing and consuming renewable energy. Incentivizing production helps feed in more green energy to the grid, although making its consumption cheaper helps offset grey energy.

NRGcoin rewards prosumers for their renewable energy. Although this mechanism resembles Renewable Energy Certificates (RECs), there are some important differences. RECs bring income to a given producer located anywhere in the electricity network for energy they generated in the past. With RECs, one is purchasing the environmental benefits of renewable energy, but not the energy itself, for which they need to pay in addition. In contrast, paying with NRGcoin ensures that the consumer is buying the actual green energy recently produced in their local district. Thus, NRGcoin helps bring the consumption of green energy closer to its production both in space and in time. Furthermore, RECs and carbon credits are issued by centralized authorities, which can at any time change the issuance rate of these instruments or stop them altogether. These events can result in market shocks, which discourage renewables investments. In contrast, NRGcoins are issued at a pre-determined rate by a decentralized smart contract, which no single party can alter or terminate. This brings more confidence to prosumers for the stability of returns for the renewable generators in which they invest.

9.3.3.4 Why Does NRGcoin Need Blockchain Technology?

A robust and future-proof mechanism should not rely on centralized institutions. Such institutions are a single point of control and also a single point of failure. NRGcoin aims to offer an alternative to the traditional centralized support policies by relying on Blockchain technology for decentralization.

In order for NRGcoin to reduce the risk of policy change facing prosumers, the mechanism needs to provide solid guarantees that the tariff will not be altered in time. An effective way to achieve that is to rely on immutable smart contracts, rather than on software running on centralized servers. In this way prosumers need not trust any organization or institution for their green energy award payment, as they can rely on the guarantees of Blockchain technology itself. Thus, NRGcoin makes use of smart contracts that run on a Blockchain, which gives users transparency in the reward mechanism, and reliability that it will execute exactly as written and identically for everyone.

In addition to the above necessity to use Blockchain, there is another reason. Traditional policies and reward systems always rely on funds outside of the energy market system—the state governments or the individual utility companies. With NRGcoin, currency is minted from within the system, by the very end users, thus reducing the involvement of centralized institutions and even the burden on the governments. The Blockchain technology has given us so far the only feasible way to mint currency in a decentralized system. NRGcoin builds on top of this innovation and thus it could not have existed before this technology came to light in 2008 (Nakamoto, 2008).

Lastly, Blockchain allows for direct transfer of value between users, whereas the smart contract reduces friction and operational costs of utilities. This also allows for micro-payments of energy, speeding up the cash flow for large stakeholders.

9.4 Practical Implementation

NRGcoin comprises three main components: (i) a smart contract, (ii) a currency market, and (iii) gateway devices. The smart contract is envisioned as a piece of open-source software running on a Blockchain. The currency market is an exchange platform that allows users to buy and sell NRGcoins for existing currencies (e.g., Euro, Dollar, Pound, Bitcoin, etc.). The gateway devices are deployed in residential homes, integrated with the local electric installations and connected to the Internet. The gateway devices measure electricity inflows and outflows as well as communicate with both the smart contract and the currency market.

9.4.1 Deployment

There are several stages involved in the deployment of the NRGcoin system. At stage one, the system will be deployed in pilot locations with trusted environment in order to test the real-life operations of the mechanism and ensure its robustness before any large-scale rollout on a real Blockchain. For example, the Blockchain and the currency markets can be emulated or provided as a service, depending on the requirements of the pilot. Any prosumer or consumer in the respective pilot location can choose whether or not to participate in the NRGcoin mechanism after weighing the benefits and drawbacks presented in Section 9.2.

At stage two, the project can be adopted as a support policy by existing utility companies or energy service companies (ESCOs) and thus up-scaled to a larger user base. The implementation of the Blockchain and currency market can vary based on the operation of the given institutions and relevant regulations of the respective country. For example, the NRGcoin mechanism can be running on a public or on a permissioned Blockchain, whereas the NRGcoin currency can be traded on a locally hosted exchange platform, or on an existing FOREX market. The customers of the respective utility may have the choice to opt-in or opt-out of this new support policy.

In the long run, it is envisioned that the NRGcoin mechanism is implemented as a Decentralized Autonomous Organization (DAO). The NRGcoin smart contract will be hosted on a fully decentralized public Blockchain run by the gateway devices themselves. The public Blockchain is envisioned to use Proof of Stake or any other low-energy consensus protocol.

Anyone can buy or build a gateway device from simple components, flash an open-source firmware on that device, integrate it with one's electrical system and thus become part of the NRGcoin ecosystem and DAO. These gateways will run autonomous trading agents to buy and sell NRGcoins on behalf of the homeowners. The NRGcoin currency can be traded on existing exchange markets—whether centralized (such as Kraken, Bitstamp, etc.), or decentralized (such as Bisq)—using Application Programming Interfaces (APIs). The gateways will then use these coins to pay their green and grey electricity consumption directly to the NRGcoin DAO, which in turn, will reward prosumers according to the mechanism described above.

The NRGcoin DAO will operate as an autonomous, decentralized, and independent utility. It will supply electricity to its "customers" and bill them in real time. It will fulfill its responsibilities as a balancing party by trading on the wholesale electricity market using autonomous trading agents, and will pay any necessary taxes and grid fees according to the regulatory requirements of the country. The NRGcoin DAO will have neither any CEO nor any ruling party, but it will be run entirely by transparent code and Blockchain-type consensus among all users. Nevertheless, the NRGcoin DAO can still employ people to carry out important, though not central tasks, which are typical to traditional utilities—infrastructure and code maintenance, legal pursuit in case of nonpayment and marketing activities to attract more customers. However, unlike traditional utilities, the NRGcoin DAO will be comprized entirely by code running on customers' devices. Thus, it will greatly reduce its operational costs and hence, offer lower electricity retail prices to the end customers.

Lastly, as NRGcoin is designed to boost the local green energy production and consumption, it can operate not only on traditional grids, but also on microgrids (in developing countries). If the microgrid creates its own grid infrastructure and becomes independent from the DSO, this change can be reflected in the code of the smart contract using majority consensus. Thus, the budget initially ring-fenced for paying grid costs to the DSO can be redirected toward maintenance of the local microgrid instead and thus, contribute to the growth of the local renewable energy economy.

9.4.2 Challenges

Given that NRGcoin is a novel concept, it may face not only technical and economic challenges, but also social and regulatory challenges.

Like other Blockchain-based solutions, NRGcoin faces technical challenges such as hardware and Blockchain scalability. Such scalability is different from that of NM or FiT as NRGcoin begins with offering unbalanced incentives as the number of prosumers starts to rise. Other challenges are standardization as well as privacy and security (see also the chapter on Blockchain 101 in the book).

Unlike other Blockchain-based solutions, however, NRGcoin also faces challenges regarding transparent and reliable alignment/interaction between electricity grids and information and communications technology (ICT). In this regard, it is important to guarantee that the NRGcoin rewarding system can handle asynchronous reports of gateway devices. Otherwise, it would need to require "perfect" matching among all ICT components, which is generally not feasible.

NRGcoin's economic challenges mostly relate to the costs and the benefits that it brings to the electricity grid. On one hand, NRGcoin must consider not only capital investment costs (Capex) and operation and maintenance costs (Opex), but also costs related to other services that may be required to guarantee the operation of the local grid (Ueckerdt, Hirth, Luderer, & Edenhofer, 2013). These costs could be estimated with current metrics such as System Levelized Cost of Electricity (LCOE) (Ueckerdt et al., 2013). Briefly, System LCOE aims to also take into account integration costs that occur at the system level (e.g., balancing services) (Ueckerdt et al., 2013). On the other hand, for a better understanding of the benefits that NRGcoin brings to the electricity grid, assessments based on System Value should also be conducted (International Energy Agency, 2016b). Unlike System LCOE, System Value tries to capture the net effects on the system by interplaying positive and negative effects arising from the integration of renewables (International Energy Agency, 2016b).

Another major challenge is the social adoption of NRGcoin, which will require full engagement with all participants so that they are motivated and confident about using NRGcoin. The main concerns that must be addressed to achieve customer adoption mostly relate to data privacy and security (International Energy Agency, 2015b).

Finally, the main regulatory barriers that NRGcoin faces will be either energy-related or market-related. Energy-related regulations dictate whether consumers are allowed to become full prosumers (i.e., self-consumption is allowed) as well as whether they can freely inject their excess electricity to meet demand from their neighbors. Market-related regulations pertain to the rights and obligations that must be fulfilled by NRGcoin participants to compete and collaborate with other stakeholders in the current electricity grid. For instance, in order to be compliant with current regulations that promote the idea of open markets in Europe, NRGcoin prosumers and consumers should be able to freely join and leave the NRGcoin mechanism (Sokolowski, 2016).

9.4.3 Possible Mitigation Strategies

To potentially replicate the NRGcoin concept in different market conditions in different countries, one should be aware of these challenges outlined above and some mitigation strategies. The following are a few mitigation strategies that can help address the above challenges:

- Implementation of standardized technology: Regardless of the deployment stage, NRGcoin must always make use of standardized procedures and specifications that ensure interoperability, security, and data privacy for all stakeholders (International Energy Agency, 2015b).
- Systemic analysis: NRGcoin developers must present a good economic case for its adoption in the sector. It is therefore desirable to conduct a systemic analysis of its economic value. To this end, different metrics such as system LCOE or System Value can be applied (Ueckerdt et al., 2013; International Energy Agency, 2016b).
- Transparency: In the interest of social acceptance, NRGcoin needs to establish a clear proposition. For instance, consumers and prosumers need to be sure that their information regarding consumption and production will be used exclusively in the NRGcoin rewarding mechanism.

9.5 Conclusions

Blockchain is a promising technology with enormous potential to help transform climate and renewable energy finance. The rising popularity of Blockchain means increasing number of projects and companies will apply this technology to address a variety of challenges. However, it is sometimes unclear what the added value of each project is or whether the use of Blockchain is really needed in the first place. In this chapter, therefore, we propose four key questions that every innovative project in this field should be able to answer. Doing so will dissipate the fog caused by the rapid growth of Blockchain uses in the renewable energy sector: boundaries between individual contributions will become clearer and new opportunities for collaboration and co-creation more visible. We also demonstrated how the NRGcoin system described in this chapter answers those key questions. We argued that such understanding is mandatory to effectively position and highlight the value of any given Blockchain-based solution.

Our work focuses on renewable energy and how Blockchain technology can help us meet climate change mitigation targets. The traditional incentive mechanisms for residential green energy production and consumption were a vital stimulus in the race to meet these targets. NM and FiT were successful in stimulating the initial penetration of decentralized microgeneration units. However, their continued operations will probably be counter-productive in the long run. These two flagship policies focus only on prosumers but they neither consider the issue of grid stability, nor provide adequate incentives for consumption of green energy. With renewable generation reaching critical mass, we need more robust and future-proof mechanisms that not only incentivize prosumers, but also have a broader scope and align the interests of all stakeholders. We believe that the Blockchain technology is a crucial infrastructure for such mechanisms.

NRGcoin is a decentralized Blockchain-based alternative to the traditional support policies and reward programs. It can bring more stable revenue streams to prosumers, lower the risk of policy changes for investors, and automatically scale as the number of prosumers grow. It will promote supply-demand balance on the grid, lower the cost of green energy to end consumers and hence, offset the need for grey energy. Likewise, NRGcoin can also be combined with **load shifting** and **storage capabilities** to provide more benefits to DSOs, retailers, and consumers (Razo-Zapata et al., 2016). As preliminary results indicate (Razo-Zapata et al., 2016), retailers, and DSOs benefit from grey energy demand reduction during peak times, which can be down by 50%. Additionally, consumers benefit from low prices for green energy (Razo-Zapata et al., 2016). In this way, NRGcoin incentivizes the local energy production and consumption and, effectively, boosts the local renewable energy economy.

Finally, the NRGcoin concept provides a clear example of a novel mechanism in which public sector (e.g., DSOs) and private sector businesses (e.g., retailers) could work together to reduce greenhouse gas emissions (International Energy Agency, 2016c). A more sustainable public-private partnership in the renewable energy sector will be conducive to leveraging more investment for accelerating global energy transitions.

References

Bertoldi, P., Rezessy, S., & Oikonomou, V. (2013). Rewarding energy savings rather than energy efficiency: Exploring the concept of a feed-in tariff for energy savings. *Energy Policy*, *56*, 526–535.

De Boeck, L., Van Asch, S., De Bruecker, P., & Audenaert, A. (2016). Comparison of support policies for residential photovoltaic systems in the major EU markets through investment profitability. *Renewable Energy*, *87*(2016), 42–53.

Butler, L., & Neuhoff, K. (2008). Comparison of feed-in tariff, quota and auction mechanisms to support wind power development. *Renewable Energy*, *33*(8), 1854–1867.

Cerd, E., & del Ro, P. (2015). Different interpretations of the cost-effectiveness of renewable electricity support: Some analytical results. *Energy*, *90*(Part 1), 286–298.

Couture, T. D., Cory, K., Kreycik, C., & Williams, E. (2010). A policymaker's guide to feed-in tariff policy design *(No. NREL/TP-6A2-44849)*. United States: National Renewable Energy Lab (NREL), Golden, CO.

European Commission. (2011). *Energy roadmap 2050: Communication from the commission to theeuropean parliament, the council, the European economic and social committee and the committee of the regions.*

European Commission. (2013). *European Commission guidance for the design of renewables support schemes. Delivering the internal market in electricity and making the most of public intervention.* Brussels: The European Commission.

Greer, C., Wollman, D.A., Prochaska, D.E., Boynton, P.A., Mazer, J.A., Nguyen, C.T.... Hefner Jr, A.R. (2014). *NIST framework and roadmap for smart grid interoperability standards, release 3.0*, Tech. Rep. NIST SP-1108r3, The National Institute of Standards and Technology, United States.

International Energy Agency. (2015a). Energy technology systems analysis programme (IEA−ETSAP), International Renewable Energy Agency (IRENA), *Renewable energy integration in power grids*, Tech. Rep. Technology Brief E15, IEA-ETSAP.

International Energy Agency. (2015b). *Technology roadmap: How 2 guide for smart grids in distribution networks*. OECD/IEA, Tech. Rep.

International Energy Agency. (2016a). *Renewable energy − Market analysis and forecasts to 2021*, Tech. Rep. Medium-Term Market Report 2016, IEA.

International Energy Agency. (2016b). *Next generation wind and solar power—from cost to value*, Tech. Rep., 2016.

International Energy Agency. (2016c). *Energy, climate change and environment*. OECD/IEA, Tech. Rep.

International Energy Agency. (2017). *Market-based instruments for energy efficiency*, Tech. Rep.

Jacobs, D., Marzolf, N., Paredes, J. R., Rickerson, W., Flynn, H., Becker-Birck, C., & Solano-Peralta, M. (2013). Analysis of renewable energy incentives in the latin america and caribbean region: The feed-in tariff case. *Energy Policy*, *60*, 601−610.

Mihaylov, M., Jurado, S., Avellana, N., Van Moffaert, K., de Abril, I. M., & Nowe, A. (2014a). *NRGcoin: Virtual currency for trading of renewable energy in smart grids. European energy market (EEM), 11th International Conference on the* (pp. 1−6). IEEE.

Mihaylov, M., Jurado, S., Van Moffaert, K., Avellana, N., Nowe, A. (2014b). *NRG-X-Change: A novel mechanism for trading of renewable energy in smart grids*, In: Proceedings of the 3rd International Conference on Smart Grids and Green IT Systems, pp. 101−106.

Nakamoto, S. (2008). *Bitcoin: A peer-to-peer electronic cash system*.

Phatak, D., Robucci, R., Banerjee, N., (2014). Beyond a usage threshold, NO form of energy is sustainable or green, In: Proceedings of the 3rd International Conference on Smart Grids and Green IT Systems, pp. 185−190.

PwC. (2017). *Transitioning to a low carbon energy system*, 2017.

Rathmann, M., de Jager, D., De Lovinfosse, I.D., Breitschopf, B., Burgers, J., Weores, B. (2011). *Towards triple-a policies: More renewable energy at lower cost*, Re-Shaping D16 Report.

Razo-Zapata, I.S., Mihaylov, M., Nowe, A. (2016). *Analysing the impact of storage and load shifting on grey energy demand reduction*, SMARTGREENS 2016, VEHITS 2016: Smart Cities, Green Technologies, and Intelligent Transport Systems, pp. 27−48.

REN21. (2016). *Renewables 2016 global status report*, Tech. rep., Renewable Energy Policy Network for 21st Century.

Rickerson, W., Couture, T., Barbose, G.L., Jacobs, D., Parkinson, G., Chessin, E.... Barrett, H. (2014). *Residential prosumers: Drivers and policy options (re-prosumers)*, Tech. rep., International Energy Agency (IEA)- Renewable Energy Technology Deployment.

Sokolowski, M. (2016). *Regulations in the European electricity sector*. Routledge Studies in Energy Law and Regulations, Routledge.

Ueckerdt, F., Hirth, L., Luderer, G., & Edenhofer, O. (2013). System LCOE: What are the costs of variable renewables? *Energy*, *63*, 61−75. [(Online].)Available: http://www.sciencedirect.com/science/article/pii/S0360544213009390.

U.S. Energy Information Administration. (2017). *Short-term energy outlook*, July 2017.

INTERLUDE III

Blockchain for Smoother International Climate Finance Transfers

> If we stop thinking of the poor as victims or as a burden and start recognizing them as resilient and creative entrepreneurs and value-conscious consumers, a whole new world of opportunity will open up.
>
> ***C.K. Prahalad (1941–2010), late University of Michigan Professor, The Fortune at the Bottom of the Pyramid (2004)***

The preceding section has explored the tremendous potential of Blockchain technology to transform the supply and demand sides of mature energy markets. In this section, let us turn to the needs of developing countries. Likewise, the openness, transparency, and immutability of data stored on a Blockchain, distributed across millions of users on the network, will deliver transformative opportunities to the poor in developing countries. At least it can efficiently address the three "gaps" in international climate finance mentioned in the Editor's Prologue.

First, Blockchain is set to accelerate financial inclusion for the poor, i.e., "banking the unbanked." Anyone with Internet access, e-mail address, or phone number is eligible for an "online bank account"—a "digital wallet." There is no requirement for legal identification, maintenance fees, or minimum balance. Blockchain, therefore, expands the coverage of microfinance facility to billions more people in developing countries, which can channel climate finance to local communities more directly. Second, Blockchain can track every single penny of climate funding for governments and donors in almost real time. The project and funding data on a customized Blockchain will be visible to all taxpayers to ensure the accountability of aid-recipient countries. Third, smart contracts will enhance the monitoring and evaluation of donor-funded climate projects in developing countries. In the spirit of results-based climate finance, climate funds cannot be disbursed until the smart contracts are triggered by the attainment of agreed project milestones. It can therefore streamline lots of procedures within international climate funds or financial institutions. In Chapter 10, Blockchain—Powering and Empowering the Poor in Developing Countries, Dr. Jane Thomason, together with her 12-person team experienced in applying Blockchain to poverty alleviation, will discuss a series of practical use cases. They include tracking climate finance, program results tracking, climate adaptation, financial inclusion, and digital identity.

As with the peer-to-peer energy transaction system introduced in the preceding section, Blockchain also enables peer-to-peer donations and, more importantly, secure voting systems. A secure and credible voting system will democratize the currently top-down decision-making or assessment processes of many climate change project proposals within international climate funds. More stakeholders such as (voluntary) climate policy experts and even residents in remote villages can be involved in selecting the most suitable project proposals for the communities. In Chapter 11, Disintermediating the Green Climate Fund, Dr. Tim Reutemann will introduce his concept of liquid democracy to "disintermediate" the Green Climate Fund and increase its legitimacy. In Chapter 12, COCOA—Crowd Collaboration for Climate Adaptation, a team of three innovators, Dr. Cristián Retamal González, Dr. Iván Razo-Zapata, and Dr. Gustavo Arciniegas López will present their proposal for Crowd Collaboration for Climate Adaptation (COCOA). It connects donors with beneficiaries to transfer financial resources via smart contracts and allows stakeholders to share climate change expertise and local know-how on an open platform as part of the process of assessing the feasibility of climate change project proposals.

The application of Blockchain is not confined to public climate finance but private climate finance too! The defining features of Blockchain technology and smart contracts can rewire relevant financial infrastructure and thus boost confidence of private investors in various green investment projects. Blockchain will mobilize substantially more climate finance if it is combined with other digital technologies such as artificial intelligence. In Chapter 13, Using Smart Algorithms, Machine Learning, and Blockchain Technology to Streamline and Accelerate Dealflow in Climate Finance, Neil Salisbury and Jenya Khvatsky of CleanTek Market will introduce their marketplace that provides a smart algorithm-supported matchmaking platform for institutional investors and green project developers. It marries machine learning, a subset of artificial intelligence, and a package of Blockchain services to enhance dealflows, that is, to direct private capitals to the best-matched green investment projects. In which case, the poor is really being empowered to be resilient and creative entrepreneurs in their climate-friendly community businesses.

Integrated with green finance concept, Blockchain could turn natural resource management into profitable businesses or investments. Water is one of them. A Blockchain-based water management system would make a huge difference in regions which are increasingly arid because of climate change. In Chapter 14, Addressing Water Sustainability With Blockchain Technology and Green Finance, Anna Poberezhna will analyze the global problem of water scarcity and propose an innovative concept of water credit trading on a Blockchain network based on the principles of circular economy.

CHAPTER 10

Blockchain—Powering and Empowering the Poor in Developing Countries

Jane Thomason[1], Mira Ahmad[2], Pascale Bronder[2], Edward Hoyt[2], Steven Pocock[1], Julien Bouteloupe[3], Katrina Donaghy[4], David Huysman[5], Tony Willenberg[5], Ben Joakim[6], Loretta Joseph[7], David Martin[8] and David Shrier[9]

[1]Abt Associates, Australia [2]Abt Associates, United States [3]Doshup Ltd, United Kingdom [4]Civic Ledger, Australia [5]Neocapita, Romania [6]Disberse, United Kingdom [7]Australian Digital Currency Commerce Association[1] [8]Power Ledger, Australia [9]University of Oxford[2], United Kingdom

10.1 Introduction

This chapter focuses on the relevance of Blockchain and climate finance for the poor in developing countries. Climate change has a disproportionate impact on the world's poor. The Global Humanitarian Forum's Human Impact Report estimated that climatic disasters impact 375 million people every year, and will displace up to 750 million people by 2050 (Global Humanitarian Forum, 2009; Strauss et al., 2015). Impoverished communities are more dependent on climate-sensitive sectors, such as agriculture and farming; thus, climate change poses an extreme threat to their food security, livelihoods, and health. A study by the United Nations Food and Agriculture Organisation showed that more than 22% of the damage caused by climate disasters in developing countries between 2003 and 2013 occurred in the agriculture sector (Food and Agriculture Organization of the UN, The impact of disasters on agriculture and food security The United Nations Research Report, 2015). Climate change affects rainfall, temperature, and water availability for agriculture in vulnerable areas, impacting productivity, agricultural practices, and land availability (Best Climate Practices, 2015). Climate change also impacts marine environments, including coral reef systems that have suffered bleaching due to warming seas. Land-use changes, particularly deforestation in tropical regions, where forests tend to be rich in biodiversity, are responsible for roughly 18% of human-driven carbon dioxide emissions (Wikiwand, 2017). Disadvantaged communities also have increased exposure and vulnerability to health threats due to climate change (Commission on Social Determinants of Health, 2008).

---

[1] http://adca.asn.au/

[2] https://www.sbs.ox.ac.uk/community/people/david-l-shrier

Transforming Climate Finance and Green Investment with Blockchains.
DOI: https://doi.org/10.1016/B978-0-12-814447-3.00010-0

Developing countries will require tens of billions of dollars in new funding annually to adapt to climate changes. Forty eight of the Least Developed Countries[3] have drawn up national adaptation plans that identify priority actions estimated at least US$2 billion per year (Tenzing et al., 2015). Financial Technology (FinTech) and Blockchain innovations can play a major role in implementing these adaptation plans. It is increasingly apparent that, Blockchain innovations can help build trust and transparency in international climate finance and could potentially have major implications in scaling and speeding up north-south, south-south, and peer-to-peer (P2P) climate finance transfers. This chapter will explore the potential of Blockchain to provide scalable solutions in the areas of: tracking climate finance, results tracking, climate adaptation, financial inclusion, and identity. It will discuss requirements to achieve scale, including building the ecosystem by linking Blockchain developers, and funding sources with people working on problems of poverty. The big problems of our time need bold, collaborative solutions.

10.2 Where Does Blockchain Fit in?

In 2016, the World Economic Forum released a report[4] calling Blockchain technology a "mega-trend"[4] that will shape society in the next decade, predicting that Blockchains could store as much as 10% of global Gross Domestic Product (GDP) by 2027 (Thomason, 2017a). It has spawned a flood of activity with that $1.4 billion was invested globally in Blockchain start-ups representing the most rapid scale-up of a technology ever seen (PwC, 2016).

In climate finance and green investment for local adaptation, Blockchain can strengthen transparency in the reporting, monitor effectiveness, simplify settlements, and reduce intermediaries and long funding delays. Access to climate funding has been described as "excruciatingly painful" by India and Namibia (Lo, 2016). Blockchain offers solutions for identity, health records, financial services for the unbanked, microfinancing, financial transparency, off-grid solar power, water trading, carbon trading, supply chain, provenance of goods, and land registry.

10.3 Immutable Identity on Blockchain

Blockchain's ability to create an immutable identity makes it a game changer for the global poor. Identity is central to everything—it is a fundamental human right and a prerequisite for financial inclusion, access to services, formal employment and movement of people. Currently, 2.4 billion people worldwide do not have an identity (Dahan and Gelb, 2015). Blockchain provides a game-changing opportunity to confer a permanent, immutable record of identity where no state identity exists. This will improve accuracy and transparency for

[3] https://www.un.org/development/desa/dpad/least-developed-country-category/ldcs-at-a-glance.html

[4] https://www.weforum.org/reports/the-future-of-financial-infrastructure-an-ambitious-look-at-how-blockchain-can-reshape-financial-services

carbon trading, beneficiary identity, and cash transfers for climate finance and financial inclusion more generally.

"Digital identity" is either (1) the credentials issued by a government to an individual (making the distinction between credentials, which are by decree and revocable, and identity, which is intrinsic and irrevocable), or (2) at best, a set of attributes assignable to an individual (fingerprint, iris scan, behavioral traces) that in aggregate form a unique identity. Digital identity in the context of climate change and Blockchain can be so much more. Identity data held immutably on a Blockchain remains in the ownership of the individual and individuals give permissions on who can see what data for what purpose and for how long. It is a transformative concept that will have many benefits.

Digital identity provides portability and recognition. Blockchain enables a decentralized, irrevocable, unique identity, and then creates a personae that can exist for one-time-use or for a longer duration. The result is the protection of sensitive identity attributes and increased identity transferability and utility.

Digital identity can deliver better governance. The immutable nature and individual ownership of digital identity that Blockchain provides would assist in the mitigation of legitimate privacy concerns that developed nations currently face in existing state-based, digital identity programs. Availing a digital identity for the people should be the prerequisite for developing any climate or development financing platform on which to conduct transactions. "Self-sovereign" identity means the identity owner has the ability to *govern* their data: to decide who can possess it, how it is used, and most importantly, how it is reclaimed (the "right to be forgotten") (Allen, 2016). With the advent of Blockchain-based smart contracts, it is possible to have self-sovereign digital ID for things as much as for people.

To reach "the last mile," innovative ways of recording identity for remote and low-infrastructure populations are needed. Case Study One provides an example of an innovative device that is being tested to provide secure identity to populations without electricity and internet.

Case Study One

IDBox[5], developed by Doshup, is a simple, low-cost solution identity being developed and tested in Papua New Guinea, in partnership with the Central Bank of Papua New Guinea and Abt Associates[6]. It enables the hashing of a digital fingerprint onto a Blockchain, creating an immutable identity. It has been developed for low-resource, low-infrastructure settings and can be used with basic analog phones. IDBox does not require electricity or the internet. The use case trial is taking place in July and August of 2017. The IDBox is durable and portable and has great potential in emerging markets, hard to reach populations and in humanitarian settings. http://www.idbox.io/about.html

[5] http://www.idbox.io/

[6] http://www.abtassociates.com/Blockchain.aspx

Digital identity can also introduce greater transparency. The intersection of inexpensive sensors enabled by the Internet of Things[7], with the capabilities that the Blockchain unlocks could expand dramatically the economic management of the environment. Every tree and every iceberg (and every factory) could have a digital ID. Once measured, it can be tracked and managed. For example, just as carbon credits became a useful tool for regulating carbon output through economic forces, we have oxygen credits (for oxygen-producing entities), carbon sink credits (for extracting carbon from the atmosphere), and so on. New identity systems offer unprecedented opportunities for understanding, control, management, and conservation of critical environmental resources.

10.4 Blockchain and Tracking Financial Flows to the Poor

Climate finance is the flow of donor and government funds toward activities that reduce or mitigate greenhouse gas emissions or help communities adapt to climate change's impacts, and can catalyze other investment flows. Climate finance and adaptation will play an important role in helping developing countries meet the UN's Sustainable Development Goals[8] (United Nations, 2017).

Each year, over one trillion dollars flows from individuals, governments, and businesses to address the challenges of poverty and crisis across the world (Carraro, 2017). The distribution and tracking of global development and humanitarian aid funds remains complex, opaque, and hugely inefficient. Transfers can take weeks to arrive and 5–10% losses are not uncommon (The World Bank, 2017).

Lack of transparency is also a key issue as there is an inability to trace the flow of funds from end to end. The UN estimates that up to 30% of official development assistance is lost due to fraud and corruption (Jenny, 2012). The result is less funds, leading to reduced impact for those who need it most. On the ground, organizations face multiple barriers to ensuring full transparency of fund distribution—weak infrastructure, capacity limitations, and cash-based systems all contribute to the challenge for international agencies and local organizations to be fully accountable to both their donors and the communities and individuals they work with.

The Climate Funds Update[9], a joint initiative of the Heinrich Böll Stiftung (HBF)[10] and the Overseas Development Institute (ODI)[11], monitors dedicated climate change funds from the stage when donors pledge funding, through to the actual disbursement of financing for

[7] https://www.forbes.com/sites/jacobmorgan/2014/05/13/simple-explanation-internet-things-that-anyone-can-understand/#20bd6c5b1d09

[8] http://www.un.org/sustainabledevelopment/sustainable-development-goals/

[9] http://www.climatefundsupdate.org/

[10] http://www.boell.org/

[11] http://www.odi.org.uk/

projects, in an effort to increase the transparency of climate finance flow (Climate Finance Fundamentals, 2017). Blockchain applications to track financial flows and report results, and smart contracts that target beneficiaries have potential to speed up financial flows and vastly improve accuracy and transparency of reporting.

Applications are being tested to improve the flow and targeting of donor funds. The Case Study Two highlights Disberse[12], a Blockchain platform to track flow of donor funds.

Case Study Two

Disberse[12] is a fund-management platform for the global development sector, built on Blockchain technology. It drives the transparent, efficient, and effective flow and delivery of development and humanitarian aid. It enables donors, governments, and Non-Governmental Organizations (NGOs) to transfer and trace funds through the whole chain, from donor to beneficiary, via intermediaries. It has potential to be used for social cash transfers to enable mobile money transfers and voucher schemes for the most vulnerable. It could be employed by any development and humanitarian projects that involve the transfer of funds between two or more stakeholders (Disberse, 2017).

Swaziland: Disberse[12] has partnered with the UK charity Positive Women[13] which raises funds in the UK from individuals and foundations, and then distributes the funds to partner NGOs in Swaziland. This pilot has a school fees component (developed in conjunction with schools, government, and the families) and a wider personal-development support program. The pilot, tested between the UK and Swaziland, sent funds to four Swazi schools via a local NGO. Disberse[12] enabled Positive Women to reduce their transfer fees by 2.5% and trace the flow of funds down the chain, resulting in zero losses at the points of delivery. Thanks to the cost savings, Positive Women were able to fund an additional three students' school fees for a year.

Start Network: Disberse[12] have partnered with the Start Network[14] who manage a $100 million humanitarian response fund from institutional donors such as Department for International Development (DFID), Swedish International Development Cooperation Agency (SIDA), and Irish Aid. Start Network will roll out a full cross border, end-to-end pilot in late 2017, and will be the first organization to fully track the flow of funds though the chain, from donor through to beneficiary.

www.disberse.com/

10.5 Blockchain and Tracking Results

Results-based climate finance pairs measurable changes in environmental factors with payments as incentives (or disincentives for negative environmental changes). Currently cash transfers for international climate finance are slow, cumbersome and expensive and

[12] http://www.disberse.com/

[13] http://www.positivewomen.org/

[14] https://startnetwork.org/

rely on intermediaries. In the future, cash transfers could be driven by a network of interconnected Blockchains. With Blockchain, reliance on humans and institutions will be exchanged for reliance on code that can be checked and calculated from the data itself. Given the interconnectedness of the variables that make up natural environmental systems, it is inevitable that greater amount of Blockchain technology will be used to automate the measurement of each link and remunerate those responsible for positive contributions to the individual links, which benefits the environment as a whole.

Blockchain smart contracts offer a tamper-proof and zero-cost administrative mechanism for connecting positive (or negative) environmental changes or outcomes to financial incentives/disincentives. This is infinitely more efficient than the current global system of aid administration we have in place today.

For example, a measurable reduction in CO_2 or methane measured by an IoT based network of atmosphere monitoring sensors placed around a village, might "trigger" the release of cryptocurrency or record a commitment to pay to the village account—an amount based on the observed measurement in the environment. As soon as the sensor readings are transmitted via the Internet to the smart contract, the smart contract will apportion payment for the demonstrable reductions in greenhouse gases at the local level. Case Study Three provides an example of an application that aims to track climate finance and results.

Case Study Three

StoneBlock[15] is a secure decentralized platform tying funds to agreed-upon actions to ensure mutual accountability. It provides transparent and tamper-proof "metering" service of development aid flows at a range of scales: a project, programe, portfolio, and national, regional, and international. Applied to the many millions of dollars that target poverty, these platforms will be able to significantly enhance targeting and accountability. Stoneblock provides a general registration service or a "digital registry" for citizen information in the custody of a government, on a Blockchain. Information currently in the custody of the government, but belonging to the citizen, can now be rendered on the Blockchain and tied to a national ID. Rather than try to provide a use case specific workflow support system to manage the various and numerous e-government instances, Stoneblock cuts across all use cases providing an identity based "data layer" that all e-government systems can take advantage of. The basis of all storage is a public-private key pair, meaning that Stoneblock is the first platform that provides a unique kind of data layer to existing applications, one that is tethered tightly to the identity of the actors involved in storing the record. Stoneblock, effectively solves two e-government problems such as identity of the owner of the citizen information, and ensures that information is recorded in a tamper-proof manner. Stoneblock

(Continued)

[15] http://neocapita.com/

Case Study Three (Continued)

is being piloted now in a number of government agencies in a number of use cases in Somalia, Armenia, and Samoa Neocapita (2017).

http://neocapita.com/

10.6 Blockchain and Renewable Energy

The production and delivery of modern forms of energy, and especially electricity, represents one of the major challenges the world faces today—especially for the developing countries. The provision of access to affordable electricity through the introduction of off-grid renewable energy technologies, especially based on solar photovoltaic, has the potential to dramatically improve the lives of the more than 1.1 billion people who lack access worldwide, including some 650 million in Sub-Saharan Africa, 350 million in Asia, and about 20 million in Latin America and the Caribbean ("Energy Access Database", 2017).

Access to affordable electricity can have a transformational impact on poor rural communities by impacting a range of sectors including: education, economically important activities such as cottage industries and commerce, communications, access to national and international media and information, security, and, of course, recreation. In short, in the hyper-connected world of today, access to affordable electricity is critical.

Blockchain can play a useful role in the process of introducing renewable technologies suited to the needs and capacity to pay of the rural poor by reducing transaction costs, increasing the transparency and traceability of customer payment data, and increasing trust between parties. Furthermore, the growing traction in Blockchain and cryptocurrencies open the doors to autonomous market players and services to disrupt the entire energy market using a P2P business model. Developing countries represent one of the most interesting use cases for decentralized energy through a P2P infrastructure and using new, low-cost renewable energies.

Imagine a world where everyone could be a reseller and a buyer of renewable energy and where the cost of the energy was automatically calculated based on the number of people in a neighborhood producing and consuming electricity. A combination of Blockchain, new technologies which significantly reduce fabrication costs, and the massive adoption of mobile phones in developing countries makes this feasible. For example, the price of solar panels has dropped over 80% over the last decade (Whitwam, 2016). It is now cheaper to produce and buy solar energy than fossil energies (Griffin, 2017). Solar panels can now be connected to the Blockchain to enable consumers in developing countries to benefit from distributed generation (i.e., LO3 Energy Brooklyn Microgrid[16]).

With Blockchain, someone from a village in a developing country can buy small solar panels and plug them to a mini grid network of cables to produce electricity for their local community. Another person from the same village can use their phone to buy electricity from the solar panel as a pay-as-you-go service, or the provider of electricity to villagers on a pay-as-you-go basis and collect payment from the end user (and disconnect the solar unit in the event of a nonpayment). The incentive of buying solar panels for people in remote areas is compelling as they can immediately generate electricity and get paid instantly by the consumer. More and more people from the same village can buy solar panels and plug them into the Blockchain network. The smart contracts on the Blockchain allow participants in that system—consumers and producers—to buy and sell solar energy from each other, using digital tokens that can be redeemed for a local cryptocurrency. As the Blockchain is immutable and highly secured through a distributed infrastructure, this process is automated and cannot be corrupted by an individual or institution.

Clean energy tech start-ups are developing solutions and applications that focus on renewable energy, reducing carbon emissions and developing smart grids that will optimize energy consumption (Climate Change, 2017). Blockchain can assist these start-ups in offering a decentralized energy solution that is reliable and effective. As such, Blockchain has a significant chance to help combat climate change as it enables the creation, usage, and exchange of renewable decentralized energy in a trustworthy and reliable manner. This makes renewable energy more attractive to consumers, organizations, and societies.

Solar start-ups Azuri Technologies, Off-grid Electric, and Mobisol are producing low-cost solar panel solutions for off-grid areas in rural Africa (Azuri Technologies, 2017). Affordable solar panels offer a tangible solution that bring clean energy to markets where once kerosene was the only option. This smart PayGo system, makes solar technology affordable a fraction of the price of kerosene (Burger, 2017). PayGO energy access business model when households pay off the solar panels, they move from renting to owning an asset. This can transform the lives of off-grid rural citizens making them owners of cutting-edge technology and building a healthier safer home environment.

Case Study Four

Power Ledger[17] has developed a Blockchain-enabled P2P energy-trading platform that allows for the secure and transparent allocation of value from distributed energy resources, including

(*Continued*)

[16] http://lo3energy.com/projects/

[17] https://powerledger.io/

Case Study Four (Continued)

roof-top solar PV, to the owners of the PV systems, even if they are not the consumers of energy.

Where individuals produce more energy than they consume, the Blockchain-based trading platform supports the sale of excess energy to other consumers within a mini grid.

The Blockchain-based system allows for the tracking of energy generation and consumption in close to real time, meaning the owners of roof-top PV systems can get a secure return on their investment in the same moment the energy they produce is consumed.

Relying on smart-meter data, the trading platform can distribute returns to the owners of roof-top PV systems, or where the PV systems are provided under microfinancing terms, to the original finance provider until the initial cost of the installation is paid back.

Once the amount of the microfinancing has been returned, the full value of the energy generated by the PV system accrues to the owner, building economic capacity and, potentially, the opportunity to install more PV and increase both the return to the individual as well as the amount of energy available in developing communities.

The system is currently being deployed in embedded networks (multitenanted buildings) in modern electricity systems in Australia, but the opportunities for the application of the technology in emerging economies, particularly those undergoing rapid urbanization and electrification, are immense.

In this energy-trading scenario, the Blockchain provides the "source of truth" that enables the pairing of a physical transaction (in this case energy) with a financial transaction giving security and certainty to the providers of finance as well as economic opportunity to those who can sell their excess energy to energy-hungry consumers in their community Power (2017).

https://powerledger.io/

A newly developed Blockchain-enabled P2P trading platform—Water Ledger—is creating a way to solve transparency problems in Global Water Markets.

Case Study Five

The ability to trade water has been essential in maintaining irrigation sector incomes during drought, and is expected to be even more important under future climate change scenarios. Although water resource management is complex, it doesn't need to be complicated.

Australia is widely acknowledged as a global leader in water resources management, and more specifically in the application of water markets and trading as a key tool to share water between competing uses including urban, industry, environment, and agriculture. As governments globally continue to reform their water resources management in the face of

(*Continued*)

Case Study Five (Continued)

climate change and greater competition between users, many are turning to water markets and water trading. China is just beginning to embrace water trading (Wei, 2016). The European Union has created water trading protocols, Spain needs a water trading platform and opportunities for a simple trading platform exist in California's Central Valley, Nevada groundwater, and Texas water markets.

In early 2017, Civic Ledger[18] successfully secured a grant from the Australian Government to undertake a Feasibility Study to solve transparency problems in Australia's water markets to boost confidence with the view to increasing the volume and reliability of water trading. As a result of the Feasibility Study, Civic Ledger[18] is now developing a Blockchain-enabled –P2P trading platform–Water Ledger–that allows for the secure and transparent trading of water entitlements and allocations within water markets, increase participation by irrigators, and increase the overall allocative efficiency of the water resource. Water Ledger will provide substantial value for money for governments, contribute to increased revenue for irrigators, and keep farming communities strong while conserving additional water for the environment Civic (2017).

www.civicledger.com

10.7 Blockchain and Mobilization of Capital

There remains a gap between the intermediaries in the climate finance "system" and the entrepreneurs in emerging markets who face an enormous challenge in accessing capital. Sustainability will require engagement with local financial institutions by helping them identify investment opportunities in renewable energy, and supporting them in the development of new financial products and services to channel capital to the most attractive segments of the renewable energy space.

Achieving lower costs and lower risk is especially important for the expansion of renewable energy in areas of the world that still are largely excluded from access to modern forms of energy and financial services. This is because the ability to mobilize capital for these sectors will increasingly depend on the involvement of local financial institutions providing capital to local entrepreneurs, as well as emerging international companies in the energy access space.

International banks, investors, and development finance institutions can play a role in expanding the reach of renewable energy into regions previously lacking access to electricity. Many have done so, but their reach is constrained by their need to make investments and issue loans on a scale that is far larger than what businesses require (Dalberg Global Development Advisors, 2010), and typically not in local currencies.

[18] http://www.civicledger.com/

The mobilization of capital to support businesses in the energy access industry will require the expansion of a variety of investment vehicles, modernization, and innovation on the part of local financial institutions, rapid evolution in the sophistication and capabilities of local entrepreneurs, and a change in the approach to regulation of the energy sector on the part of governments. Fueled by capital deployed by international banks, impact investors, and development finance institutions, some major corporations such as Engie (Engie, 2017), have attracted individual investors through crowd financing platforms such as Bettervest (Bettervest, 2017).

However, the rapid growth of the energy access industry will only result in a transformational impact for the world's unelectrified population once local capital is mobilized to these regions. This necessitates the participation of local financial institutions, entrepreneurs, and pension funds. The process that should take place will look similar to the way that cellular phones have become ubiquitous throughout emerging markets as well as in industrial economies. Although increasingly aware of the opportunities in clean energy technologies, local banks and other financial institutions have yet to fully engage in the sector.

In order for it to be possible to harness Blockchain technology to address the needs of businesses, lenders, and financiers in the energy access space, several key shifts must occur. Local financial institutions must position themselves to engage in the sector, they must come to see energy access as a viable and potentially profitable business opportunity, and they must understand the technologies that underpin the business models of emerging energy access companies. Many of these companies are expanding their businesses in Sub-Saharan Africa and globally (The Chicago Council on Global Affairs, 2015). Financial institutions must also develop their own versions of the mobile money payment systems, similar to M-Pesa[19] (introduced by Safaricom in 2007). Developing mobile payments will increasingly be viewed as essential to remaining competitive in the banking industry, which even in emerging markets can be extremely dynamic and highly competitive.

10.8 Faster Cheaper Remittances

Blockchain's unique characteristics allow financial institutions to tailor their products and services to promote ease of use for the unbanked and underbanked. A Philippine company, Coins.ph[20], offers a good example of Blockchain's potential for this use case. Situated in the country ranked third in the world for receiving remittances (totaling about USD$28 billion a year) (The World Bank, 2016), Coins.ph provides Filipino users a mobile, Blockchain-based platform to allow them to send money at a more affordable and faster

[19] https://www.safaricom.co.ke/personal/m-pesa

[20] https://coins.ph/

rate. Blockchain allowed Coins.ph to build an application to facilitate fund transfers without reliance on existing bank infrastructures and to be more agile in their services at a more affordable price (Coins.pH, 2016). Current remittance processes are slow and expensive. Digital currency remittances can move funds from remitter or donor to recipient almost instantaneously with low transaction costs (Holmes, 2016). CASHAA[21] is already operating in India and Nigeria and transmits remittances at no cost to the consumer. Recent experience from Bangladesh shows that women actually prefer to receive payments by phone, rather than cash (Klapper, 2017). In order to be possible, this will rely on the confirmation of the identity of the recipient, emphasizing the criticality of solving the identity piece. With Blockchain-enabled digital currencies and identity—financial inclusion can be provided to the two billion unbanked[22] in the world. Blockchain would reduce the transaction costs for remittances and climate transfers, giving the unbanked access to financial systems, and ensuring that climate funds intended for the poor actually reach them.

Mobilization of domestic financing and private financial flows will be also critically important. The private sector is a primary source of employment, influence, and ideas. Blockchain enables new forms of finance to address global climate finance problems, including crowdfunding and dynamic funding mechanisms from private finance markets. Tax mechanisms and incentives allow Blockchain to encourage the private sector to invest in renewable energy; for example, through tax credit schemes where credit flows back to investor. A second example is the mobilization of global efforts to tap into pension funds, and advocacy with governments to bring tax relief to pension funds that pursue social investment into climate adaptation transparently recorded on the Blockchain. To mobilize private money will also necessitate efforts to create better measurement of social return. Blockchain enables that transparency. More funding is needed in venture capital to support the development and testing of new technologies for climate adaptation, that have longer time horizons, lower financial returns but higher social returns financing.

10.9 Crowdfunding Through Digital Currencies and Initial Coin Offerings

Blockchain-based cryptocurrencies are increasingly in circulation. Initial coin offerings (ICOs), token sales, and crowdfunding are rapidly growing in popularity and scale. Blockchain start-up Golem[23], for example, raised \$8.2 million in less than 20 min based on a white paper (Thomason, 2017b). The term white paper originated in Britain and was initially an authoritative report on a complex issue (Anthony, 2017). The term has been adopted by tech companies as a marketing presentation. Over \$327 million was raised via

[21] https://cashaa.com/

[22] https://www.weforum.org/agenda/2016/05/2-billion-people-worldwide-are-unbanked-heres-how-to-change-this

[23] https://golem.network/

ICOs in 2017 (http://www.coindesk.com/ico-investments-pass-vc-funding-in-Blockchain-market-first/). This new fundraising phenomenon is fueled by a synergistic convergence of Blockchain technology, new wealth, clever entrepreneurs, and cryptoinvestors backing Blockchain-fueled ideas. Start-ups are raising funds by creating their own cryptocurrencies and offering discounted rates on digital assets before they hit the cryptocurrency exchanges. Some of the ICO's are focusing on social value, so this presents potential for raising finance for clean energy as an example. The Blockchain community is experimenting with ideas around tokenizing social value; so, it is likely that this phenomenon may be a means of raising more climate finance in the future.

10.10 What Still Needs to Be Done to Take This to Scale?

There are several things that will be required for Blockchain to scale in developing countries.

Too few governments and international development practitioners are familiar with Blockchain and its potential. If Blockchain technology is to diffuse without a significant lag, then it needs to be socialized with development professionals and how it applies to use cases. Institutional resistance is to be expected by institutions that will be disrupted by this technology. Middle men will be disintermediated and resist the change and those who benefit from the lack of transparency in financial flow will be unlikely to embrace Blockchain.

The role and uptake of governments remains crucial, as the process of developing bankable projects has high upfront costs, often untenable by small local entities and communities. Public intervention will thus be more effective where it targets local capacity building and strives to overcome impediments of scale (e.g., setting up pooled finance facilities) so that private investments can be fully mobilized.

There remain several regulatory and operational barriers, such as the high-energy consumption of Blockchain processes (Energerati, 2017) and the governance of personal data. The cost model for "last mile" users needs to be developed as current transaction based models will be expensive for poor consumers. There also remain infrastructure barriers, with low-power settings, poor Internet, and in some developing areas, low penetration of mobile phones. There are also not enough developers worldwide to keep up with the scale of Blockchain expansion.

Funding to create seed investments in early stage start-ups is insufficient and difficult to access. There is scant funding to support developers and teams to get prototypes to the point at which they can commercially scale. Current VC funding models do not support early stage start-ups. Taking successful use cases to scale will likely require new forms of financial and institutional partnerships.

There have been some disconnected efforts to connect technology developers and the international development community. However, the ecosystem remains nascent and there is a need to unite the Blockchain systems in advanced economies with the people that work with the intractable problems of poverty and inequality. This includes, developers, platforms, people who know problems, government, and finance (VC, impact investing, local finance donors, ICOs, etc.).

10.11 Conclusion

Blockchain technology offers a path forward in the quest to address two of our times' most pressing issues: climate change and the growing levels of poverty. Blockchain's potential for social impact is yet to be realized, however, with insufficient use cases in developing countries. The focus of the global community needs to move from the underlying technology to exploring use cases—and looking for solutions that can scale. There are many practical questions to be answered in settings with little Internet access and electricity. There is a need to reach out even more to connect innovative ideas, with the local private sector, and social finance organizations to encourage innovation, entrepreneurship, and action for pressing development challenges. Blockchain represents a strategic ally, not an alternative route, toward a greener, inclusive, and resilient economy.

References

Allen, C. *The path to self-sovereign identity, CoinDesk.* (2016). <http://www.coindesk.com/path-self-sovereign-identity/> Accessed 20.07.17.

Anthony, J. *Origin of white paper, Klariti.com.* (2017). <http://klariti.com/white-papers/origin-of-white-papers/> Accessed 19.07.17.

Azuri Technologies. (2017). <http://www.azuri-technologies.com/> Accessed 06.7.17.

Best Climate Practices. *How does climate change affect water and agriculture?* (2015). <http://www.bestclimatepractices.org/how-does-climate-change-affect-water-and-agriculture/> Accessed 20.07.17.

Bettervest. (n.d.). <https://www.bettervest.com/en/> Accessed 20.07.17.

Burger, A. (n.d.). *Easy energy with mobile pay, Eniday.com.* <https://www.eniday.com/en/technology_en/pay-as-you-go-energy-with-mobile-payments/> Accessed 19.07.17.

Carraro, C. *Towards a greener finance landscape, International Center for Climate Governance.* <http://www.carlocarraro.org/en/topics/finance/towards-a-greener-finance-landscape/> Accessed 20.07.17.

Civic Ledger. (n.d.). <http://www.civicledger.com/> Accessed 06.07.17.

Climate change. (n.d.). *The new economy, COP23 series—How can Blockchain help combat climate change?* <http://climatechange-theneweconomy.com/cop23-series-can-Blockchain-help-combat-climate-change/> Accessed 19.07.17.

Climate finance fundamentals. (n.d.). <http://www.climatefundsupdate.org/resources/finance-fundamentals> Accessed 06.07.17.

Coins. (n.d.). pH. <https://coins.ph/> Accessed 06.07.16.

Commission on Social Determinants of Health. *Closing the gap in a generation: Health equity through action on the social determinants of health (Final Report of the Commission on Social Determinants of Health, World Health Organization).* <http://apps.who.int/iris/bitstream/10665/43943/1/9789241563703_eng.pdf.>. Accessed 20.07.17.

Dahan, M., & Gelb, A. *The identity target in the post-2015 development agenda, The World Bank.* (2015). < http://www.worldbank.org/en/topic/ict/brief/the-identity-target-in-the-post-2015-development-agenda-connections-note-19 > Accessed 20.07.17.

Dalberg Global Development Advisors. *The growing role of the development finance institutions in international development policy.* (2010). < https://www.deginvest.de/DEG-Documents-in-English/Download-Center/EDFI_The-growing-role-of-DFIs-in-International-Development-Policy_2010.pdf >. Accessed 20.07.17.

Disberse. (n.d.). < http://www.disberse.com/ > Accessed 06.07.17.

Energerati. (n.d.). *The downside of energy Blockchains? They consume lots of power.* < http://www.engerati.com/article/downside-energy-Blockchain-carbon-footprint > Accessed 19.07.17.

Engic. (n.d.). < http://www.engie.com/en/ > Accessed 20.07.17.

Food and Agriculture Organization of the UN. The impact of disasters on agriculture and food security (The United Nations Research Report). (2015). < http://www.fao.org/3/a-i5128e.pdf >. Accessed 20.07.17.

Global Humanitarian Forum. *The anatomy of a silent crises (Human Impact Research Report—Climate Change).* (2009). < http://www.ghf-ge.org/human-impact-report.pdf >. Accessed 20.07.17.

Griffin, A. *Solar and wind power cheaper than fossil fuels for the first time, independent.* (2017). < http://www.independent.co.uk/environment/solar-and-wind-power-cheaper-than-fossil-fuels-for-the-first-time-a7509251.html > Accessed 20.07.17.

Holmes, B. *Bitcoin, remittance, service, brave new coin.* (2016). < https://bravenewcoin.com/news/global-remittances-slow-while-regulation-restricts-access/ > Accessed 20.07.17.

Jenny, R. (n.d.). *30 percent of aid lost to corruption- Ban Ki-moon, Devex, July 12, 2012/July 5, 2017.* < https://www.devex.com/news/30-percent-of-aid-lost-to-corruption-ban-ki-moon-78643 >. Accessed 20.07.17.

Klapper, L. (n.d.). *Why women workers and managers prefer electronic wage payments in Bangladesh, Better Than Cash.* < https://www.betterthancash.org/news/blogs-stories/why-women-workers-and-managers-prefer-electronic-wage-payments-bangladesh > Accessed 08.03.17.

Lo, J. *Why can't poor countries access the climate finance they were promised? The Guardian.* (2016). < https://www.theguardian.com/global-development-professionals-network/2016/feb/15/small-island-states-green-climate-fund > Accessed 20.07.17.

Neocapita. (n.d.). < http://mdanderson.libanswers.com/faq/26219 > Accessed 06.07.17.

Power Ledger. (n.d.). < https://powerledger.io/ > Accessed 06.07.17.

PwC. (n.d.). < https://www.cryptocoinsnews.com/pwc-expert-1-4-billion-invested-Blockchain-2016/ >. Accessed 20.07.17.

Strauss, B., Kulp, S., & Levermann, A. *Mapping choices: Carbon, climate, and rising seas: Our global legacy (Climate Central Research Report).* (2015). < http://sealevel.climatecentral.org/uploads/research/Global-Mapping-Choices-Report.pdf >. Accessed 20.07.17.

Tenzing, J., Gaspar-Martins, G., & Jallow, B.P. *LDC perspectives on the future of the Least Developed Countries Fund (LDC Paper Series).* (2015). < https://ldcclimate.files.wordpress.com/2012/05/ldc-perspectives-on-the-ldcf.pdf >. Accessed 20.07.17.

The Chicago Council on Global Affairs. *AFRICA IS NOW: The opportunity for mid-sized US companies.* (2015). < https://www.lw.com/mediaCoverage/africa-is-now-the-opportunity-for-mid-sized-US-companies > Accessed 20.07.17.

The World Bank. *Remittances to developing countries edge up slightly in 2015.* (2016). < http://www.worldbank.org/en/news/press-release/2016/04/13/remittances-to-developing-countries-edge-up-slightly-in-2015 > Accessed 20.07.17.

The World Bank. *Remittance prices worldwide: An analysis of trends in cost of remittance services.* (2017). < https://remittanceprices.worldbank.org/sites/default/files/rpw_report_june_2017.pdf >. Accessed 20.07.17.

Thomason, J. *Opinion: Embracing Blockchain solutions to fight global poverty, Devex.* (2017a). < https://www.devex.com/news/opinion-embracing-Blockchain-solutions-to-fight-global-poverty-89564 > Accessed 18.07.17.

Thomason, J. *Changing the world crypto style – First UK conference on initial coin offerings* (2017b). < https://www.linkedin.com/pulse/changing-world-crypto-style-first-uk-conference-initial-thomason > Accessed 19.7.17.

United Nations. (n.d.). *Sustainable development doals.* < https://sustainabledevelopment.un.org/sdgs > Accessed 19.07.17.

Wei, X. *Water trading aims at better use of resource, China Daily*. (2016). < http://www.chinadaily.com.cn/china/2016-06/29/content_25897654.htm > Accessed 21.07.17.

Whitwam, R. *We could be headed for a solar power renaissance as costs keep dropping, Extreme Tech.* (2016). < https://www.extremetech.com/extreme/241300-headed-solar-power-renaissance-costs-keep-dropping > Accessed 20.07.17.

Wikiwand. (n.d.). *Climate change and poverty.* < http://www.wikiwand.com/en/Climate_change_and_poverty > Accessed 18.07.17.

Energy Access Database. (n.d.). *International Energy Agency.* < http://www.worldenergyoutlook.org/resources/energydevelopment/energyaccessdatabase/ > Accessed 19.07.17.

Alex Sunnarborg (2017) < http://www.coindesk.com/ico-investments-pass-vc-funding-in-Blockchain-market-first/ > . Accessed 18.07.17.

CHAPTER 11

Disintermediating the Green Climate Fund

Tim Reutemann

Managing Director RedMorpheus GmbH[1], Zurich, Switzerland

11.1 The Task at Hand

Financing international climate change projects is one of the most enormous efforts in the history of human cooperation. The United Nations Framework Convention on Climate Change (UNFCCC) has brought about the most legitimate international climate agreement among 197 countries ever (UNFCCC, 2015). Part of that agreement is the creation of a pool of money and an institution to spend it for a global public good, with the mission to help poorer parts of the world to cope with climate change. But despite a well-researched and traditional task for governments, such procurement for the public good is dangerous and requires utmost care, according to the Organization for Economic Cooperation and Development (OECD) "Public procurement is one of the government activities most vulnerable to corruption" (OECD, 2016). Therefore, it is perfectly reasonable for donor countries to require extreme care and insist on a rule-driven process with little wiggle room for lower-level individuals in the climate finance institutions to mitigate abuse of funds.

Apart from the usual difficulties in public procurement, prioritization of climate change funding is extraordinarily difficult, too: Climate change interacts with every aspect of the economy and, for each sector, needs to find a balance between financing "loss and damage," "adaptation," and "mitigation." Further, the distribution of international climate finance is also prioritized toward poorer countries, as expressed by the differentiation between donor and host countries and the special status given to least developed countries and small-island developing countries (http://www.greenclimate.fund/who-we-are/about-the-fund, 2017).

A combination of these properties to make financing the response to climate change an unprecedented challenge of global governance.

[1] https://www.redmorpheus.com/

Transforming Climate Finance and Green Investment with Blockchains.
DOI: https://doi.org/10.1016/B978-0-12-814447-3.00011-2

11.2 The Second Millennium Solution

The first step toward a solution was the creation of a multilateral consensus platform, involving all nations, namely the UNFCCC and its annual Conference Of the Parties (COP) meetings in 1992 (UNFCCC, 1992). And, although there have been other efforts to finance climate change projects before, it took 15 years until the COP text first mentioned the Green Climate Fund (GCF) in 2009 (UNFCCC, 2009). The first actual projects were approved for GCF funding by the end of 2015 (http://www.greenclimate.fund/-/green-climate-fund-approves-first-8-investmen-1, 2017), but the first projects were not granted all their paper work until summer 2016 (http://www.finsmes.com/2016/08/private-equity-impact-fund-kawisafi-ventures-launches-operations.html, 2017), and so far no reports of actual financial transfers to purchase actual technologies have been issued.

This is far too slow to get anywhere near preventing catastrophic climate change (UNEP, 2016). The gap between what is scientifically required and actually feasible has become so large that the United Nations Environmental Program does not even include a scenario to achieve the 2°goal without hypothetical negative total global emissions in the second half of the century, despite such technologies still being early-stage pilots and their scale-up to predicted levels far from certain (https://qz.com/1100221/the-worlds-first-negative-emissions-plant-has-opened-in-iceland-turning-carbon-dioxide-into-stone/, 2017).

Apart from failing to fund sufficient levels of mitigation, the GCF is also struggling to fulfill its mandate to prioritize proposals by entities from developing countries and especially from least developed countries, because its current pipeline is dominated by multinational institutions such as the Multilateral Development Banks or the United Nations Development Program (https://www.worldfinance.com/markets/the-uns-climate-fund-is-failing-to-invest-in-those-most-in-need, 2017) (http://www.greenclimate.fund/what-we-do/portfolio-dashboard, 2017). Yet, the GCF is not like any other financial institution ever created before in human history. Thus, there is no reference point on how an efficient yet legitimate fund for a global public procurement by the consensus of over seven billion people should operate.

11.3 Second Millennium Legitimacy

In simple words, the UNFCCC procedure gains its legitimacy by extending national governments' legitimacy (Claude, 1966). It is also using the same bureaucratic, rule-driven hierarchical method of governance that national governments typically use (Biermann and Siebenhüner, 2009). Hierarchic approaches lose efficiency over longer chains of command, and thus the slow speed of the GCF is not surprising in light of the long chain of command from sovereign voters to the GCF. A strict and rule-driven bureaucratic management style below board level is a reasonable tool to reduce the potential for conflict over individual decisions and corruption (Peters et al., 2015). Under this management style, individual employees are required to operate mechanically and make as few autonomous decisions as possible. There is no proven and universally accepted way to do any better. In particular, as long as legitimacy is created on paper ballots, bureaucracy seems to be the only viable choice to maintain it along a chain of commands.

It is humbling to compare the complexity of the actual COP process with the naiveté of the early-stage ideas on climate change applications in the Blockchain space. Yet, the remainder of this chapter will propose another such early-stage idea, as the second millennium governance technologies have apparently reached their limits, whereas climate change is too important a problem for surrender.

11.4 Disintermediation Potential

There are some smaller issues in climate change that seem like natural use cases for Blockchains at first sight. In particular, the tracking of climate certificates (https://www.hack4climate.org/#challenges, 2017) and the Monitoring, Reporting, and Verification of emissions and their reductions are frequently targeted by Blockchain start-ups and institutions (https://unfccc.int/secretariat/employment/UserManagement/FileStorage/8QWN14FOHEGUSVIABDKPMX6JZY9L03, 2017). Although there is certainly space for improvement, the second millennium solutions have done a reasonable job on those two tasks in the past, using a central database hosted by the UNFCCC and international third-party auditing of participating companies.

But this chapter aims at a deeper issue in climate finance—**the diffusion of political legitimacy through the institutions involved**. The following points analyze the chain of paper-based intermediation steps between the will of "sovereign citizens" and funding decisions at the GCF:

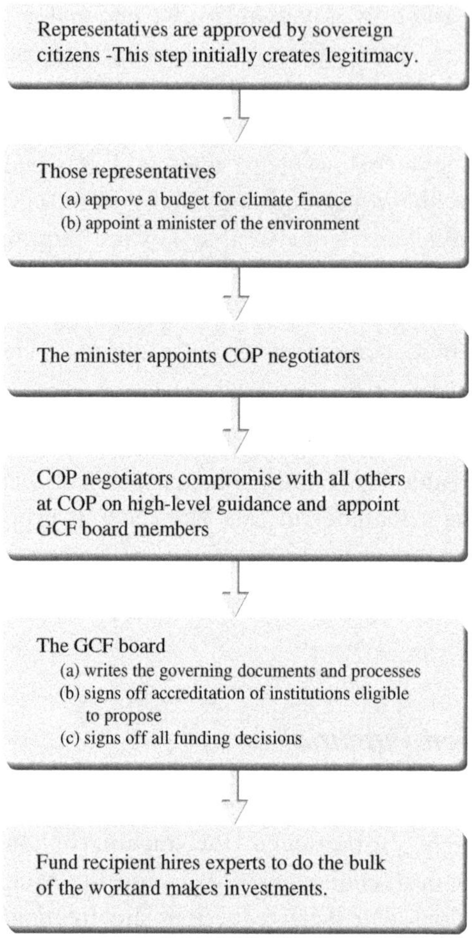

For host countries of GCF projects, the process further includes the Minister appointing a Nationally Designated Authority (NDA) who needs to approve every proposal that a project developer wants to implement in that country (http://www.greenclimate.fund/how-we-work/empowering-countries, 2017).

11.5 Decentralizing Legitimacy

There are some steps in this chain that are vital for legitimacy and operations, which therefore should not be targeted for disintermediation, namely Steps (1), (4), and (6). Step (1) creates the initial legitimacy of the decision by public approval at a national level. Step (4) forms an internationally legitimate consensus from there. Step (5) is required to inform and actually implement the decisions. But Steps (2), (3), and (5) are fair game for disintermediation, as these do not involve decisions as in Step (1), compromise between sovereigns (4) or action (6), but only appointments as in Steps (2) and (3) or the intermediated execution of decision and compromise (5).

Decentralizing legitimacy opens the door to adhocratic forms of governance and can thus dramatically reduce the burden of bureaucracy on the GCF and various stakeholders (Toffler, 1970). In an adhocratic governance form, decisions on issues like approval of funding or accreditation are made case-by-case, without a universal set of rules applicable to all cases. In a paper-based, nontransparent structure, this approach opens the door for corruption and discrimination of various forms. Yet, when using Blockchain-enabled disintermediated democratic tools, adhocracy can become a viable form of public management. When legitimacy is devolved to low-level employees or experts and their decisions directly from sovereign voters, their decisions have the same level of legitimacy as the high-level rules negotiated at the COP, and can thus operate without them. Adhocracy can dramatically increase efficiency as it allows a large number of people to make legitimate decisions as needed to achieve results, depending on their national circumstances.

11.6 Disintermediated Democracy

The concept of a decentralized and disintermediated democracy has been discussed elsewhere in depth (https://github.com/DemocracyEarth/paper, 2017; https://github.com/the-laughing-monkey/cicada-platform/blob/master/Cicada-WhitePaper-2016-10.13.GA.1.pdf 2017). As this proposal only targets *one* governmental function, namely public procurement, the platform outlined here can be significantly slimmer than what is required for a full-fledged decentralized government. Further, the set-up is simplified as donor country voters have no stake in the allocation of funds except their willingness to address climate change, whereas host country voters incur no costs, except for possible negative impacts of project implementation. Still, one core idea from the wider e-democracy literature is vital for this task, namely the use of "*liquid democracy*" as a solution to the conflict between voter fatigue due to a large numbers of small decision votes and the goal of full legitimacy of all those small decisions. Liquid democracy is a new form for collective decision making that gives voters full decisional control like a direct democracy, and also allows voters to delegate their decision power to others like in a representative democracy. It further allows to on-delegate votes.

In the current system for most democratic countries, voters delegate their vote to a regional representative (parliamentarian) once every election cycle, who collectively decide on the budget for climate finance and on-delegate their vote on the use of climate finance to the Minister of the Environment. The minister on-delegates to COP negotiators and GCF board members, who on-delegate some small decisions to GCF employees.

In a liquid democratic format, all these delegations and all decisions taken along the chain remain under direct control of the voters, allowing them to change both decisions and delegations taken anywhere in the chain.

Such an electronically enabled set-up dramatically reduces the rigidity of rules required, as the will of voters can directly interfere if decisions are taken against their interest, instead of having to wait until the end of the election cycle; and being only able to provide legally binding feedback on all decisions taken by their representative at once by voting for a different person.

Although the GCF currently uses an accreditation procedure to reduce the number of proposals from nonqualified actors, such accreditation might not be needed if funding decisions can legitimately be taken on a case-by-case basis. Therefore, only three types of decisions require yes/no voting: project eligibility, project approval, and achievement of milestones. For example, one issue that needs careful debate and compromise are projects with significant non-climate components (https://twnetwork.org/climate-change-gcf/debate-gcf-board-over-funding-development-and-climate-change-projects, 2017). Owing to limited resources, not all eligible projects will be able to gain approval; thus, some prioritization is needed along the current guidelines of the GCF, including on efficiency of the projects, geographical balance, and balance between mitigation and adaptation.

Liquid democratic voting is, in principle, direct, so every voter in every participating country has the option to vote individually on each project's eligibility. The difference to a pure direct democratic system is that in liquid democracy voters can "follow" the votes of others, or "delegate" their votes to others, thus creating a hybrid between direct and representative democracy. Liquid delegation can also be revoked at any time and differentiated by tags.

For instance, a voter can decide to follow the WWF vote on eligibility for all proposals tagged with "biodiversity" while follow the International Renewable Energy Agency (IRENA) for all proposals tagged with "solar." Further, chains of delegation are possible to allow multiple layers of engagement: a student of environmental science may ask her less-engaged family members to delegate all votes on climate finance to her. She then has the option to "forward-delegate" those votes to her favorite NGOs. Voters are also informed of all decisions made by their delegates as those decisions are registered in a public ledger. Although too busy to engage full-time, the student may spend an hour per month to read a summary of all decisions her and her families votes facilitated, and potentially revoke her delegation or communicate with the NGO in case she disagrees. This feature enables broad participation by otherwise disengaged voters through personal networks of topic-based trust and; low-entry barriers for engaging interested voters.

11.7 Voter Eligibility and Registration

Voter eligibility and registration deserves extra attention in the design of electronic democracy and has often been found as a major barrier to mass participation (https://github.com/DemocracyEarth/paper, 2017) (https://github.com/the-laughing-monkey/cicada-platform/blob/master/Cicada-WhitePaper-2016-10.13.GA.1.pdf, 2017). We propose a pragmatic approach reliant on existing second millennium technologies. Therefore, voter eligibility should be fairly easy for donor countries where voters sign up with existing unique ID solutions such as social security numbers or ID cards. In contrast, voter eligibility in host countries could be more challenging (http://www.worldbank.org/en/programs/id4d, 2017; https://www.uport.me/, 2017). Both the lack of democratic traditions and lack of access to the Internet may make broad participation difficult to achieve in many least developed countries. It is also expected that the platform will not be global from the start, but rather operate as a bilateral vehicle within a "club of the willing." To ensure maximal legitimacy, it is proposed that the platform start from the status quo, with all host-country voting power on project eligibility being centralized at the NDA by default, whereas offering the NDA technical support for decentralization of its legitimacy to the citizens.

Voter sign-up in donor countries will be open to any citizen, at any time. See the section on "Addressing Voter Engagement" for incentives for sign-ups.

11.8 Funding Decisions

Once a project has been voted eligible and approved for funding by a global unqualified majority, provided that it has not been vetoed by the NDA, the platform could operate like a conventional crowdfunding site.

The main difference from classic crowdfunding approach is that voters do not use their own money, but decide the allocation of the available public finance. Further, the approval of milestones leading to the disbursement of further funds is required, similar to a multistage crowdfunding project. Once milestones are approved through liquid democratic voting, either traditional mechanisms for disbursements or a third millennium solution such as "Disberse" could be applied (http://www.disberse.com/).

11.9 Addressing Voter Engagement

Lack of voter engagement is a major and reoccurring point of criticism for all forms of more direct democratic governance. Thus, this section discusses a pathway to growth and engagement.

The platform envisioned in this chapter needs to cross a threshold in terms of public participation if it is to generate real legitimacy; and can only aspire to develop the checks and balances necessary to avoid fraud and corruption once a diverse set of actors participates in it. The second step—after coding a functional prototype—is thus to create an incentive for broad participation.

Citizens, climate change experts, and project developers who could get paid for, say, project assessment work through the platform are eligible to vote. Assuming an initial donor provides sufficient seed funding for a few climate change projects, estimated in the lower two-digit million dollars, stakeholders with commercial interest in the approval of projects are expected to be the first voters to sign up. This creates a competition among project developers to convince larger numbers of voters to delegate votes to them. Also, NGOs like the WWF have an incentive to reach out to their members asking for sign-up and vote following, as they can take on multiple roles in the process, either as evaluators or project developers. Although the GCF so far operates only with governmental donors and control, the Paris Agreement also stipulates the private sector and other "nonstate actors" to contribute to climate finance (UNFCCC, 2015). The ideas presented here not only work for donor nations with their citizens as voters, but also any other "donor group," such as Google with voting rights for their users, or Starbucks with voting rights for all its customers. Such donor groups would not be able to gain extra votes on project eligibility, but could decide on project approval and milestone acceptance, given that the funds to pay for those projects come entirely from their own donations.

If every donor group controls the budget they provide (b); and each of the individual voters in that group controls an equal share, as the total number (n) of active voters increases, the budget controlled per person (b/n) decreases. Therefore, the incentive to participate is very high initially; however, as more voters join the platform, the marginal reward for convincing an additional voter decreases. This effect is counter-balanced by the expectation of more public contributions (increase in b) as the voter base grows.

11.10 Where Does the Project Stand?

The project at this stage is primarily an idea developed in collaboration with a climate professional and a group of developers during the "Consensys Blockchain for Social Impact Hackathon," called the "Liquid Climate Fund." Although no actual code has been written, a possible architecture of the platform has been drafted and a number of open-source projects have been identified that solve parts of the problem; for example, the sovereign app by the

Democracy Earth Foundation or CarbonX. At this point, numerous features are being examined:

-Proof of Identity (By leveraging national ID if possible)

This step is aspirational and only required if individual voters are to gain direct access to voting in the Liquid Climate Fund. An immediately available solution would be to provide voting power to NDA's initially and offer further support to disintermediate the process for individuals.

-Smart contracts for e-voting on project eligibility

Project eligibility includes both acceptance of a project as relevant for climate change; and acceptance of the project proponent as eligible and capable. Thisincorporatesthe function currently fulfilled by the accreditation process. It is expected that such votes will be delegated to experts (similar to the current employees of the secretariat), but leave them more room for individual and ad-hoc decisions. Eligibility voting should be a relatively simple contract, as it is basically a yes/no decision.

-Smart contracts for e-voting on budgeting

While significant stakeholder consultations will be necessary to define the details, the current idea is to allow donor country voters to allocate the funds provided by their country to individual projects on an even per-capita basis. The Blockchain solution will be similar to a traditional crypto-currency, which can be transferred to an escrow account created for every approved project. Projects can launch as soon as their requested funding is allocated to them.

-Smart contracts for e-voting on milestone verification and payoutsTo access funds in the escrow, milestones need to be approved with reference to data collected from various remote sensing or Internet of Things devices. Approval will be a simple yes/no decision again, where only voters who allocated funding to the project are eligible to vote.

-Access to platform in under-developed countries wherethe Internet has not been popularized, for example by SMS.

The most difficult problem is the decentralization of voting power in developing countries. Yet, the project can start without this step by accepting the legitimacy of the existing NDAs. It will have to be left to these existing government institutions to define which form of ID is accepted for voting, which confines the experiment to a "club of the willing." Further, we expect voters in donor countries to be more likely to support projects in countries that democratize decision making, which thereby creates an incentive for NDAs to seek innovative ways of low-tech voter engagement. Proposals for voter engagement technologies could also be financed through the platform.

Beyond those points, we will continue to publish further ideas and thought experiments on the authors blog at www.medium.com (@tim.reutemann) and the code on the open-source code sharing site, Github (LiquidClimateFund).

11.11 Outlook

Climate finance is an excellent place for first experiments with global democratic consensus mechanisms. And the most vital component of it is the creation and conservation of legitimacy. But initially, electronic democratization of climate finance will remain for a "club of the willing." Most importantly, lots of experiments on the rules, user interfaces, and incentives, etc., are required before any serious proposals for a USD$100 billion apparatus can be made. Believably, creating a show-case for functional electronic democracy could be a door opener for wider efforts to turn the tide on the global recession of democracy with new technologies (https://www.eiu.com/topic/democracy-index, 2017).

References

Biermann, F., & Siebenhüner, B. (Eds.), (2009). *Managers of global change: The influence of international environmental bureaucracies*. MIT Press.

Claude, I. (1966). Collective legitimization as a political function of the United Nations. *International Organization*, *20*(3), 367–379. Available from https://doi.org/10.1017/S0020818300012832.

<http://www.disberse.com/>.

<http://www.finsmes.com/2016/08/private-equity-impact-fund-kawisafi-ventures-launches-operations.html> Accessed on 15.10.17.

<http://www.greenclimate.fund/-/green-climate-fund-approves-first-8-investmen-1> Accessed on 15.10.17.

<http://www.greenclimate.fund/how-we-work/empowering-countries> Accessed on 15.10.17.

<http://www.greenclimate.fund/what-we-do/portfolio-dashboard> Accessed on 15.10.17.

<http://www.greenclimate.fund/who-we-are/about-the-fund> Accessed on 15.10.17.

<http://www.worldbank.org/en/programs/id4d> Accessed on 15.10.17.

<https://github.com/DemocracyEarth/paper> Accessed on 15.10.17.

<https://github.com/the-laughing-monkey/cicada-platform/blob/master/Cicada-WhitePaper-2016-10.13.GA.1.pdf> Accessed on 15.10.17.

<https://qz.com/1100221/the-worlds-first-negative-emissions-plant-has-opened-in-iceland-turning-carbon-dioxide-into-stone/> Accessed on 15.10.17.

<https://twnetwork.org/climate-change-gcf/debate-gcf-board-over-funding-development-and-climate-change-projects> Accessed on 15.10.17.

<https://unfccc.int/secretariat/employment/UserManagement/FileStorage/8QWN14FOHEGUSVIABDKPMX6JZY9L03> Accessed on 15.10.17.

<https://www.eiu.com/topic/democracy-index> Accessed on 15.10.17.

<https://www.hack4climate.org/#challenges> Accessed on15.10.17.

<https://www.uport.me/> Accessed on 15.10.17.

<https://www.worldfinance.com/markets/the-uns-climate-fund-is-failing-to-invest-in-those-most-in-need> Accessed on 15.10.17.

OECD, (2016). *Preventing corruption in public procurement.* <http://www.oecd.org/gov/ethics/Corruption-in-Public-Procurement-Brochure.pdf> Accessed on 15.10.17.

Peters, B. G., von Maravić, P., & Schröter, E. (Eds.), (2015). *Politics of representative bureaucracy: Power, legitimacy and performance*. Edward Elgar Publishing.

Toffler, A. (1970). Future shock. <https://archive.org/stream/FutureShock-Toffler/Future-Shock_-_Toffler_djvu.txt>

UNEP, (2016). *Emission gap report.* <http://wedocs.unep.org/bitstream/handle/20.500.11822/10016/emission_gap_report_2016.pdf?sequence=1&isAllowed=y> Accessed on 15.10.17.

UNFCCC, (2015). *Paris climate accord.* <https://unfccc.int/files/meetings/paris_nov_2015/application/pdf/paris_agreement_english_.pdf> Accessed on 15.10.17.

UNFCCC, (2009). *Copenhagen accord.* <https://unfccc.int/resource/docs/2009/cop15/eng/l07.pdf> Accessed on 15.10.17.

UNFCCC, (1992). *Kyoto protocol.* <https://unfccc.int/resource/docs/convkp/kpeng.pdf> Accessed on 15.10.17.

Further Reading

Stern Review on the Economics of Climate Change, (2006).

CHAPTER 12

COCOA—Crowd Collaboration for Climate Adaptation

Gustavo Arciniegas López[1,2], Iván Razo-Zapata[2,3] and Cristián Retamal González[2,4]

[1]Geo-Col GIS and Collaborative Planning, Delft, South Holland, The Netherlands [2]COCOA Collaborative Innovation, Delft, South Holland, The Netherlands [3]Luxembourg Institute of Science & Technology, Esch-sur-Alzette, Luxembourg [4]Amanhã Ventures, Denver, CO, United States

12.1 Introduction

All over the world, there is a need to adapt to climate change by helping local communities to build resilience. It usually takes local communities to organize themselves in civil societies that propose climate adaptation initiatives, so as to request funding for such initiatives. Adaptation initiatives, however, often face several bureaucratic processes to access financial resources. For instance, a local community—in any given location—that organizes to design a certain measure to improve their resilience to climate change, and wants to apply for funds from the international climate funds, will most likely need to secure governmental endorsements, necessitating completion of huge amounts of paperwork with information and other bureaucratic actions, which are usually far from the community's actual know-how.

The main problem is the lack of trusted entities to support such initiatives. The funding entities may change funding programs as a result of governmental or corporate policies. Consequently, accessing financial resources becomes a crucial yet challenging issue for organized communities or civil societies in general. These resources, nonetheless, constitute key elements for supporting local communities in building resilience to climate change. An example of this situation is the set of adaptation measures on water availability promoted in the Coquimbo region in Chile, where the International Research Institute for Climate and Society at Columbia University has provided pivotal resources for improving the resilience of communities in Chile's Elqui Valley[1].

[1] For more details please visit https://vimeo.com/98403077

Transforming Climate Finance and Green Investment with Blockchains.
DOI: https://doi.org/10.1016/B978-0-12-814447-3.00012-4

Blockchain technology becomes an interesting instrument to tackle this problem as it offers three main features: a transparent decentralized ledger (TDL), fast and trustless transactions (FTTs), and no need for central third parties (NCP) (Business Insider, 2016).

The TDL offered by Blockchain is important as it supports tracking the transfers of financial resources, whereas FTTs are essential to rapidly mobilizing resources from funders to beneficiaries. In this context, "trustless" means that funders and beneficiaries are not required to know each other. Likewise, the fact that no central third parties are required guarantees that resources can transparently flow from funders to beneficiaries without any bureaucratic formality.

This chapter presents the concept of COCOA^2 platform, which aims to facilitate the transfers of financial resources to support projects for adaptation to climate change. COCOA^2 comprises two layers: technical knowledge and financial resources. They rely on three main modules: knowledge-sharing technologies, multicriteria analysis (MCA), and smart contracts, which are (software-based) forms of engagement that use the Blockchain's TDL to rule the transfers of values among parties.

On one hand, knowledge-sharing technologies and MCA are important tools for understanding the impacts of local initiatives and tracking their progress. On the other hand, by exploiting the aforementioned Blockchain features (TDL, FTT, and NCP), COCOA's^2 smart contracts can facilitate the transfers of financial resources (e.g., fiat money and/or cryptocurrencies)—or other types of assets—from funders to beneficiaries. TDL offered by Blockchain makes it possible for COCOA's^2 smart contracts to track and verify transactions between funders and beneficiaries. Likewise, via FTTs, smart contracts can help efficiently disburse financial resources for projects that require prompt actions (e.g., timely adapting to forecasted floods). Finally, NCP guarantees that smart contracts-aided transfers of resources address the interest of both beneficiaries and funders, rather than simply following the interest of governmental or corporate policy regimes.

12.2 Challenges Facing Adaptation Finance

12.2.1 Adaptation Ideas Could Come From Unforeseen Origins

Climate change impacts are not homogeneous across territories and depend crucially on the characteristics of each natural ecosystem, and on both the vulnerability and resilience of the local communities inhabiting those territories. In such context, climate adaptation measures depend imperatively on the local knowledge of the territory, its geography, ecosystem services, and the priorities and features of the local communities directly affected by climate change impacts.

[2] https://www.climatecolab.org/contests/2012/enabling-adaptation/c/proposal/1304127

It is understood that planning for climate change adaptation fundamentally requires the development of participatory processes that consider the understanding, concerns, and enlightenment of local communities, since great adaptation initiatives could come from simple measures and local know-how (CARE, 2014).

12.2.2 Origination of Adaptation Measures and Traditional Flows of Climate Finance

As the international climate change negotiations in the UNFCCC processes have historically been government- or state-based, climate change adaptation measures tend to be a derivative of governmental or corporate top-down policies, instead of bottom-up initiatives developed with active involvement of organized communities in civil societies. This explains the global climate finance flows within the framework of the international climate governance regime being traditionally subject to bureaucratic processes pivotal to stakeholders in donor countries, funds managers, as well as accredited entities implementing agencies or, recipient governments as intermediaries, while the local communities (i.e., ultimate beneficiaries) are sometimes not in the equation.

Such bureaucratic structures of the international climate governance regime and its financial ecosystems typically impede the origination of many adaptation initiatives directly from organized communities. They do not regularly participate in the established dynamics on the international climate finance landscape. In addition, the bulk of climate finance flows has always been directed toward mitigation initiatives, and to a much lesser extent toward building resilience through climate adaptation (CPI, 2015).

On another front, there have been very limited endeavors studying how Blockchain can support climate finance flows, putting together supply and demand in this sphere (Unfccc. int, 2017). Current Blockchain-based solutions (including Carboncoin[3], NRGcoin[4], and Solarcoin[5]) address some issues but still come short of full engagement with local communities.

12.2.3 The Sources of the Problem

The lack of technical knowledge and existing bureaucracy in the current climate finance architecture may be quite restrictive for *organized communities* to implement climate adaptation initiatives in their territories.

First, the selection or prioritization of adaptation projects for funding is subject to governmental discretion, justified with technical arguments and metrics, and usually also impacted by unavoidable political calculations. Consequently, many interesting climate

[3] http://carboncoin.cc/
[4] http://nrgcoin.org/
[5] https://solarcoin.org/

adaptation ideas originated at local community levels have never passed the proposal assessment stage by conventional international climate funds.

Second, funding entities are characterized by cumbersome bureaucracies and procedures, lengthy approval processes and complicated structures designed to safeguard the integrity of the funds. It could take years for an international climate fund to complete a project funding cycle.

12.2.4 COCOA: Crowd Collaboration for Climate Adaptation

Against the current backdrop, where adaptation is of utmost priority to keep pace with climate change and community activism is crucial to sustainable communities, COCOA^2 is based on the ethos that great adaptation initiatives can come from any local sources.

With massive penetration of social networks into people's lives, community activism is much more feasible than it was decades ago. The COCOA^2 idea has been developed with the understanding that effective climate change adaptation planning requires technology-based and evidence-driven participatory processes which integrate otherwise out-of-reach technical knowledge into the development priorities of local communities. Such participatory processes will no longer confine adaptation planning to government officials and associated personnel in their bureaucracies.

On the COCOA^2 platform, this web-based network of stakeholders shares common motivation, interest, and necessities, without more procedures than needed. Crowd communities focus on developing their ideas to be turned into finance-ready climate adaptation projects, but not on maintaining the governance structures of any organizations framing the projects.

12.3 The COCOA Platform

The COCOA^2 is a web-based platform featuring three main components:

1. Smart contracts to ensure transparency and trust in the transfers of financial resources;
2. Knowledge-sharing technology (i.e., webinars and short videos) to allow efficient promotion of adaptation initiatives (i.e., projects); and
3. Proposal evaluation methodologies based upon MCA to assess the impacts (positive and negative) and potential benefits of promoted projects.

12.3.1 Overall Design

Figure 12.1 presents the overall architecture that supports the operation of the COCOA^2 platform. For explanatory purposes, we organize the three main components by two layers:

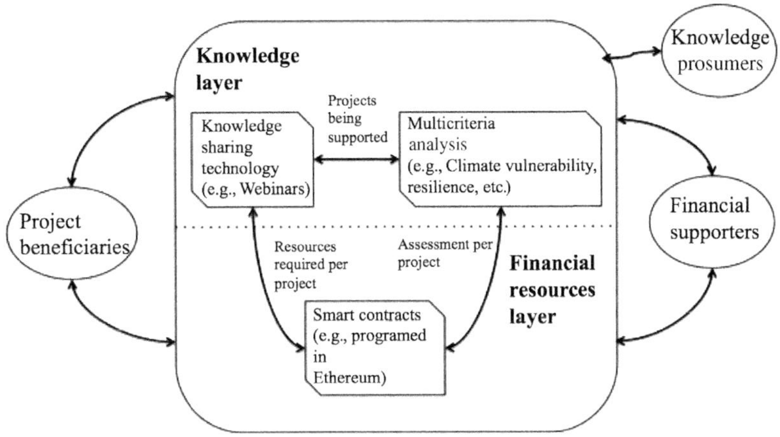

Figure 12.1
COCOA's[2] main components.

a *knowledge layer* and a *financial resources layer*. Likewise, we foresee at least three types of actors that will interact with the COCOA[2] platform, namely *project beneficiaries*, *knowledge prosumers*, and *financial supporters*.

On one hand, the knowledge layer supports the exchange of information between the knowledge-sharing technologies module and the MCA module. The former provides details about the projects being supported and/or promoted, whereas the latter performs assessments of the possible impacts of each project, and whether they are on track to achieve their goals. On the other hand, the financial layer contains only the smart contracts module that exchanges information with the modules on the knowledge layer. The communication among the three modules is facilitated via web-based service technology (though the knowledge layer itself may or may not need to be on a Blockchain network).

Project beneficiaries interact seamlessly with both layers since they provide (and potentially use) knowledge about projects and make use of financial resources. Similarly, *financial supporters* not only contribute financial resources but can also generate and consume knowledge. For example, they can learn about the unique attributes characterizing local initiatives in different countries.

In contrast, *knowledge prosumers* provide and use knowledge mostly regarding project designs as well as monitoring and evaluation. That is, climate change experts can contribute their knowledge to several initiatives without financially supporting them. In some cases, they can also be financially rewarded by project beneficiaries or funders for their expertise. Arrangement of this kind is freely determined by all stakeholders and then prescribed within the smart contract ruling the project.

The next paragraphs elaborate on the way each component supports the operation of the COCOA^2 platform.

12.3.2 Knowledge Sharing (Stage Performing)

This module deals with two main objectives. The first objective is called *bidding*. It allows communities to promote adaptation initiatives for which financial resources are required. At this stage, they clearly define the *goals* and *milestones* as well as exchange information with potential supporters through webinars. Together, the goals, milestones, and relevant information being exchanged will feed into both the project proposal assessment, funding disbursement, and the subsequent monitoring and evaluation processes.

The second objective is *two-way knowledge sharing*. It offers an interactive space in which communities and experts can mutually exchange (technical) knowledge. In this way, communities can share experiences on successful (and not-so-successful) adaptation projects with local and international supporters. Likewise, further technical know-how or perspectives can be crowd-sourced among experts across the world to either refine the proposals, suggest improvements with best practices, or better assessment approaches regarding specific elements of the projects. In this sense, for instance, the *makers* or do-it-yourself culture have proven to be very successful in sharing knowledge for solving different kinds of problems (Makezine, 2016). All the information about the projects is registered in this module and allows the other modules to access it via web services.

12.3.3 Multicriteria Analysis

The multicriteria analysis (MCA) module provides assessments of possible impacts or performances of projects proposed by participants (Arciniegas Lopez, 2012). MCA is a tool to address and formalize decision problems with conflicting objectives. It compares a set of alternatives against one or more criteria to determine the anticipated performance of each alternative options, which allows for ranking and prioritization of a set of project proposals (Belton & Stewart, 2002).

On the COCOA^2 platform, MCA methods are used to evaluate projects with regard to *sustainability* and *economic impact*. The sustainability aspect is based on four of the UN's 17 Sustainable Development Goals. Relevant to COCOA's^2 sustainability criterion are:

- Goal 6: "Ensure access to water and sanitation for all";
- Goal 7: "Ensure access to affordable, reliable, sustainable and modern energy for all"; Goal 11: "Make cities inclusive, safe, resilient and sustainable"; and
- Goal 13: "Take urgent action to combat climate change and its impacts" (United Nations, 2015).

The MCA module will assess to what extent each project will accomplish each of the four Sustainable Development Goals.

As for economic impact, the MCA module evaluates each project on the basis of its effects on local economies as a function of gross domestic product per capita. Relevant to COCOA's^2 economic impact criterion is Goal 8: "Promote inclusive and sustainable economic growth, employment and decent work for all."

It is very important to select robust criteria for scoring proposals, assessing project performance, and the right approach for standardization, weighting, and aggregation. Aggregation of the individual criteria scores is conducted to determine the total score of a particular project. This is performed in two steps: (i) the weighted sum of the criteria scores is calculated; and (ii) the result is rescaled to maintain the 0–10 range of the scores. A similar approach is applied to all goals.

For each project, an application undertakes an initial assessment on the basis of the two aspects with underlying criteria. The projects must follow a common template and format such that all of them are described using the same representation scheme. A *suitability scoreboard*, which is contained in the COCOA^2 knowledge-sharing module, contains suitability scores of all projects against each criterion. When a new project is uploaded, a new tuple (row) is added to the table. The table is subsequently populated with the results of the initial assessment. The suitability scores are calculated as a function of the information contained in each uploaded project.

12.3.3.1 Visualization of Impacts: Dynamic Bar Charts and Weight Sliders

As mocked up in Fig. 12.2, the MCA module of COCOA^2 features an interface that juxtapositions the scores of uploaded project proposals against selected criteria on a visualization panel. Projects which have gone through MCA are visualized in bar charts on the visualization panel. Customized criteria can be used to rank projects in an efficient way. Projects can be compared by a single criterion or aggregated total score.

The interface also features sliders with which criteria weights can be adjusted for exploring "what if?" scenarios (see Fig. 12.3). The dynamic charts will be redrawn in real time to provide instant response to a variety of project assessment scenarios.

12.3.4 Smart Contracts

The financial resources layer supports the operation of smart contracts, which guarantee the integrity of financial transactions between funders and beneficiaries. The overall process is conducted in several steps:

1. The supporters choose a project to be funded based on (a) the beneficiary's bidding performance; and (b) the MCA's forecasted impacts.

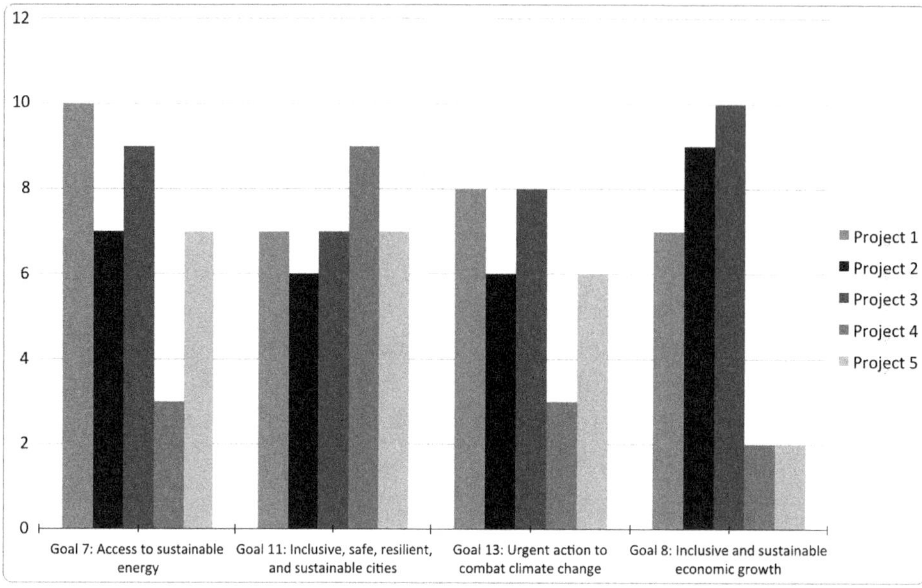

Figure 12.2
Visualization of project assessments.

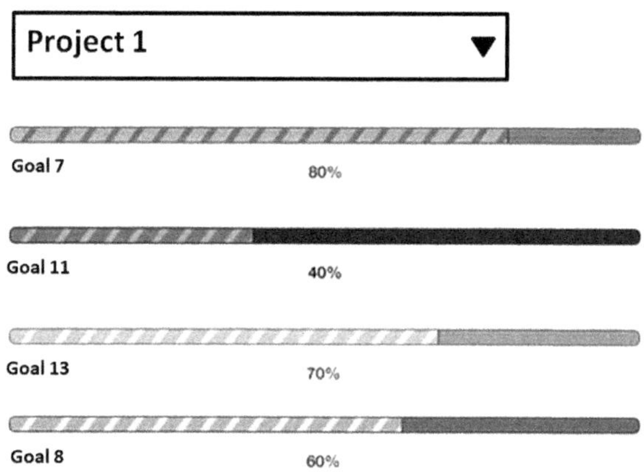

Figure 12.3
Visualization of interactive modification of weights.

2. The funders select a funding program. COCOA^2 initially supports two funding program and several payment methods (including the use of crypto-currency—Euro, Dollar, Bitcoin, Ether, etc.). Funding programs can either be (a) 100% transfer of resources up-front; or (b) progressive transfers of resources based on milestones achieved by the project. If the payment method is in fiat currency, COCOA^2 will convert it (via an

exchange platform such as bisq (https://bitsquare.io/)) to a cryptocurrency chosen by the beneficiaries.

3. A predefined smart contract is invoked to govern the interparty transfers of resources. COCOA's^2 smart contracts can be written on Blockchain protocol such as Ethereum (https://www.ethereum.org/) or Counterparty (https://counterparty.io/).
4. On beneficiaries' side, they are able to receive funding convertible to local currency or other resources needed for their projects.
5. If funding is progressive, based on the progress of a project itself, there are two possible outcomes:
   a. In case the project is progressing, the funding continues normally. To keep track of progress, for instance, knowledge-sharing technologies such as webinars can be used. Likewise, depending on the project, Internet of Things (IoT) technology can be deployed to perform (remote) tracking of project progress. For instance, there are already several IoT devices able to report on different environmental measurements (e.g., thingful.net).
   b. In case milestones are not achieved (i.e., progress is doubtful), the smart contract can suspend the funding until the project is on track again.
6. Once the funding program completes, the smart contract stops transferring resources to the beneficiaries. The knowledge layer, nonetheless, stores all relevant data about the funded projects.

To be a successful adaptation finance accelerator, the COCOA^2 also faces different technical and socio-economic challenges. The next section elaborates on each of the identified challenges with some strategies to address them.

12.4 Challenges

As COCOA^2 is a first-of-its-kind web-based multistakeholder funding platform, not only does it face technical challenges around the maturity of Blockchainized adaptation finance management, but also scalability challenges before its widespread adoption internationally.

12.4.1 Technical Challenges

Interoperability issue: The technicality of COCOA^2 as the first Blockchainized adaptation finance management platform is not mature yet. The primary technical challenge is the *interoperability* between the elements within the knowledge and financial resources layers. It requires seamless coordination, alignment, and synchronization between smart contracts and the MCA, as well as knowledge-sharing technologies.

Project impact assessment: Along this line, another technical challenge for COCOA^2 lies on its capacity to *thoroughly assess* the possible project impact and performance using the

MCA methodology. In the same vein, the MCA module should be connected with certain acute technologies able to accurately and credibly *track* the performances of every component of projects being supported, which may determine whether or not the smart contract keeps on transferring resources.

12.4.2 Scalability Challenges

Participation of practitioners: Local communities need clear and transparent approaches to assess and manage the climate risks of their habitats in order to increase their climate resilience and avoid catastrophe. Similarly, they need free and open channels to pitch their adaptation ideas to potential funders. A main challenge of COCOA^2 is to *attract the attention and participation of practitioners* on this platform. To this end, COCOA^2 aims to gather a group of pioneers keen to develop climate adaptation ideas wherever necessary on the globe.

Such a challenge is pertinent to the idea of triggering direct and indirect *network effects* (De de Reuver, Sørensen & Basole, 2016). On one hand, direct network effects increase the base of users, such as project beneficiaries, financial supporters, and knowledge prosumers. On the other hand, even harder, indirect network effects increase the number of social actors who do not directly benefit from COCOA^2 but still perceive the value of this platform. Indirect network effects, for instance, attract the attention of other social actors, such as science communicators or journalists who can report on successful stories on COCOA^2.

Governance system: Another challenge lies on *governance*. On COCOA^2, although the platform is initially "governed" by the COCOA's^2 management team (i.e., a centralized organization), it will be more attractive for participants to gradually "govern" their interactions on their own (i.e., a distributed organization) in the spirit of Blockchain technology. The transition from a centralized towards a distributed scheme requires a prolonged multistakeholder negotiation on the most suitable protocol probably in the UNFCCC process.

Longevity of the platform: During such negotiation, it is also important to study how to enhance a value cocreation process from which all participants can benefit. Obviously, beneficiaries will mostly receive economic value (i.e., financial resource) from supporters, whereas supporters and knowledge prosumers should perceive some forms of social and altruistic value (Holbrook, 2006). COCOA's^2 *longevity* also depends on the sustainability of financial resources and the continuity of selfless contributions from a community of volunteers to improve the platform.

User base: The "financial resources" layer, which relies on smart contacts, is one of the most important elements within COCOA^2. Smart contracts offer interesting features such as immutability—meaning once an agreement takes place, there is no way to change it and the smart contract will behave as agreed. But smart contracts may not be powerful enough to streamline the transfers of resources from funders to beneficiaries unless there is a considerable user base so that beneficiaries can easily exchange their digital resources for fiat money or other valuable resources (i.e., maximizing direct and indirect network effects as explained in the previous section). Unsurprisingly, owing to its novelty, smart contract is neither widespread nor easily understood by all people as yet.

12.4.3 Possible Strategies to Address Technical Challenges

Interoperability issue: As regards technical challenges, it is highly desirable that COCOA's[2] deployment uses standardized protocols and technologies throughout different layers. In this way, interoperability issue on the platform is diminished while maintaining some degree of openness as well.

Project performance tracking: To track the progress of funded projects, two approaches can be applied.

1. Self-reporting from project beneficiaries via Webinars – This requires project beneficiaries to coordinate with all supporters to meet and report status of project components on a periodical basis.
2. The potential of COCOA[2] will not be fully unleashed unless this Blockchain platform combines with other acute technologies; and as explained in Section 12.3.4, automatic monitoring can also be achieved via either remote sensing tools or IoT devices. But these are conditional upon COCOA[2] participants ensuring that the project goals and milestones with indicators are precisely defined against MCA criteria. Remote sensing can be done using tools such as high-resolution spatial images from the communities involved (Arciniegas Lopez, 2012) and even verified with data collected from space satellites before triggering a smart contract. IoT applications can be made possible in partnerships with web services that provide measurements collected by IoT devices (e.g., https://thingful.net). In this case, data authenticity must be ensured via distributed techniques such as collective attestation (Ambrosin et al., 2016). This technique allows validating the integrity of IoT devices (i.e., indirectly the authenticity of the transmitted data) by relying on a network of neighboring devices which work together to guarantee that devices on the network are uncompromised (Ambrosin et al., 2016). The costs incurred by the IoT-based approach, however, should be taken into account as part of the process of establishing smart contracts (i.e., some budgets should be ring-fenced to cover such costs).

Lastly, the openness of the architecture is also important to its applicability. In the long term, COCOA's[2] main *software architecture* should become open-source for external collaborators who wish to either alter minor elements or customize some of the functionality to suit their purposes. It would be enormously attractive if the platform allows collaborators to design their own smart contracts and connect their project components to IoT devices which can continuously track the project progress.

12.4.4 Possible Strategies to Address Scalability Challenges

Active participation of practitioners on the network: To directly engage beneficiaries and supporters, COCOA^2 needs to be promoted via well-known channels within climate change policy communities such as the Nazca Portal (https//climateaction.unfccc.int/), the WeAdapt initiative (https://www.weadapt.org/), and social media. Social media can go extra miles to spread the words about COCOA^2-supported projects.

Robust governance system: COCOA^2 needs to clearly define its value proposition, that is, how participants benefit from working together. To this aim, it is possible to introduce a credibility rating system to better harness the social or altruistic value (Holbrook, 2006) on top of any economic value to be received.

Financial sustainability: To ensure the longevity and financial sustainability of the COCOA^2 platform, one may consider charging small fees for each project that gets funded; or securing long-term sponsorships from public and private organizations that may find COCOA^2 valuable.

In brief, this section has presented several challenges to be addressed with some initial strategies. The challenges above are not exhaustive as others may appear along the way.

12.5 Corollary: COCOA'S Contribution to the Climate Efforts

Despite the infancy of the concept, COCOA^2 is indeed an innovative idea that proposes to combine Blockchain technology, smart contracts, MCA, and Internet-based instances—such as webinars or others—to potentially accelerate global climate change adaptation efforts. The COCOA^2 idea has not been tested yet. Thus, it is not possible to appraise its effectiveness in supporting a transparent, fair, and decentralized development of adaptation initiatives, and more importantly, the coherent, unobstructed, and trustful allocation of climate funding.

In this context, the most valuable contributions of COCOA^2 would be:

1. The possibility of sharing knowledge and adaptation experiences at virtual gathering sessions—also in real time through webinars—where climate adaptation *pioneers* community could freely and openly exchange views and solutions to climate impact problems;
2. The option of pitching climate adaptation ideas to individuals located globally could liberalize the traditionally locked-up climate policy fortress to the crowd for reducing

the vulnerability of a community through concerted efforts and, ultimately, democratize adaptation program design processes; and

3. The use of Blockchain-based features (including TDL, FTT, and NCP) to transfer (financial) resources to smart contracts to turbo-boost the implementation of climate adaptation solutions, the pace of which have lagged far behind that of climate change.

All in all, COCOA's[2] ethos and innovative approach will demonstrate to the world that community knowledge, engagement, and innovation are the overlooked cornerstone yet the most powerful force to tackle the climate challenge.

References

Ambrosin, M., Conti, M., Ibrahim, A., Neven, G., Sadeghi, A. R., & Schunter, M. (2016). *SANA: secure and scalable aggregate network attestation. Proceedings of the 2016 ACM SIGSAC Conference on Computer and Communications Security* (pp. 731–742). ACM.

Arciniegas Lopez, G. A. (2012). *Map-based decision support tools for collaborative land use planning. PhD Thesis*. VU University Amsterdam.

Belton, V., & Stewart, T. J. (2002). *Multiple criteria decision analysis: an integrated approach.*. New York: Springer.

Business insider. (2016). *There is a 'game changer' technology on Wall Street and people keep confusing it with bitcoin*. [online]. Available at: http://www.businessinsider.com/what-is-blockchain-2016-3/#blockchains-are-ledgers-like-excel-spreadsheets-but-they-accept-inputs-from-lots-of-different-parties-the-ledger-can-only-be-changed-when-there-is-a-consensus-among-the-group-that-makes-them-more-secure-and-it-means-theres-no-need-for-a-central-authority-to-approve-transactions-1. [Accessed 15 July 2017].

CARE. (2014). *Participatory monitoring, evaluation, reflection and learning for community-based adaptation: A revised manual for local practitioners*. London: CARE International.

De Reuver, M., Sørensen, C., & Basole, R. C. (2016). The digital platform: A research agenda. *Journal of Information Technology*, 1–12.

Holbrook, M. B. (2006). Consumption experience, customer value, and subjective personal introspection: An illustrative photographic essay. *Journal of Business Research*, *59*(6), 714–725.

Makezine. (2016). Monitor Light Pollution with Photodiodes: Measure sky glow—and fight for your right to the night. http://makezine.com/projects/surveying-light-pollution/. [Accessed 1 July 2017].

United Nations. (2015). Transforming our world: the 2030 Agenda for Sustainable Development. Resolution adopted by the General Assembly on 25 September 2015.

Unfccc.int. (2017). Blockchain Technology Can Boost Climate Action—UNFCCC Recognizes Potential [online]. Available at: http://newsroom.unfccc.int/climate-action/how-blockchain-technology-could-boost-climate-action/. [Accessed 15 July 2017].

Further Reading

Buchner, B., Trabacchi, C., Mazza, F., Abramskiehn D., Wang D. (2015). *Global Landscape of Climate Finance 2015*, Climate Policy Initiative.

Raid, H. (2016). Ecosystem- and community-based adaptation: Learning from community-based natural resource management. *Climate and Development*, *8*(1)), 4–9. Available from https://doi.org/10.1080/17565529.2015.1034233.

Swan, M. (2015). *Blockchain: Blueprint for a new economy*. "O'Reilly Media, Inc.".

CHAPTER 13

Using Smart Algorithms, Machine Learning, and Blockchain Technology to Streamline and Accelerate Dealflow in Climate Finance

Neil Salisbury[1] and Jenya Khvatsky[2]

[1]*Founder & Director of CleanTek Market, Melbourne, Victoria, Australia*
[2]*CEO of CleanTek Market, Melbourne, Victoria, Australia*

We are in the midst of a clean technology revolution. There is now widespread acceptance that the global economy is transitioning away from a traditional fossil-based model to one based on sustainable use of resources and energy.

To achieve an orderly transition, the scale of financing required can be counted in the trillions—not billions—of dollars. A widely-quoted report by the International Energy Agency estimated that US$1 trillion a year to 2050 is required to finance this transition (IEA, 2015). Yet, the flows of climate finance has averaged US$360 billion per annum over the first half of this decade, peaking at US$391 billion in 2014, to be valued at US$5.5 trillion (CPI, 2015). The current scale of the market is impressive—and yet it is some 60% below what is required. This is a clear market failure (Fig. 13.1).

So what is going on? And, how can we improve it?

The massive and persistent structural under-investment in climate change actions is due to a number of market inefficiencies. These include a lack of clear local, regional, and global regulatory signals; poor access to risk capital for new initiatives; poor access to specialist expertise; low visibility and the small scale of many initiatives.

The Paris Agreement attempts to orchestrate the global cacophony of regulations governing climate change business into a harmonized plan of action. But the Agreement ultimately pushes the action down to individual countries, where actual adherence continues to be patchy—closer to jazz than symphony. Europe and China are leading the way. Canada is back. The USA is back-tracking. Australia is still vacillating.

Transforming Climate Finance and Green Investment with Blockchains.
DOI: https://doi.org/10.1016/B978-0-12-814447-3.00013-6

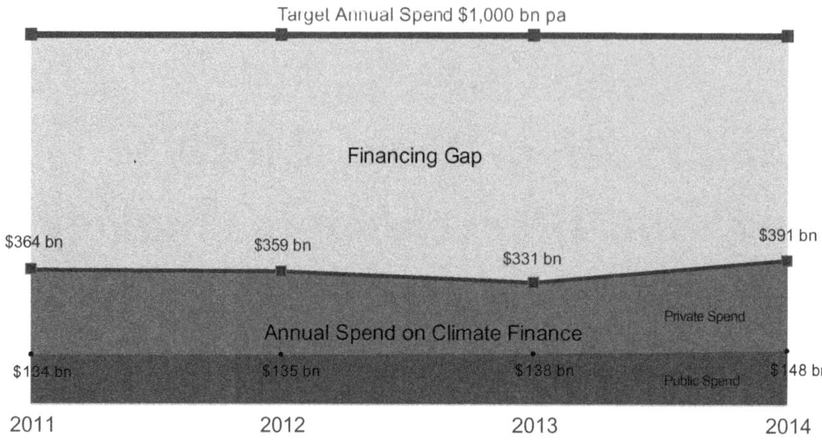

Figure 13.1
Climate finance flows. *Source: Global Landscape of Climate Finance 2012–15, CPI.*

Although it is comforting to blame the politicians, it is also true that the private sector is simply not positioned to deploy an additional US$600 billion per year to plug the climate finance gap.

Here are some reasons:

Visibility

Dealflow origination and transaction is mired in the twentieth century processes. Transactions still largely depend on physical networks of developers, finders and investors. "Luck" is a key factor in a project developer or a start-up finding an investor with an aligned focus on technology type, technology readiness, location, and spend, who, critically, also has funds to invest at that point in time. If this is a challenge in the mature economies of Europe, North America, and Australasia, it is an impossibility in most developing economies. Yet, this is where clean energy, air, and water matter most. It is also where such projects make financial sense without government subsidies.

Scale

Deploying US$600 billion per year is something large financial institutions do well. They collectively deploy US$8–10 trillion per year on a managed portfolio worth over US$300 trillion, which is growing at 3–4% p.a. These organizations have a minimum deal size in the tens of millions to justify their high transaction costs and overheads.

Nonetheless, many climate mitigation and adaptation measures, and their underlying technologies, are best deployed at relatively small scale, often costing substantially less than US$1 million. Examples include local- or community-scale projects around water efficiency; and small-scale renewable energy generation and storage. These projects are the under-rated backbone of the global clean technology transformation.

Whereas large financial institutions like pension funds try to participate in such projects and technologies through private equity and venture capital firms, these "more nimble" investors also have their limitations—minimum spend (again), relatively narrow focus, etc.—and represent considerable value leakage in their management fees.

Deep Tech[1] and Information Asymmetry

Encouragingly, the rise of fintech is disrupting all financial services—including deal origination and transaction. For example, the rise of crowdfunding for equity and debt is simplifying and democratizing the investment process. Blockchain technology is improving the security of transactions and opening the path toward their full automation.

There are, however, reasons that climate mitigation and adaptation activities, and cleantech in particular, are generally considered too challenging to benefit from the fintech revolution. Emerging "deep" technologies (e.g., storage, advanced biofuels, etc.) require specialized technical and commercial expertise to adequately evaluate them prior to investment. Such information asymmetry is the main reason why crowdfunding has not yet penetrated into this space.

It is with such barriers in mind that CleanTek Market[2] was founded.

13.1 CleanTek Market[2]

CleanTek Market[2] was founded in 2014 with the goal of *accelerating the development and deployment of clean technologies in order to facilitate the transition toward a sustainable low-carbon economy*. CleanTek Market's[2] on-line marketplace, or "Platform," has been designed specifically to address the key barriers which are, believably, restricting climate finance flows: visibility, scale, and information asymmetry. The Platform is developing tools to deploy smart algorithms, machine learning, and Blockchain technologies to connect clean technologies and projects with multiple stakeholders: investors, end users, advisors, government agencies, and other market intermediaries—channelling technology, finance, and expertise to where they are most needed (Fig. 13.2).

[1] Deep Tech is having profound impact on segments including autonomous systems, robotics, smart home/cities, medical devices, clean tech, energy efficiency, and many more developing or emerging application areas. TechWorks defines Deep Tech as technology that is based on tangible engineering innovation or scientific advances and discoveries. Deep Tech is often set apart by its profound enabling power, the differentiation it can create, and its potential to catalyse change. Deep Tech can span across many technological areas and affect diverse applications. On the technological front, these can include processing and computing architecture innovations, advances in semiconductors and electronic systems, power electronics, vision and speech algorithms and techniques, artificial intelligence and machine learning, haptics, and more (https://www.techworks.org.uk/about/what-is-deep-tech).

[2] https://www.cleantekmarket.com/

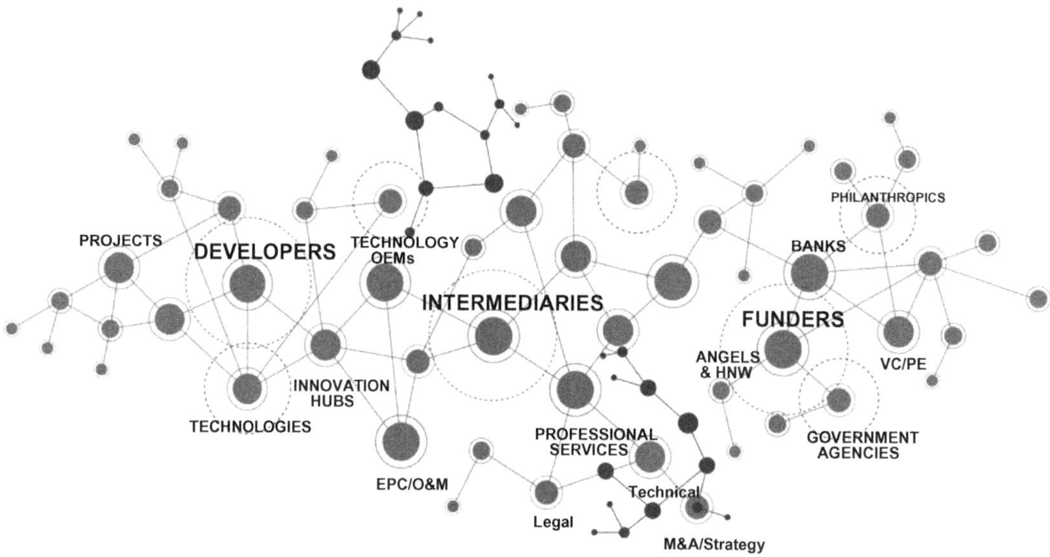

Figure 13.2
Ecosystem.

By creating an efficient on-line ecosystem, we enable any organization active in the cleantech sector to raise its visibility to others on the Platform. Our algorithms then match organizations with each other, along with market intelligence, tenders, events, etc., in accordance with their specified criteria.

13.1.1 Tiered Validation Process

We address the challenge of *information asymmetry* by creating a **tiered validation process** that aims to rate the "quality" of dealflow on the Platform. The first tier algorithmically scores each deal using over 20 criteria, which are combined to indicate its overall "quality." The second tier uses discrete dealflow channels. It is managed by industry-trusted organizations that add their expertise and reputation to validate the deal. The third tier uses pre-approved agents on the Platform to review and "qualify" specific deals. Eventually, all organizations on the platform will be similarly rated.

13.1.2 Investment Aggregation Tools

We address the challenge of *scale* by creating **investment tools** that can aggregate small projects into larger investible legal entities, and thereby enable institutional investors to efficiently deploy tens or hundreds of millions of dollars at a time would dramatically increase the flow of capital into the cleantech and wider climate change project portfolios.

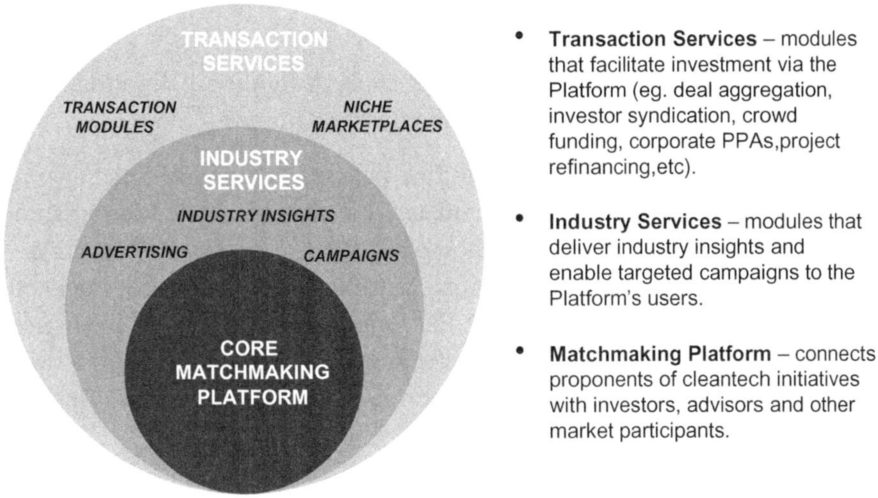

Figure 13.3
Roadmap.

13.1.3 Matchmaking Platform

At its core, CleanTek Market[2] is a **matchmaking platform** that connects users based on their specified criteria. CleanTek Market[2] is rolling out dozens of service modules broadly grouped by their functionality: Matchmaking, Industry Services, and Transaction Services (Fig. 13.3).

So, what do we mean by deploying "smart algorithms," "machine learning," and "Blockchain technology"? How can they facilitate greater investment in the climate change projects?

13.2 Smart Algorithms

Smart algorithms refer to the algorithms we have developed to efficiently match users in accordance with their specified criteria. These algorithms represent the base layer of CleanTek Market's[2] technology. Superficially, these matching algorithms are similar to those used by other matchmaking platforms for dating, employment, or real estate, etc. Setting CleanTek Market's[2] algorithms apart from them is its capacity to support the higher-order data management requirements of the Industry and Transaction Services being rolled out. These algorithms are being continuously refined to improve their effectiveness in channeling information to users, or streamlining deal origination between "buyers" and "sellers."

13.3 Machine Learning

Machine learning, a subset of Artificial Intelligence, is the ability of algorithms to acquire new knowledge or insights without being explicitly programmed to do so.

On the Platform, machine learning is used to find patterns in individual and aggregated user behavior to then predict what information and activities can be curated for a particular user. As the user base expands into the tens of thousands, we will enable our matching algorithms to start drawing inferences and insights from the users' profiles and activities to further enhance their user experience on the Platform by predicting their likely needs. Data collected on the user patterns on the site will be applied to determine the "trustworthiness" of a user based on their activities on the Platform in comparison to others. This would then be incorporated into the user's overall rating.

For example:

A newly enrolled developer of solar projects in Southeast Asia could automatically be channeled potential projects in the region, related technologies under development, grants, and industry intelligence that could support their activities based on ongoing interactions among similar organizations.

As the solar developer engages with the Platform, our algorithms will increasingly tailor content to the developer's explicit and implicit preferences, based on their activities on the Platform. This could include the services the developer may need to develop projects (e.g., advisory, insurance, etc.), raise finance (grant, equity, debt, etc.), or peripheral intelligence about the project (e.g., regional issues, technology, etc.). Counterparties will be guided by our algorithms to assess the developer's risk profile, thereby assisting with their risk management, compliance, and fraud protection activities.

13.4 Blockchain

Blockchain refers to the use of a decentralized, secure database to support critical steps in a transaction, namely authentication, contracting, and payment. CleanTek Market[2] is developing a private, Turing-complete Blockchain to support its Transaction Services for clean energy and climate finance.

Authentication

Organizations wishing to participate in CleanTek Market's[2] Transaction Services portals will be required to verify their digital identity. This encrypted digital identity, or *Passport*, could be created on our Platform or by trusted third parties. In both cases, the identity attributes will be hashed and stored on the Blockchain for algorithmic verification of the organization's identity at each transaction. The *Passport* will be the primary source of identification and authentication for any Platform-based interactions and transactions—thereby minimizing the risk of fraudulent behavior. The *Passport* could also be used to verify organization's identity on third-party services.

Contracting

CleanTek Market[2] is working with its legal partners to create "standard contracts" for specific transactions (e.g., power purchase agreements, Special Purpose Vehicle structuring). These standardized contracts will be translated into "**smart contracts**" by encapsulating their key terms into a Blockchain-distributed ledger, and enabling automated execution and settlement upon the contractual terms being satisfied. Whereas the legal status of such smart contracts is still being debated in the society, its use will provide documented evidence of parties agreeing to use the Blockchain facility for execution of contracts. Over time, all Platform-based transactions will be executed via smart contracts.

Settlement and Payment

Using the automated escrow function inherent in Blockchain design, parties will be paid per terms of contract as each phase of the contract is executed—thereby eliminating the need for a trusted third party for administering payments. Payment will be in proprietary cryptocurrency hard-linked to the major fiat currencies.

Working seamlessly together, our algorithms streamline the process of identifying and evaluating dealflows, whereas our Blockchain-backed applications will streamline the transaction processes. With the technology being highly scalable, we are able to host and run multiple concurrent processes to efficiently deliver a range of transaction services. This light-touch frictionless approach—using scalable algorithmic and Blockchain tools to match, authenticate, contract, and settle—will significantly reduce the risks and costs associated with deal origination and transactions.

In practical terms, these tools are being encapsulated into function-specific Portals on the Platform. A Portal is a "walled" space on the Platform set up for a particular purpose (e.g., dealflow aggregation, or a niche clearinghouse) with its own customized participation criteria. So, for example, a portal set up for running a clearinghouse for energy procurement (e.g., Corporate power purchase agreements) would have multiple providers using the Platform's clearinghouse tools to offer their own branded versions of energy procurement services.

Section 13.5 will outline some examples on how it will actually work.

13.5 Example 1—Deal Aggregation

One of the key barriers in mobilizing material flows of climate finance is the relatively small scale of many transactions. Specifically, most financial institutions have a minimum spend in the tens of millions, reflecting the cost of processing the deal. But the vast majority of cleantech deals are less than that threshold, leaving a critical sector of the market without easy access to finance. Blockchain can enable the aggregation of smaller deals into investment-ready legal entities, or special purpose vehicles (SPVs). It will work as follows:

An organization active in the corporate finance space (e.g., investment manager, law firm, private equity firm . . .) would set up its own branded aggregation facility on the Deal Aggregation Portal—becoming the Manager of that facility. The Manager would create one or more SPVs, specifying key selection criteria for each (i.e., deal type, deal size, location . . .) as well as its terms (i.e., timing, fees . . .).

Part of each SPV offer materials will be a standardized SPV agreement, encapsulated on CleanTek Market's[2] *proprietary Blockchain which will be used for execution, on a "take it or leave it" basis.*

The Manager's branding SPV criteria and contractual terms would differentiate it from similar SPV facilities on the Portal. The Manager could engineer the SPVs to have specific risk profiles—balancing higher and lower risk deals. Also, it is likely that the Manager would already have the capital to fund the SPV, or it could use the Platform to find investors.

Organizations looking for funding to support their company or project would gain access to the Portal by setting up a Passport (enabling them access to all Transaction Services) and electing to access the Aggregation Portal. These organizations would be algorithmically matched with one or more SPVs on the Portal, the terms of which they could review before deciding whether to participate.

All contracts would be executed on-line via our smart contracting facility. Payments, in both directions, would be made in accordance with the milestones stipulated in the individual contracts.

It is important to note that this algorithmic aggregation and Blockchain-execution approach would also be used to syndicate investors to tackle larger deals, and on both sides of crowdfunding campaigns.

13.6 Example 2—Corporate PPAs

As their name suggests, a PPA is a contract where a generator of electricity (the seller) agrees to sell electricity to one or more off-takers (the buyer). A PPA is one of the key contracts project developers require to secure finance for their projects. Traditionally, PPAs have involved a single buyer and a single seller. However, with the traditional utilities business being disrupted and disaggregated, it is increasingly difficult for developers to secure "bankable" PPAs in many mature energy markets.

Corporate PPAs are increasingly becoming popular ways of funding utility-scale renewable energy projects. Initially, they featured a large corporate off-taker (e.g., Google) entering into an off-take agreement with planned projects, thereby making the project bankable. The next wave of corporate PPAs is featuring multiple off-takers seeking one or more projects. Interestingly, with an energy retailer acting as an intermediary, the generators and the off-takers need not be physically connected.

CleanTeck Market's algorithms can readily match off-takers with projects, grouping both sides by output, geography, technology, commercial operation, etc.

As with the previous example, our solution is to create a bespoke Portal for matching multiple projects with off-takers. It will work as follows:

An organization active in the energy trading space (e.g., existing energy retailer) would set up its own branded Corporate PPA Procurement facility on the PPA Marketplace Portal—becoming the Manager of that facility. The Manager would create one or more Procurement Processes, specifying key selection criteria for each (i.e., annual output, unit cost, location, timing . . .) as well as its terms (i.e., penalties, fees, . . .).

Project Developers looking for off-takers would gain access to the Portal by setting up a Passport (enabling them access to all Transaction Services) and electing to access the PPA Marketplace Portal.

Corporate off-takers looking for multi-year electricity supply would likewise gain access to the Portal by setting up a Passport and electing to access the PPA Marketplace Portal.

Projects and Corporates would be algorithmically matched together in accordance with their criteria, and across one or more procurement processes. Each process may have their own terms, which both the buyer and seller could review before agreeing to participate.

A PPA process would conclude when supply and demand is satisfied through the build-up of buyers and sellers. All contracts would then be executed on-line via our smart contracting facility. Payments would be made in accordance with the milestones stipulated in the individual contracts.

It is important to note that this clearinghouse functionality and Blockchain execution approach could also be used to run similar multiparty procurement processes, including demand-side management (i.e., load shedding at times of high system load), project refinancing, environmental bonds, and emission reduction purchase agreements (ERPAs).

13.7 Outlook

We believe that the market failure evident in the bullish climate finance flows can be overcome by providing the private sector with technology-driven tools to enable the efficient and scalable dealflow origination and transaction. This, in turn, would drive increased innovation, development, and deployment of clean technologies. CleanTek Market's[2] algorithms are designed to streamline the processes of originating dealflows. This "*Tinder for cleantech*" approach, backed by machine learning and Blockchain technology, will facilitate the private sector to efficiently deploy the additional US$600 billion per annum necessary to fill the global climate finance gap and hence, fuel the decarbonization of the global economy.

CHAPTER 14

Addressing Water Sustainability With Blockchain Technology and Green Finance

Anna Poberezhna

Smart4Tech, London, United Kingdom

14.1 "Green" as a Principle Concept: Intro, Facts, and Recent Numbers

Innovative businesses have the power not just to generate profit, but to solve critical environmental problems, addressing broader challenges and contributing to communities.

Sustainability or a green paradigm is much more complex than dealing with climate change, resource depletion, erosion of territories, generation and accumulation of waste. Addressing climate change is only one of many constituents of what is known as "green philosophy" or "inclusive sustainability" today, which contributes to the environmental and ecosystem changes.

Green business strategy has been associated for a long time with a soiled "ecology science," low-cost corporate social responsibility (CSR) policies, sunk costs, and part of the marketing.

In the past a short-termism and stakeholders' vested interests defined companies' performance, investment rates of return, and market behavior. For a long time, enterprises were focused on traditional capitalization-weighted indices, including return on investment (ROI) metrics with no or little connection to nonfinancial ones and their correlation towards resilience. Not until recently have principles of sustainability become an actual part of a company's business strategy and practices.

The green corporate philosophy began its active advancement in 1950s with the development of the ESG (environmental, social, governance) paradigm. It was followed by "The Sullivan Principles," code of conduct for American-owned companies in South Africa, in the late 1970s followed by a number of international treaties (for instance, including the Montreal Protocol, Kyoto Protocol, Paris Agreement) along with international initiatives (United Nations Global Compact, Financial Stability Board—Task Force on Climate-related Financial Disclosures, Carbon Disclosure Project, B Corporation, and others). The fundamental task behind them was to deploy the concept of sustainable development within global players and especially at an enterprise level.

Transforming Climate Finance and Green Investment with Blockchains.
DOI: https://doi.org/10.1016/B978-0-12-814447-3.00014-8

It is crucial to state that sustainable business is not about gender diversity or about more resource efficient operations; it is about creation and implementation of practices for the restorative or circular economy. Green strategy is about the value chain, the products or services, and the people, not the company. Respectively, the green finance sector is about assets and services not financiers and those raising finance.

Darwin stated 158 years ago: "in the long history of humankind (and animal kind, too) those who learned to collaborate and improvise most effectively have prevailed." Both sustainability and ESG trends became more prevalent recently. Due to the novelty, it is quite difficult to measure sustainability and reflect the impacts in quantitative data.

A difference in ESG metrics and absence of one universal green or sustainability standard to which organizations would be accountable and comply with adds to the challenge.

Yet it has become clear that sustainability-business strategy and high ESG performance translate into higher financial performance and greater investor interest. Comparing the development of the ESG market to the tech innovation world, it is at the stage of "proof of concept," or "early adopters."

The size of the green bond market is expected to exceed US$150 billion in 2017 after hitting a record in 2016 of US$92.2 billion, doubling its size from US$42.2 billion in 2015. Each year new countries and enterprises pioneer green bond issuance with corporates leading the way. A positive signal in the United Kingdom this year came in the issuance of water bonds by Anglian Water, the water utility company, raising £250 million, breaking records not only for the water utilities industry, but for the overall green finance sector in the United Kingdom.

14.2 Thinking Circular: Applying Systems Thinking

Living today in an era of technological advancement, mixed with neo-Marxism and advanced capitalism, a good economist should know some ecology and a good ecologist should know some economics. The current system is characterized by fragmented globalization and new operating models that move away from the linear "produce–consume–dispose" model. For a long time a traditional, open-ended economy was based on growing consumption, and heavily reliant on conventional energy (oil and gas). In contrast, a circular economy is grounded on systems thinking, where people, processes, and products are aligned towards sustainable transformation. The model advocates continuity of quality and quantity, where in the heart of the processes are strategic innovative solutions that will facilitate change and connect processes into a closed-loop cycle.

Digital transformation has become an integral part of the systems and processes optimization, which has been actively developed over the past 50 years. New emerging technologies, including Blockchain, internet of things (IoT), artificial intelligence (AI), and data science are rapidly replacing legacy systems and accelerating progress.

14.3 Blockchain for Commerce

The key challenge for Blockchain is to be considered as an innovative integrative solution that can provide a value proposition across a number of business processes, rather than a simply disruptive technology. The advantage of Blockchain is its stimulus to create new operational business models for different types of organizations and to digitize assets.

Blockchain is a foundational technology which sits as a building block underneath various business processes, enabling industries and companies to structure their interactions and processes in fundamentally new ways. Within the water utility sector, Blockchain solutions can automate back-end, administrative and legal processes, whether it is billing and customer relationship management (CRM) or digitization of water assets and trading.

Blockchain technology can help to build a trusted ecosystem with one version of the truth. Instead of needing separate systems for each company that needs to run processes and store data, Blockchains provide a single shared logical system within which companies can operate, in collaboration with their counterparts. This creates a trusted environment between participants where, traditionally, many manual checks and reconciliation processes would be required.

From the financial point of view, it is all about the revenues and costs. An additional challenge is to create a circular economy operation with new revenue models, that would be economically and ecologically resilient.

An existing example of a Blockchain application is wholesale energy, run on multiple Blockchains to provide required levels security, immutability, and scalability. This application reduces the cost of running wholesale energy trading back offices.

Similarly, in a water trading application, Blockchain will provide not just a secure decentralized database for the private ecosystem, but ensure security and data consistency flows to all counterparties.

The Blockchain technology can help to prevent speculation in water trading, which is crucial in such a sensitive sector. This is achieved by creating greater transparency to auditors and regulators through recording trades and actions on the chain.

One of the requirements to enable water trading, is to create a new digital identity for organizations (that is, farmers) and digitize the water assets (for instance, entitlements, quotas, certificates, allocation). By storing these on a shared ledger, counterparty data management—a process which typically costs millions across an industry sector—can become a trivial issue.

14.4 Blockchain Applications for the Water Industry

14.4.1 Recordkeeping

Application of Blockchain can help water sector companies to manage balancing and settlement more efficiently, in comparison to the existing applied methods. All information, including consumption and transactional records can be automatically stored on an immutable ledger.

Blockchain and smart contracts can potentially help water sector companies and government bodies to access real-time data regarding market shares, consumption patterns, management of utility bills of consumers, and other possibilities. Managing records on a Blockchain platform can significantly reduce the cost of recordkeeping.

14.4.2 Reporting Repository/Compliance Reporting/Audit

Blockchain could be used to keep track of the steps required by regulation. Recording actions and their outputs immutably in a Blockchain would create an audit trail for regulators and streamline compliance. Such a change could reduce dramatically the time and effort (and therefore cost) that financial institutions spend on regulatory reporting, as well as improving quality, accuracy, and confidence in the process. Blockchain provides immutability, immediacy, and transparency of information, where stakeholders can be part of the real-time process instead of being recipients of posthoc reports and analytics.

The unique design of some Blockchain platforms, particularly Interbit, can provide powerful, yet lightweight systems with the following core characteristics:

- Effortless scalability across multiple networks.
- True segregation of data for industry grade privacy.
- Flexible redundancy across permissioned nodes.

14.4.3 Data Reconciliation

By creating a distributed data repository using a network of computers, each of which would hold a copy of the ledger, rather than a centralized authority, could enable regulators to monitor and control the status of assigned allotments, water licenses, and the state of transactions, validating availability of resources in real-time, rather than following the conventional regulatory reporting process.

14.4.4 Bond Issuance on Blockchain

Blockchain technology will simplify green bond issuance processes, by providing a shared space for the various agencies involved in the verification of requirements, the issuance and compliance of the securities, and the eventual trading of the bonds themselves.

14.5 Digitizing Water

The demand for the most precious resource, fresh water, is expected to surpass supply by 40% within the next two decades. In many recent reports, including World Bank's "High and Dry: Climate Change, Water and the Economy," it highlights that the price of water scarcity, exacerbated by climate change, could cost some regions a significant amount of their Gross Domestic Product (GDP), spur migration, and multiply the risk of conflict. Both scientists and economists agree that under a business-as-usual scenario, high water stress by 2050 puts at risk 45% of global GDP, primarily affecting water scarce areas, such as the Middle East, the Sahel in Africa, Central Africa, and East Asia.

Water as a resource is pivotal in the water–energy–food and water–energy–waste nexus, where agricultural use accounts for 70% of global water withdrawal, industry 20%, and municipal use 10%. Energy production accounts for 15% of global water withdrawals, and may potentially increase by 20% by 2035 (UNESCO, 2014).

In order to drive sustainable development and growth, it is crucial to address aspects of water security that are deeply interrelated with food and energy security. Water finance requires investments in water infrastructure, information technology, institutions, and communities.

Across regional levels, there are multiple problems in the water sector. These include flooding, droughts, scarcity, aged infrastructure, water quality, inefficient water management, and poorly controlled consumption. However, the most challenging task is to establish the right metrics and incentives for pricing water, which will encourage more efficient water management, enhance better water quality and decrease pollution, and stimulate more investment in water infrastructure.

In comparison to other sustainable development goals (SDGs), the water sector lacks investment, innovation, and collaboration. Despite the fact that water is the number one global strategic resource, investment into the sector is much less in comparison with other similar strategic resources. Low profit margins and returns, regionalism of problems, aging or the absence of infrastructure, quasi-ownership, complex multilayered regulation, interrelated political and business interests, high entry barriers—all these factors often result in a decreasing number of water utility projects and delayed advancement of the industry, in general. According to the World Bank's "The Private Participation in Infrastructure" 2016 report, investment in the water sector in emerging markets in 2016 was US$1.9 billion, which is 53% lower than the 5-year average of US$4.06 billion. Water utility projects attracted only US$669 million of private investment in 2016, an 88% drop from the previous 5-year average.

Given that food security is highly dependent on water resources and the fact that agricultural production will double by 2050 due to the increased population, water finance is crucial not only for water projects, but for all related, especially for water or food-dependent, sectors of global economies. Following the OECD Water, Growth and Finance Policy Perspectives 2016 Report (OECD, 2016); The four lines of action combined that can incentivise investment in water security and sustainable growth are:

- Maximising value from existing water security investments.
- Selecting investment pathways that reduce water risks at least cost over time.
- Ensuring synergies and complementarities with investments in other sectors.
- Upscaling financing through attractive risk-return allocation.

The basic economic and environmental problem is of multiplying demand, against decreasing, scarce resources. This systemic problem requires a holistic and innovative approach, which could be applied in both private and public sectors and also repeated on a cross-border level. Given the complexity of the water challenge, a holistic, innovative solution is required, which in turn would integrate green finance, innovative technology, and aspects of water stewardship.

The Water Credits project, being developed by Smart4tech[1], an innovative strategy and technology startup, is a combination of green finance, circular economy, and technological thinking.

It encompasses a monetary incentive program, underlined by technology for water consumers, based on a number of metrics, including water usage performance, innovation, and community involvement, and nonwater related metrics. The main goal of the project is to improve water efficiency, improve resilience, and advance innovation in the sector. Taking into account the complexity of the industry, the project brings together a number of public and private organizations.

It will be an important strategy to achieve several SDGs concerning industry, innovation and infrastructure, sustainable cities and communities, responsible consumption and production, climate action, and partnerships, for the goals (SDGs 6, 9, 11, 12, 13, and 17). The scheme develops a strategic platform for collaboration with public–private incentives, local tax programmes with further development of green finance instruments, including water bonds.

---

1 http://smart4.tech/

The key purpose of Smart Water Credits, a financial incentive, is to improve efficiency of water use and incentivize public and private sectors (urban and industrial water users), using aspects of behavioral economics as incentives for compliance, innovation, and partnerships. As part of the concept there is a technology solution. A private Blockchain is expected to be used to optimize existing and future recordkeeping, smoothen the audit process, and compliance reporting, where required, with a focus on disclosure of environmental impacts (sustainability indices, CSR, ESG, and international ratings indices, including MSCI, Trucost, Thompson Reuters, S&P Dow Jones), providing multiple decision makers with the unified data they need to change market behavior.

Using the Blockchain technology as an independent verifier and a trusted source of shared information will help to provide necessary transparency and verified accountability to the market. An ability to earn, accumulate, exchange, and trade credits will be recorded, and verified on the shared ledger. It could also be part of the preventive mechanism against misleading communication or speculation (Fig. 14.1).

The ability to trade water has been essential in maintaining irrigation sector incomes during drought, and is expected to be even more important under future climate change scenarios. Although water resource management is complicated, it does not need to be complex.

Australia is widely acknowledged as a global leader in water resources management, and more specifically in the application of water markets and trading as a key tool to share water between competing uses including urban uses, industry, the environment, and

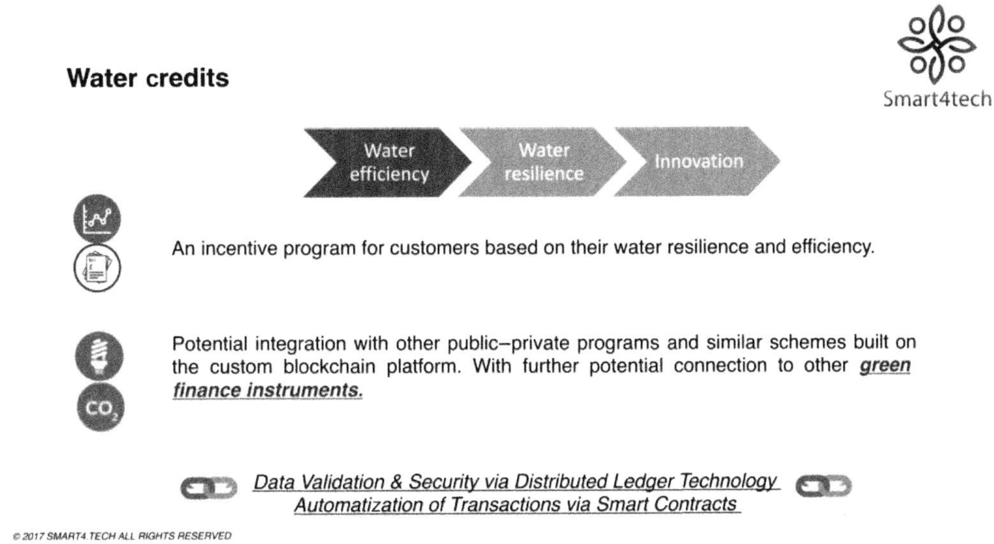

Figure 14.1
Water finance innovation.

agriculture. As governments, globally, continue to reform their water resources management in the face of climate change and greater completion between users, many are turning to water markets and water trading. China is just beginning to embrace water trading. The European Union has created protocols and Spain needs a water trading platform. In the US, opportunities for a simple trading platform exist in California's Central Valley, Nevada groundwater, and Texas water markets.

Additionally to the Water Credits, an application of Blockchain to existing water trading can be part of the integral solution. In early 2017 Civic Ledger, a startup, undertook a feasibility study to solve transparency problems in Australia's water markets, to boost confidence, with a view to increasing the volume and reliability of water trading. As a result of the feasibility study, Civic Ledger considered development of a Blockchain enabled peer-to-peer trading platform—water ledger—that would allow secure and transparent trading of water entitlements and allocations within water markets, increasing participation by irrigators, and boosting the overall allocative efficiency of the water resource. Water Ledger was expected to provide substantial value for money for governments, contribute to increased revenue for irrigators, and keep farming communities strong and providing additional water for the environment.

To achieve maximum results in the shortest possible time, collaboration and application of integrative solutions is essential, both in the public and private sectors. The concept of Smart Water Credits and strategic Blockchain integration by Smart4tech[1] as well as the water trading could be a perfect example of innovative thinking and prospects for cooperation, which are aimed to address the systemic problem of water scarcity.

Reference

OECD. (2016). *Water, growth and finance. Policy perspectives*. OECD.

Further Reading

United Nations report warns rising energy demand will stress fresh water resources, © UNESCO," [Online]. Available: <http://en.unesco.org/news/united-nations-report-warns-rising-energy-demand-will-stress-fresh-water-resources> Accessed 31.07.17.

World Bank. (2016). *High and dry: Climate change, water, and the economy*. Washington, DC: World Bank.

UN Water: <http://www.unwater.org/water-facts/water-food-and-energy/> Accessed 31.07.17.

World Water Council.

United Nations Water 2012. Facts and figures. Managing water under uncertainty and risk. [Online]. Available: <http://unesdoc.unesco.org/images/0021/002154/215492e.pdf> Accessed 31.07.17.

<https://media.anglianwater.co.uk/anglian-water-makes-financial-history-with-first-public-utilities-green-bond/> Accessed 31.07.17.

INTERLUDE IV

SECTION 4

Blockchain for Fraud-Free Emissions Management

In God we trust; all others must bring data.

Anonymous

Efficient mobilization of private climate finance relies on a trustworthy mechanism. To achieve the goals of the Paris Agreement, this mechanism may be a cross-border trading of carbon allowances or credits. Article 6 provides a vehicle for such trading by allowing voluntary cooperation involving "Internationally Transferred Mitigation Outcomes" (ITMOs). Talks around ITMO may be hollow unless the accounting rules for emissions reductions achieved in different carbon markets are harmonized and the risk of double counting of carbon credits minimized.

A common carbon accounting system necessitates reliable processes on which to base the greenhouse gas inventories (both national and international registers) as well as the tracking, subtraction, and addition of certified emission reductions. It means we all must bring credible data. Nonetheless, greenhouse gas "monitoring" and "reporting" are dependent on trust in an intricate web of operators or suppliers' data which culminate into emissions data reported largely based on mathematical calculations. Independent, third-party "verification" may be helpful but unable to fully guarantee reliable monitoring and reporting of emissions data. A burning question is who can or should verify the verifiers?

A complex problem in nature, the entire monitoring, reporting and verification system of emissions management should be revisited and solidified with the help of emerging digital technologies if we are to make this potential cross-border mechanism a success in curbing global emissions. It is the prominent features of Blockchain technology which will make it a good fit for this mission. Blockchain can save a huge amount of manpower and financial resources currently devoted to paperwork or administrative procedures and hence enhance considerably the efficiency of many emissions trading schemes.

This technology, if deployed in emissions trading schemes, has tremendous potential to disrupt the current transaction system. Spring 2017 was an iconic season for Blockchain in emissions management. In March, IBM China and the Shanghai-based Energy Blockchain Lab announced the world's first Blockchain-based carbon asset management platform based on the open-source, openly governed Hyperledger Fabric. It can streamline and accelerate the issuance of carbon credits in China's emissions trading scheme. In the same month, the Russian Carbon Fund and Aera Group (France) pioneered the first international carbon credit transaction using Blockchain technology in DAO IPCI.

Speaking of carbon pricing, Dr. Delton B. Chen will first present a case for expanding the central bank remit to developing a Central Bank Digital Currency for rewarding climate mitigation actions in Chapter 15, Central Banks and Blockchains: The Case for Managing Climate Risk With a Positive Carbon Price. The digital currency system will demonstrate a new model for costing externalities and pricing systemic climate risk. In Chapter 16, Carbon Deposits—Using Soil and Blockchains to Achieve Net-Zero Emissions, Edward Dodge will introduce an innovative form of carbon finance—the "Carbon Deposit" system that uses Blockchain to account for and price the amount of carbon to be sequestered in soils by farmers paid to practice regenerative agriculture. With regard to emissions trading, Anton Galenovich, Alexey Shadrin, and Sergey Lonshakov will introduce the concept and design of their decentralized public Blockchain ecosystem for carbon markets in Chapter 17, Blockchain Ecosystem for Carbon Markets, Environmental Assets, Rights, and Liabilities: Concept Design, and Implementation. Probably, a constructive contribution to developing new market mechanism or even redeveloping the clean development mechanism, Chapter 18, How a Blockchain Network Can Ensure Compliance With Clean Development Mechanism Methodology and Reduce Uncertainty About Achieving Intended Nationally Determined Contributions, by Steven Dunkel will recommend a new concept for transferring digital carbon credits among different clean development mechanism projects over distributed ledgers.

Linking or unifying emission trading schemes (ETS) can, arguably, result in a single carbon price. By leveling the playing field for businesses, at least in theory, linking ETS could increase the cost-effectiveness of curbing global emissions as unlimited abatement opportunities abound for exploration while avert carbon leakage. Notwithstanding these great benefits, linking ETS of different jurisdictions with different regulations and of different development stages will prove challenging. Mutual trust among states is not apparent yet. The first and foremost hot potato is for all states to agree on a price band, which is nearly impossible. In Chapter 19, Networked Carbon Markets: Permissionless Innovation With Distributed Ledgers?, Adrian Jackson, Ashley Lloyd, Justin Macinante, and Markus Hüwener will propose an alternative, bottom-up solution based on Distributed Ledger Technology. Their proposal would enable the "networking" of carbon markets without "unifying" their legal and regulatory frameworks, but data sharing and innovative transaction management which enhance the interoperability of carbon market infrastructure.

CHAPTER 15

Central Banks and Blockchains: The Case for Managing Climate Risk with a Positive Carbon Price

Delton B. Chen

Center for Regenerative Community Solutions[1] 501(c)(3), Basking Ridge, NJ, United States

15.1 Introduction

15.1.1 The Paris Agreement

The main ambition of the celebrated Paris Agreement—of the Twenty-First Conference of Parties (COP21) under the United Nations Framework Convention on Climate Change (UNFCCC)—is to hold global warming to well below 2°C above preindustrial levels and pursue a limit of 1.5°C of warming, recognizing that this would "*... significantly reduce the risks and impacts of climate change*" (UNFCCC, 2015; Article 2). Managing climate risk is clearly a major component of the Paris Agreement in terms of the intended actions, but what is meant by "risk"? A standard definition of risk is that it is "*the effect of uncertainty on objectives*" (ISO, 2009) and so the Paris Agreement should be concerned with improving the *certainty* that unwanted levels of climate change will be avoided.

Parties to the Paris Agreement have agreed to define their voluntary mitigation actions as Nationally Determined Contributions (NDCs) on a five-year pledging cycle. The NDCs and new technologies are helping to drive a low-carbon transition, but the Paris Agreement as a whole lacks a binding mechanism to enforce the 1.5/2.0°C limits (or any other limit). According to a group called "Mission 2020", to achieve the 2.0°C ambition, global carbon emissions will need to peak by 2020 and then decline, otherwise "*... the temperature goals set in Paris become almost unattainable*" (Figures et al., 2017).

15.1.2 Likelihood of a Climate Catastrophe

If the world's greenhouse gas (GHG) emissions were to cease completely in 2017, the GHGs already in the atmosphere would produce 1.3°C (0.9–2.3°C) of "committed

[1] www.crcsolutions.org/.

Transforming Climate Finance and Green Investment with Blockchains.
DOI: https://doi.org/10.1016/B978-0-12-814447-3.00015-X

warming" by 2100 (Mauritsen & Pincus, 2017). When future GHG emissions are extrapolated, it becomes apparent that the 1.5/2.0°C limits will be difficult to achieve. Raftery, Zimmer, Frierson, Startz, and Liu (2017), who undertook a probabilistic assessment of future emissions, conclude that the 1.5°C and 2.0°C limits only have a 1% and a 5% chance of being met, respectively. William Nordhaus, who is well-known for developing the *Dynamic Integrated model of Climate and the Economy* (DICE), similarly found that "*...a limit of 2°C appears to be infeasible with reasonably accessible technologies...*" (Nordhaus, 2016, p. 3).

Nordhaus (2016, p.3) does not appear optimistic about limiting warming to 2.5°C, which he described as "*...technically feasible but would require extreme virtually universal global policy measures.*" Garrett (2012), who developed a lumped-parameter model of the economy based on global energy requirements, deduced that future carbon dioxide (CO_2) emissions could be so difficult to mitigate, that future emissions could actually "*...push civilization towards an accelerating decline.*" In this chapter we will consider how central banks can address the climate crisis with new remits that specifically reduce the physical and socioeconomic uncertainties of climate change.

15.1.3 Central Bank Narrative on Climate Risk

The Prudential Regulation Authority (2015), which is part of the Bank of England, and the Financial Stability Board (2017), which is an international body promoting financial stability, have considered three causes/types of climate-related financial risk, namely: (1) physical causes that arise from climate and weather-related events; (2) transitional causes that arise from the process of adjusting to a lower-carbon economy; and (3) liability causes that arise from parties who have suffered loss or damage.

Aglietta and Espagne (2016) coin the term "climate systemic risk" in reference to the combined financial and physical risks, and they recommend that collective insurance against these risks be implemented as "*...equivalent of a value that society attributes to mitigation activities*" (p. 5). Some important questions arise from their discussion, such as: should central banks use Quantitative Easing (QE) to finance a low-carbon transition?

Appeals for climate-friendly QE, often called "green QE," have been made by numerous scholars, including: Sir David King (Harvey, 2012), Werner and Lucas (2012), Ferron and Morel (2014), Sirkis et al. (2015), and others. Matikainen, Campiglio, and Zenghelis (2017) assessed the QE programs of the European Central Bank (ECB) and the Bank of England to find that their asset purchases have actually favored emission-intensive sectors of the economy. This finding leads to another key question: should central banks remain "market neutral" with respect to the assets that they purchase?

If new remits are to be given to central banks to generate climate finance, such as with QE, then these remits will require clarity regarding policy objectives and tools. This clarity is needed because, as Volz (2017, p. 19) points out, the central bankers might not be prepared for the "*...demands and expectations that central banks should now become a jack of all trades and also solve the world's environmental problems.*" The adoption of new central bank remits appears plausible if the role of central banks is assumed to be evolving (e.g., Goodhart, 2011). Given that mandates of central banks are currently compatible with managing "financial systemic risk", the feasibility of new remits for "climate systemic risk" appears to be mainly a political question (Volz, 2017, p. 20). In the following discussion, a case is presented for new central bank remits that provide a long-term macroprudential response to climate change.

15.2 Managing Climate Risk with a Positive Carbon Price

15.2.1 Positive Carbon Price

The proposal to manage climate systemic risk with a positive carbon price is explained here in terms of (i) what it is, (ii) why it is needed, and (iii) how it should be implemented. These descriptions are based on a global climate policy developed by Chen, van der Beek, and Cloud (2017), called *Global 4C Risk Mitigation* or "Global 4C." The tool that is recommended for creating the positive carbon price is a Central Bank Digital Currency (CBDC), and the generic name of the proposed CBDC is *Complementary Currencies for Climate Change* (4C).

15.2.1.1 What Is Recommended as the Positive Carbon Price?

The positive carbon price is a financial reward for each metric tonne of carbon dioxide equivalent (CO_2-e) that is mitigated relative to certain emissions baselines and under long-lived contracts that define the required standard of service. The reward is, therefore, a commercial offer that can incentivize *voluntary* mitigation, and the aggregate of this mitigation should limit the climate systemic risk over the long term. Under the Global 4C policy, any carbon that is abated or sequestered and then rewarded with 4C will not be traded in carbon markets. This implies that all of the carbon in the 4C stocktake is "retired" and cannot be used to offset other carbon emissions.

4C is a proposed parallel currency that can be traded with other currencies. Central banks will require remits to influence the 4C price with QE and currency trading, and the main objective of this currency trading is to set the price of 4C to reflect the average market price of limiting the climate systemic risk. When the 4C is correctly priced to the climate systemic risk, it will incentivize sufficient climate mitigation for an agreed risk limit to be achieved. The central banks will employ a trading strategy that involves "telegraphing" to the market their future trading intentions, and this can be used to generate private demand for 4C. This trading strategy is described below and in the "Avoiding Catastrophe" storyline (Section 15.3).

15.2.1.2 Why Should There Be a Positive Carbon Price?

The positive carbon price is recommended as preventative insurance against the possibility of dangerous-to-catastrophic climate change. It is a global price that can respond quickly and proportionally to man-made climate forcing and natural climate feedbacks. The operational objective is to stay below specific temperature changes over a rolling 100-year planning horizon and with a limited probability of failure (e.g., remaining below 4°C of warming with 3% chance of failure). The 100-year planning horizon is not a precise time period, but it is not arbitrary either because human activities could drive 2.0–4.9°C of global warming within 100 years[2] (Raftery et al., 2017).

15.2.1.3 How Does the Positive Carbon Price Work?

The positive carbon price works by offering in the marketplace a global reward for mitigating carbon emissions. By issuing the global reward as a digital parallel currency over the Internet—called 4C—the reward should bypass most political, social, and financial barriers that would otherwise inhibit the establishment of many low-carbon projects.

It is proposed here that central banks be given remits to use QE and currency trading—as much as is needed—to set 4C prices and influence the market's expectations for future 4C prices. These remits should be used to manage long-term "bull" and "bear" markets in private 4C trading: this is a core macroeconomic strategy of the approach. It should be made clear that 4C prices will actually mirror the cost of preventing dangerous-to-catastrophic climate change, and that the scheduled 4C price is a public safety announcement based on prices. The "bear" market in 4C trading will occur after the risk has peaked and is reducing. The utility of 4C during the "bull" and "bear" markets is unchanged, although the central banks will need to adjust their currency trading strategies to manage these phases of the 4C market.

15.2.2 New Model for Externalities

The standard model for assessing the external cost of carbon emissions, is as follows:

$$\text{External Cost} = \text{Social Cost of Carbon(SCC)} \tag{15.1}$$

The *Social Cost of Carbon* (SCC) is the time-discounted average economic welfare loss per tonne of CO_2-e emissions (IAWG, 2013); and the SCC is used to guide the price of carbon taxes to reach a welfare maximum. The new model, which was first proposed by Chen *et al.* (2017), identifies two types of external cost, as follows:

[2] About 60% of the equilibration surface temperature occurs 25–50 years after the CO_2 is emitted (Hansen et al., 2013); and atmospheric CO_2 concentrations adjust about 100 years after the CO_2 is emitted (IPCC, 2013).

$$\text{Total External Cost} = \text{Social Cost of Carbon(SCC)} + \text{Risk Cost of Carbon(RCC)} \quad (15.2)$$

The *Risk Cost of Carbon* (RCC) in Eq. 15.2 is defined as the average cost of climate mitigation services (per tonne of CO_2-e) that will be sufficient to avoid specific levels of global warming and in accord with politically agreed probabilities of failure. For example, the cost of not passing 4°C of global warming by 2120 with a 3% chance of failure can be used to define the RCC. The SCC and RCC are summed in Eq. 15.2 to aid our understanding that carbon emissions are creating two kinds of external cost. These costs associate with climate damages and systemic risk, respectively.

The possible existence of the RCC as a second external cost is explained by a market hypothesis—called the *Holistic Market Hypothesis* (HMH)—that was developed by Chen *et al.* (2017) based on a thermodynamic interpretation of carbon pricing and social networks. A critical issue is that carbon lock-in is a feature of the economy that produces dangerous carbon emissions (Unruh, 2002), and that a thermodynamic analysis of the economy reveals that carbon lock-in could be much worse than previously anticipated (Garrett, 2012). The HMH is a hypothesis that the carbon lock-in can be most effectively addressed with a combination of conventional carbon taxes and a global carbon reward that is delivered as a parallel currency (i.e., 4C). To be effective, 4C should be priced to mirror the RCC in Eq. 15.2, and 4C should be convenient to trade with national fiat currencies. It is recommended that 4C not be used to trade for goods and services, because this option will unnecessarily complicate the policy and its implementation. When holders of 4C decide to spend their 4C, they will first have to exchange their 4C for a local currency.

The RCC is the cost of limiting climate risk and avoiding climate "tipping points" (Lenton, 2012); and it should be estimated with comprehensive risk assessments that take into account the total cost of mitigation and all administrative costs, including the hidden costs of monitoring for carbon leakage and for managing bad actors. The application of Eq. 15.2 is illustrated in Fig. 15.1 in terms of a single market actor. Fig. 15.1 shows that the SCC is internalized with taxes linked to emissions, and that the RCC is internalized with rewards linked to mitigation. The SCC and RCC are interactive costs. Next, we will use a thought experiment to examine the utility of a positive carbon price created with a parallel currency.

15.3 The Avoiding Catastrophe Storyline

15.3.1 Hypothetical

A hypothetical 100-year storyline called "Avoiding Catastrophe" is presented as a thought-experiment and to illustrate how a positive carbon price could potentially "move the trillions"[3] of new climate finance. In this storyline an international political agreement is

[3] Sirkis et al. (2015)

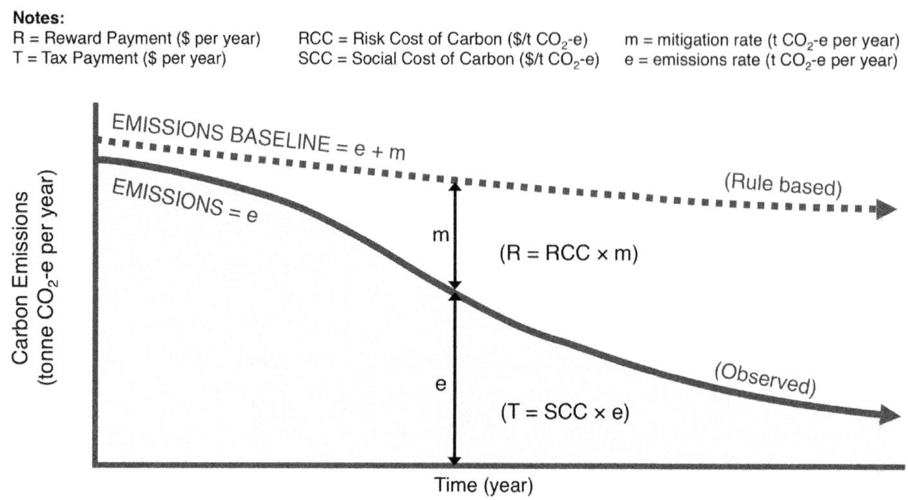

Figure 15.1
Illustrative example of complementary carbon pricing for a single market actor, including idealized representations of (e) carbon emissions and (T) tax payments for the internalization of the SCC; and (m) carbon mitigation and (R) reward payments for the internalization of the RRC.

established under the UNFCCC to prevent global warming from exceeding 2°C with a 33% chance of failure, and 4°C with a 3% chance of failure, and with both risk limits being addressed concurrently over a rolling 100-year time horizon.

The amount of additional finance that will be needed for a timely low-carbon transition could exceed USD$1 trillion per year based on published assessments (e.g., Global Commission on Economy and Climate, 2014; IEA, 2015). Of particular interest is whether 4C trading can help resolve the "Tragedy of the Horizon" conundrum that Mark Carney, the governor of the Bank of England, described in a speech:

> "*That is, climate change is a tragedy of the horizon which imposes a cost on future generations that the current one has no direct incentive to fix*" (Carney, 2016).

In this hypothetical it is assumed that all G20 nations and most non-G20 nations are supporting the Global 4C policy as active partners and beneficiaries of 4C rewards, and that central bank remits are internationally coordinated such that financial markets and the public can trust 4C as a store of value.

The "Avoiding Catastrophe" storyline begins in the year 2020 with the issuance of 4C to actors who abate carbon emissions or undertake direct air-capture and sequestration of carbon. At the beginning of each year, a risk assessment is conducted to estimate the RCC

Notes:
The 4C currency has a unit of account of 100 kg of CO_2-e mitigation, and its value is managed over time to equal the RCC. RCC has units of $USD/1000 kg of CO_2-e mitigation.

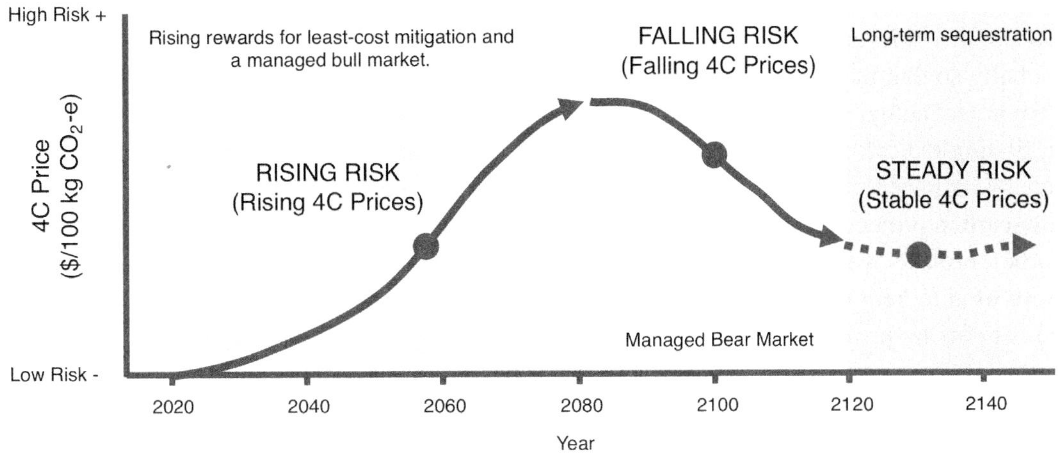

Figure 15.2
The "100-year advance 4C price alert" in the "Avoiding Catastrophe" hypothetical storyline. *Adapted from Chen* et al. *(2017).*

over the following 100 years for the objectives of avoiding greater than 2°C with 33% chance of failure, and greater than 4°C with 3% chance of failure. The RCC time-series is known as the "100-Year Advance 4C Price Alert" (see Fig. 15.2), and it is advertised in the mainstream press so that markets are aware of the 4C reward price (for each 100 kg of CO_2-e mitigated) and the yield on 4C holdings.

15.3.1.1 The 4C Supply-Demand Storyline

In the storyline, the central banks of the G20 and other participating nations coordinate their QE and 4C currency trading to guarantee the prices shown in the "100-Year Advance 4C Price Alert" (Fig. 15.2), and they accumulate their 4C purchases in holding accounts. The monetary inflation created by QE is dispersed evenly across the world economy such that firms and citizens are minimally impacted by the QE.

During the first 60 years, the required yield on 4C is greater than the average interest paid on bank deposits and borrowings, and greater than the average return in stock markets (e.g., S&P 500, Dow Jones Industrial Average) and bond markets (e.g., Bloomberg Barclays US Aggregate Bond Index). Consequently, many private traders and institutional investors buy 4C during this managed currency bull market. The central banks trade 4C to help smooth the 4C price during the bull market, but by the year 2080 there is a peak in the climate risk, and for another 40 years (2080–2120) the central banks guide the 4C price to a lower price range by selling some of their 4C holdings (Fig. 15.2). Society accepts the 4C bear market as a trade-off for limiting the climate risk and stabilizing the climate over the long term. National governments accept 4C as an international trading currency and as an essential tool for ensuring international cooperation on climate change.

During the 100-year storyline (2020–2120), data within the carbon stocktake was shared globally so that market actors could optimize their investing across a wide spectrum of low-carbon technologies. The 4C rewards encouraged massive R&D and this was instrumental in finding radical new mitigation technologies, thereby increasing the long-term mitigation rate. 4C allowed market actors to qualify for larger commercial bank loans to finance new low-carbon projects, and the green/climate bond market received a huge boost because low-carbon projects were more profitable with 4C. The reported 4C earnings of each project were used to help quantify the "shade of green" of these projects, and 4C finance generated co-benefits by providing employment, reducing poverty, and protecting biodiversity. These co-benefits were especially relevant to people living in rural regions where the land was suitable for bio-sequestering carbon but economic opportunities were otherwise limited.

15.3.2 Resolving the Climate Paradox

"Resolving the Climate Paradox" is a speech presented by the Bank of England's governor, Mark Carney. Carney (2016) remarked on climate risks and the challenges of transitioning to a low-carbon economy. The hypothetical storyline is appraised against the two paradoxes[4] described by Carney (2016): (1) "Future will be past," and (2) "Success is failure."

15.3.2.1 Paradox 1: Future Will Be Past

> *"The catastrophic impacts of climate change will be felt beyond the traditional horizons of most actors including businesses and central banks. Once climate change becomes a clear and present danger to financial stability it may already be too late to stabilize the atmosphere at two degrees"* (Carney, 2016).

The "Future will be Past" paradox, as described above, is a result of the 25–50 year time lag between carbon emissions and resulting climate impacts (Hansen et al., 2013) and the time-discounting of future impacts by society. The "100-Year Advance 4C Price Alert" is a partial resolution to this paradox because it can incentivize near-term climate mitigation (Fig. 15.2) based on a socioeconomic feedback between the risk of future global warming and the 4C price. Resolution is based on the idea that a rising 4C price will generate optimism for a

[4] Jevons effect and growth vs. degrowth may be considered two additional paradoxes related to climate change (refer Chen, Cloud, & van der Beek, 2015).

low-carbon transition within the time horizon of most market actors, and will create a secular 4C "bull" market because of the positive yield on 4C investments.

15.3.2.2 Paradox 2: Success Is Failure

> *"That is, too rapid a movement towards a low-carbon economy could materially damage financial stability. A wholesale reassessment of prospects, as climate-related risks are re-evaluated, could destabilize markets, spark a pro-cyclical crystallization of losses and lead to a persistent tightening of financial conditions: a climate Minsky moment"* (Carney, 2016).

The "Success is Failure" paradox, as described above, is partly resolved because 4C finance is commensurate with the quantity of mitigation that is needed to limit the climate systemic risk, and because 4C prices are predictable and stable over many decades (Fig. 15.2). Also, the "100-Year Advance 4C Price Alert" should be designed to ensure that the financial transition is smooth by encouraging long-term investing and infrastructure planning for the low-carbon transition: potentially avoiding a "climate Minsky moment."

15.4 The FinTech Brief

Political challenges aside, the case for a positive carbon price is dependent on the technical feasibility of the 4C platform, which is a kind of CBDC. Notable examples of CBDCs are Project Ubin, developed by the Monetary Authority of Singapore (MAS, 2017), and Utility Settlement Coin (USC), developed by Clearmatics for UBS. The 4C system will require a new international authority for setting standards, notionally called the *Carbon Exchange Standard* (CES) (Chen *et al.* 2017).

A technical brief for the 4C platform is provided here for governance institutions, investors, and innovators who may be interested in collaborating on a business model for the 4C platform: a model that faces some technical challenges and political risks. A preliminary business case is to assume roughly USD$1 trillion per year equivalent of 4C issued in the marketplace as a global carbon reward (Global Commission on the Economy & Climate, 2014). An administrative fee of just a few percent of the 4C supply may be sufficient to finance the entire 4C platform—and this will provide the CES with political autonomy and independence from fiscal budgets.

15.4.1 The Global 4C Platform

The CES should target all geographic locations and all economic sectors with the 4C platform (incl. energy, transportation, agriculture, manufacturing, buildings, land and marine management, education, etc.). The design of the 4C platform should be geared for rapid scalability and to support a global market-driven program of climate mitigation; and it should maintain scarcity of 4C by linking the 4C supply to an official carbon stocktake.

The 4C platform will outsource the Measurement Reporting and Verification (MRV) to the private sector. The platform should also leverage mitigation and enhance social and ecological co-benefits. Ideally, the 4C platform will have access to global environmental monitoring data and satellite-based observations that can support the MRV.

The 4C platform should be developed using Blockchain distributed ledgers so that the various data-intensive services can be decentralized over the Internet and mobile devices: with the aim of ensuring that the data are secure and verified. The 4C platform might use a "consortium" Blockchain network (e.g., Microsoft, 2017) so that the CES can manage permissions. The CES should be able to regulate the 4C ledger for resolving errors, thefts, and for charging demurrage fees (Section 15.4.4).

The 4C platform will require the standardization of interconnections with external databases and with real-world devices for the automation and streamlining of MRV, and to facilitate peer-to-peer (P2P), business-to-business (B2B), and machine-to-machine (M2M) communications and trade.

15.4.2 Smart Contracts

The 4C will be a P2P currency that is easily traded for other currencies. The 4C currency—with a unit of account of 100 kg of CO_2-e mitigated—will be supplied as a reward for verified carbon mitigation. 4C rewards are a kind of *seigniorage* income for mitigation actors, but 4C is rewarded conditionally and with long-lived service agreements that will be managed with "smart" contracts for administrative efficiency (e.g., Buterin, 2014). The service agreements will be separated from the 4C currency ledger, and these agreements will be private contracts that link mitigation actors to the CES. The service agreements are needed to bond actors to their mitigation claims for the approximate time that it takes carbon emissions to impact on the climate system. A maximum contract duration of 100 years is suggested, and this duration has a precedent with 99-year leases on property under common law.

Smart contracts for the service standards should be able to request MRV at suitable times for updating the carbon stocktake and for triggering contract clauses. The main aim of the 100-year service agreements (and smart contracts) is not to impose harsh penalties on actors; rather, it is to ensure the integrity of the carbon stocktake on a rolling 100-year basis. Although most low-carbon projects might only operate for just a few years to a few decades, the service agreements should contain provisions to monitor emissions and carbon storage in case of carbon leakage. If low-carbon projects outlive their owners or change hands, novation provisions should transfer the service agreements to the new owners. If low-carbon projects have no owners, then the CES or another authority should take responsibility for these agreements.

Mitigation actors who fail to meet their service agreements will be required to return the outstanding 4C or a monetary equivalent; however, actors will also be permitted to go into debt to avoid financial hardship, but actors should not be allowed to accumulate debts indefinitely. To address these issues, there will be provisions for defaulting, *force majeure*, debt collection, and debt forgiveness; and the provisions should be legally enforceable. The unrecoverable 4C debts will be reconciled using a global demurrage fee on all 4C holdings (Section 15.4.4).

15.4.3 Fees and Commissions

The CES authority should encourage an open market by offering licenses to individuals and firms who wish to do auditing and assessments for financial compensation, and by paying for these services with commissions and hourly rates. By covering all of the running costs with a percentage of the 4C rewards, the 4C platform will avoid unnecessary financial intermediaries and will be self-funding.

Although the technical details are not presented here, the basic idea is that the 4C platform will be designed to detect, discourage, and punish gaming and data counterfeiting. A virtual trust economy should also be established in which stakeholders assign comments and reputational tokens to the rules, systems, data, and other stakeholders, and help ensure that the system is transparent and accountable.

15.4.4 Demurrage Fees

Holders of a demurrage currency are charged a fee for holding the currency over time. One example is the Wörgl demurrage currency that was issued in Austria between 1932–1933, and was called the "miracle of Wörgl" because it stimulated economic activity (Hallsmith & Lietaer, 2011). A variable demurrage fee (%) will be used to eliminate an amount of 4C that corresponds to the aggregate of defaults on service agreements. This is to maintain consistency between the 4C supply and the carbon stocktake. The demurrage fee should be charged uniformly, as a percentage of all 4C holdings, to minimize its impact on 4C holders. Charging the fee will require a re-organization of the 4C ledger by the CES.

Higher demurrage fees will likely correlate with greater climate risk and a higher 4C yield. An example would be when global warming triggers forest fires and results in a *force majeure* destruction of bio-sequestration projects. The annual demurrage fee (%) will, therefore, be a vital statistic that communicates the vulnerability of low-carbon projects to changing environmental and social conditions.

15.4.5 International Business Model

For political reasons, the CES may be required to franchise the 4C platform to "climate clubs," which are clubs of nations that are grouped by mutual interests (e.g., Sirkis et al., 2015). If this approach is followed, the central banks of each climate club will be required to coordinate their 4C monetary policy under the common rules of the CES. Climate clubs will be able to define their own 4C mitigation markets (i.e., by national borders); however, it is important that the assessment rules be based on international standards, and that the various 4C (i.e., of each climate club) are managed so that their exchange rates converge on a single exchange rate over time. The various 4C originating from each climate club will be openly traded in international markets.

The 4C platform should be designed to scale-up rapidly and encourage socioecological co-benefits. For example, the 4C platform should streamline feasibility reporting, certifiable valuations, and environmental/social impact assessments. These reports should assist actors in their efforts to attract financial capital and access bank loans. The platform should also support crowdfunding and socially responsible collaborations.

15.5 Discussion and Conclusions

The 1.5/2°C global warming limits of the Paris Agreement (UNFCCC, 2015) may be unachievable if global carbon emissions do not begin to fall by 2020 because of carbon lock-in (Unruh, 2002; Mission 2020, 2017). Overshooting 1.5/2°C might inspire efforts to ratchet-up Nationally Determined Contributions (NDCs) or, alternatively, it might create pessimism and weaken the NDCs (Young, 2016).

To address the carbon lock-in problem, Chen *et al.* (2017) revised the standard model for the external cost of carbon emissions with their hypothesis for the RCC. The RCC complements the conventional SCC, as shown in Eq. 15.2 and Fig. 15.1. The new model defines complementary market-based policies for correcting the market failure: (i) the orthodox policy of applying carbon taxes that, ideally, should approach the SCC; and (ii) the new policy of offering global carbon rewards that mirror the RCC.

The hypothesis of Chen *et al.* (2017) that supports the existence of the RCC needs to be verified and validated. Scientific validations may be attempted with experiments, computer simulations and field/pilot studies. If the RCC is shown to be inherently correct, then the implications for economic assessments and climate policy will be profound. A critical issue for policy makers is that under the Tinbergen rule (Tinbergen, 1952) each unique policy objective should be addressed with one policy tool; and so an effective response to the RCC will be to "internalize" the RCC into the economy with a dedicated tool. It is argued here that the ideal tool is the CBDC, called 4C. It is essential to note that 4C will be supplied as a reward for mitigation and it is not a carbon offset. In other words, the mass of CO_2 equivalent that is rewarded with 4C will be removed from carbon markets and will not be available for trading.

The RCC is directly relevant to Article 2 of the Paris Agreement by defining a target for limiting "*risks*" (UNFCCC, 2015). Parties to the Paris Agreement may wish to assess the RCC under Article 8, which seeks to enhance knowledge and understanding of comprehensive risk management. The most relevant topics relate to Sections 4 (a), (e), and (f), as follows:

(a) "*Early warning systems*" could include the "100-Year Advance 4C Price Alert," which is a public broadcast message that communicates the risk of dangerous-to-catastrophic climate change;
(e) "*Comprehensive risk assessment and management*" could include annual assessments of the RCC; and
(f) "*Risk insurance facilities, climate risk pooling and other insurance solutions*" could include the Global 4C policy as preventative global insurance and with the insurance premium covered by QE and currency trading.

The model for the SCC and RCC offers a doorway to new central bank remits that can limit climate risk and may inspire a new global risk management strategy:

> *"Climate risk management is in its infancy and not integrated into broader risk management frameworks, but this is likely to change as it moves up policymakers' agendas"* (PwC, 2017).

For central bankers, the take-home message is to look beyond market neutrality, green QE and conventional asset purchases, and to consider a new monetary policy for 4C trading—as explained in the "Avoiding Catastrophe" storyline—to break the "Tragedy of the Horizon" conundrum and resolve Carney's (2015, 2016) climate paradoxes. Resolution appears plausible by telegraphing future 4C prices to financial markets, because the resulting 4C bull market will convert tomorrow's risk into today's profits; and this will improve inter-generational equity.

What is special about the FinTech brief for the 4C platform (refer Section 15.4), is that it suggests that central bankers do not have to be burdened by carbon auditing and MRV. This technical work can be outsourced to the private sector under the auspices of a new international institution—the CES—which is needed to assess the RCC, set standards, and manage service agreements. The 4C platform should be designed to make full use of Blockchain technologies and open-source Blockchain standards for improved interconnectivity and to accelerate mass adoption (Maupin, 2017; Tapscott & Tapscott, 2017). It is anticipated that Blockchain ledgers will improve platform scalability, decentralization, security, accountability, and transparency. A governance model for 4C was not discussed in this chapter however a skeleton of a model is provided by Chen *et al.* (2017).

There is much to consider in this proposal. Central bankers and their governing bodies are in the driver's seat for this macroprudential policy that can address the climate systemic risk. The global 4C reward should be used to invite citizens and businesses of all nations to directly participate in a global transition, rather than committing billions of people to be spectators on a moving juggernaut. Given the enormity of the climate crisis, a brave decision is needed from central bankers and political leaders to pivot on the idea—and this is why:

> *"...the lions need to learn how to roar once again"* (PwC, 2017).

References

Aglietta, M., & Espagne, E., (2016). Climate and finance systemic risks, more than an analogy? The climate fragility hypothesis. Working Paper CEPII, No 2016-10 – April, pp. 30.

Buterin, V., (2014). *A next-generation smart contract and decentralized application platform. White paper, GitHub,* <https://github.com/ethereum/wiki/wiki/White-Paper> Accessed on 14.4.16.

Carney, M., (2015). *'Breaking the tragedy of the horizon – climate change and financial stability,' speech given by the governor of the Bank of England, chairman of the Financial Stability Board, Lloyd's of London, 29 September 2015* <http://www.bankofengland.co.uk/publications/pages/speeches/2015/844.aspx> Accessed on 22.9.17.

Carney, M., (2016). *'Resolving the climate paradox,' speech given by the governor of the Bank of England, chair of the Financial Stability Board, Arthur Burns Memorial Lecture, Berlin, 22 September 2016.* <http://www.bankofengland.co.uk/publications/Pages/speeches/2016/923.aspx> Accessed on 22.9.17.

Chen, D. B., Cloud, J., & van der Beek, J. (2015). *Global 4C: World monetary union for climate change mitigation. 2015 Canberra Conference on Earth System Governance: Democracy & Resilience in the Anthropocene.* <http://www.earthsystemgovernance.org/canberra2015/wp-content/uploads/2015/11/Chen-Cloud-vdBeek-2015-Earth-System- Governance.pdf> Accessed on 28.7.17.

Chen, D. B., van der Beek, J., & Cloud, J. (2017). Climate mitigation policy as a system solution: Addressing the risk cost of carbon. *Journal of Sustainable Finance & Investment*, *7*(3), 1–42.

Ferron, C., & Morel, R., (2014). Smart unconventional monetary (SUMO) policies: Giving impetus to green investment. Appendix II – Green quantitative easing. *Climate Report: Research on the Economics of Climate Change, CDC Climate Research*, No. 46, July 2014.

Figueres, C., Schellnhuber, H. J., Whiteman, G., Rockström, J., Hobley, A., & Rahmstorf, S. (2017). Three years to safeguard our climate. *Nature*, *546*(7660), 593–595.

Financial Stability Board, (2017). *Final report. Recommendations of the task force on climate-related financial disclosures.* <https://www.fsb-tcfd.org/publications/final-recommendations-report/> Accessed on 22.9.17.

Garrett, T. J. (2012). No way out? The double-bind in seeking global prosperity alongside mitigated climate change. *Earth System Dynamics*, *3*, 1–17.

Global Commission on the Economy and Climate. (2014). *Better growth better climate: The new climate economy report* (p. 71) Washington, DC: The New Climate Economy.

Goodhart, C. A. E. (2011). The changing role of central banks. *Financial History Review*, *18*(2), 135–154.

Hallsmith, G., & Lietaer, B. (2011). *Creating wealth: Growing local economies with local currencies*. Gabriola Island, Canada: New Society Publishers.

Hansen, J., Kharecha, P., Sato, M., Masson-Delmotte, V., Ackerman, F., Beerling, D. J., ... Zachos, J. C. (2013). Assessing "Dangerous Climate Change": Required reduction of carbon emissions to protect young people, future generations and nature. *PLoS ONE*, *8*(12), e81648.

Harvey, F., (2012). *Sir David King: Quantitative easing should be aimed at green economy. The Guardian, Environment, 26 June 2012*. <https://www.theguardian.com/environment/2012/jun/26/david-king-quantitative-easing-green> Accessed on 13.8.17.

IAWG, (2013). 'Technical support document: Technical update of the social cost of carbon for regulatory impact analysis under executive order 12866,' *Interagency Working Group on Social Cost of Carbon, United States Government, Washington, DC*, May 2013, pp. 22.

IEA, (2015). *Energy technology perspectives 2015. Mobilising innovation to accelerate climate action* (p. 412) Paris: International Energy Agency.

IPCC, (2013). Climate change 2013: The physical science basis. In T. F. Stocker, D. Qin, G.-K. Plattner, M. Tignor, S. K. Allen, J. Boschung, et al. (Eds.), *Contribution of working group I to the fifth assessment report of the intergovernmental panel on climate change*. Cambridge, United Kingdom and New York, NY, USA: Cambridge University Press, 1535 pp.

ISO, (2009). *'ISO 31000:2009 Risk management – Principles and guidelines,' International Organization for Standardization, Geneva.* <http://www.iso.org/iso/home/standards/iso31000.htm> Accessed on 28.7.16.

Lenton, T. M. (2012). Arctic climate Tipping points. *Ambio*, *41*(1), 10–22.

MAS, (2017). *The future is here | Project Ubin: SGD on Distributed Ledger. Monetary Authority of Singapore, Deloitte, pp. 44.* <http://www.mas.gov.sg/Singapore-Financial-Centre/Smart-Financial-Centre/Project-Ubin.aspx> Accessed on 25.8.17.

Matikainen, S., Campiglio, E. and Zenghelis, D., (2017). The climate impact of quantitative easing. Policy Paper, May 2017. CCCEP & The Grantham Research Institute on Climate Change and the Environment.

Mauritsen, T. and Pincus, R., (2017). Committed warming inferred from observations,' *Nature Climate Change* (31 July 2017).

Maupin, J. A., (2017). *Blockchains and the G20: Building an inclusive, transparent and accountable digital economy. CIGI, Policy Brief No. 101 — March 2017.* <https://hollandfintech.com/wp-content/uploads/2017/03/CIGI-Policy-Brief-101-Blockchains-G20-March-2017.pdf> Accessed 22.9.17.

Microsoft, (2017). *Announcing support for multi-member consortium blockchain networks on Azure. Posted March 23, 2017.* <https://azure.microsoft.com/en-au/blog/multi-member-consortium-blockchain-networks-on-azure/> Accessed on 24.8.17.

Mission 2020, (2017). *2020: The climate turning point (29 pp.).* <http://www.mission2020.global/2020%20The%20Climate%20Turning%20Point.pdf> Accessed on 12.8.17.

Nordhaus, W. D., (2016). Projections and uncertainties about climate change in an era of minimal climate policies. Cowels Foundation Discussion Paper No. 2057. New Haven, CT: Yale University.

Prudential Regulation Authority, (2015). *The impact of climate change on the UK insurance sector: A climate change adaptation report by the Prudential Regulation Authority. Prudential Regulation Authority, London, Sept. 2015.* <http://www.bankofengland.co.uk/pra/Documents/supervision/activities/pradefra0915.pdf> Accessed on 22.9.17.

PwC, (2017). *Staying ahead: Central banks in a multi-mandate environment, PwC Central Bank Forum report 2016. February, 2017.* <http://www.pwc.com/vn/en/industries/assets/pwc-central-bank-forum.pdf> Accessed on 22.9.17.

Raftery, A. E., Zimmer, A., Frierson, D. M., Startz, R., & Liu, P. (2017). Less than 2°C warming by 2100 unlikely. *Nature Climate Change*. Published Online: July 31, 2017. <https://www.nature.com/articles/nclimate3352> Accessed on 22.9.17.

Sirkis, A., Hourcade, J. C., Dasgupta, D., Studart, R., Gallagher, K., Perrissin-Fabert, B., ... Aglietta, M., (2015). Moving the trillions: A debate on positive pricing of mitigation actions, Centro Brasil no Clima Ed./CIRED, Nogent-sur-Marne.

Tapscott, D., and Tapscott, A., (2017). Realizing the potential of blockchain: A Multistakeholder approach to the stewardship of blockchain and cryptocurrencies. White Paper, World Economic Forum, June 2017. pp. 46.

Tinbergen, J. (1952). *On the theory of economic policy, Contributions to Economic Analysis* (Vol. 1). Amsterdam, Netherlands: North-Holland Pub. Co.

UNFCCC, (2015). Adoption of the Paris Agreement, 21st Conference of the Parties, Paris: United Nations. Report No. FCCC/CP/2015/L.9/Rev.1 <http://unfccc.int/resource/docs/2015/cop21/eng/l09r01.pdf>. Accessed on 28.7.17.

Unruh, G. C. (2002). Escaping carbon lock-in. *Energy Policy*, *30*(4), 317–325.

Volz, U., (2017). *On the role of central banks in enhancing green finance. Enquiry working paper 17/01, February 2017. UN Environment Inquiry.* <http://eprints.soas.ac.uk/23817/1/On_the_Role_of_Central_Banks_in_Enhancing_Green_Finance(1).pdf> Accessed on 28.7.17.

Werner, R. and Lucas, C., (2012). The best use of £50bn QE? Bypass the banks and go direct to green projects, *The Guardian*. <https://www.theguardian.com/commentisfree/2012/feb/10/qe-banks-green-projects> Accessed on 28.7.17.

Young, O. R. (2016). The Paris Agreement: Destined to succeed or doomed to fail? *Politics and Governance*, *4*(3), 124–132.

CHAPTER 16

Carbon Deposits—Using Soil and Blockchains to Achieve Net-Zero Emissions

Edward Dodge

Technology Consultant, Ithaca, NY, United States

16.1 Carbon Deposit: An Innovative Concept

The carbon deposit is an innovative concept for carbon finance that leverages the power of soils to sequester carbon and achieve net-zero emissions of the greenhouse gases (GHGs) driving climate change. The *carbon-deposit system* acts like a carbon tax on the front end, but uses the funds raised to pay farmers and ranchers on an equal ton-for-ton basis to put the carbon back in the soil where it belongs to. Blockchain technology provides transparent accounting for the system with smart contracts that link a carbon dioxide source directly to a soil carbon sink that sequesters an equal amount of carbon into the soil.

The carbon deposit plan combines two popular policy proposals: the French Ministry of Agriculture's "*4 per 1000*" proposal, which seeks to improve soil carbon stocks by 0.4% per year; and a US\$40 per ton carbon tax on CO_2 emissions proposed by the US-based Climate Leadership Council (CLC). But unlike the CLC's tax proposal which would pay the funds out as dividends to citizens, in the carbon-deposit system the funds collected would be paid to farmers to deposit carbon in the soil where it provides cascading environmental benefits. The dollars follow the molecules: US\$40 per ton of CO_2 collected on emissions, translates to US\$150 per ton of carbon that is paid to farmers to practice restorative agriculture and build up healthy soil. A small portion of the funds would be used to pay for monitoring, reporting, and verification systems (MRV). The CLC estimates the CO_2 tax could reduce CO_2 emissions by 25%–30%;[1] whereas, the French Ministry of Agriculture argues that soil carbon sequestration could absorb the remaining 75% of CO_2 emissions.[2] In theory, the combination of these two measures will result in net-zero emissions and a balanced carbon cycle.

[1] Climate Leadership Council. (2017). *A winning trade*. Retrieved from https://www.clcouncil.org/wp-content/uploads/2017/02/A_Winning_Trade.pdf.

[2] French Ministry of Agriculture. (2015). Retrieved from https://www.4p1000.org/.

Transforming Climate Finance and Green Investment with Blockchains.
DOI: https://doi.org/10.1016/B978-0-12-814447-3.00016-1

Blockchain provides an effective technology platform for the accounting of this complex system. Blockchains track an asset through its lifecycle via a shared and open ledger whose contents cannot be altered but can be read by all participants. In this system, the asset is 1 ton of carbon, tracked on a Blockchain from its emission source to its sequestration sink. Auditors and regulators who certify both the emissions and sequestration would also be registered on the immutable Blockchain, which can be transparently audited.

16.2 Soil Carbon Sequestration

Carbon dioxide is the primary GHG accumulating in the atmosphere. According to the United Nations—Intergovernmental Panel on Climate Change (UN IPCC), anthropogenic GHG emissions are around 36 billion tons of CO_2 which equals 9.8 billion tons, or metric gigatons (Gt) of carbon per year.[3] Most of these emissions come from fossil fuel combustion, though a little more than one Gt are emitted from land-use changes. Of the 9.8 Gt of carbon in the atmosphere, approximately 3.2 Gt are absorbed by plants through photosynthesis, whereas approximately 2.3 Gt are absorbed by the oceans, leaving a net accumulation in the atmosphere of around 4.3 Gt of carbon per year.

The Earth's soils contain around 2400 Gt of carbon, three times the 750 Gt found in the atmosphere. The natural terrestrial carbon cycle of plant respiration and decomposition exhales carbon, whereas plant photosynthesis absorbs carbon, cycling around 120 Gt up and down annually. Likewise, the ocean's natural carbon cycle of air-sea gas exchange is around 90 Gt up and down annually. The Earth's natural carbon cycles dwarf the anthropogenic carbon emissions that are driving climate change. Human carbon emissions of less than 10 Gt are large enough to push the natural carbon cycles out of balance causing climate change; however, they are a fraction of the natural carbon cycles and can be brought back into balance.

According to the Food and Agriculture Organization of the United Nations (UN FAO), there are 4.9 billion hectares of agricultural land potentially available for carbon farming, including 1.54 billion hectares of planted crop lands and 3.36 of pasture and grasslands.[4] This is equal to 12.1 billion acres globally (1 hectare = 2.47 acres); this does not include another 4.03 billion hectares of global forests. Science and common sense farming both show us that practices that increase soil carbon also improve soil health in general, leading to improved water retention, soil microbiology, fertility, and productivity. The measure of soil carbon can be used as a proxy for soil health, generally speaking, more soil carbon

---

3 IPCC. (2014). *Climate change 2014: Synthesis report.* Contribution of Working Groups I, II, and III to the Fifth Assessment Report of the Intergovernmental Panel on Climate Change [Core Writing Team, R.K. Pachauri, & Meyer, L.A. (Eds.)]. Geneva, Switzerland: IPCC, 151 pp.

4 UN FAO. (2015). *World agriculture: Towards 2015/2030. An FAO perspective...* Retrieved from http://www.fao.org/docrep/005/y4252e/y4252e06.htm.

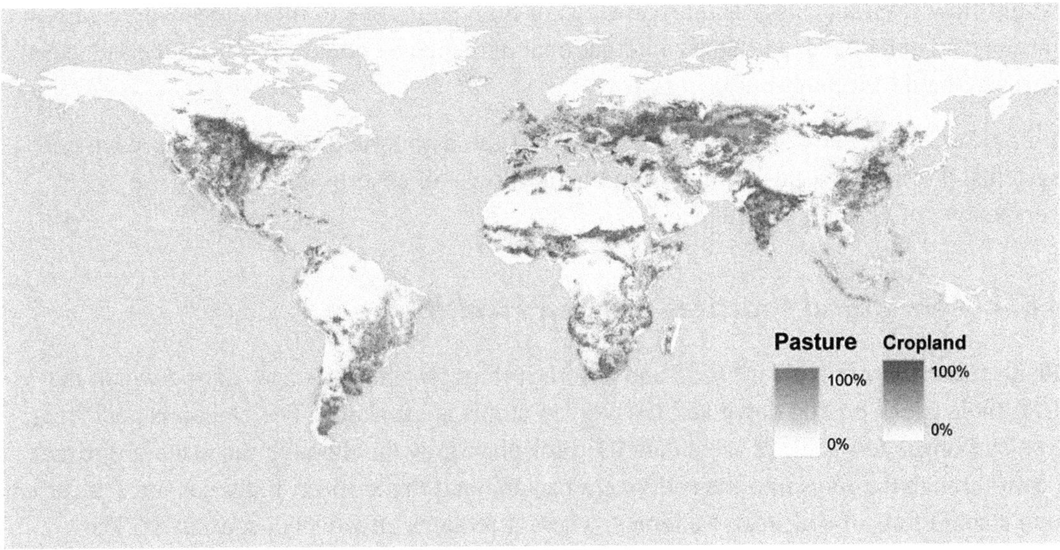

Figure 16.1
A world map of cropland and pastureland available for regenerative agriculture. *Source: Image data derived from UN FAO. (2015). World agriculture: Towards 2015/2030. An FAO perspective. Retrieved from http://www.fao.org/docrep/005/y4252e/y4252e06.htm.*

means better soil health. Research into restorative agriculture has shown consistently on small scales that it is possible to sequester a ton of carbon per hectare per year, and in practice rates range from ¼ to 20 + tons per hectare. Soil science indicates that global sequestration capacity ranges from 3 to 8 Gt per year.[5] Using 1 ton of carbon sequestration per hectare as a benchmark, spread across 4.9 billion hectares, it is reasonable to target net-zero emissions if global society can also bring emissions down by 25%–30% to manageable levels (Fig. 16.1).

16.3 Carbon-Deposit Payment to Farmers

In order to pay farmers to sequester carbon, effective systems for MRV emission reductions need to be established. These could be paid for by setting aside a small percentage of the funds raised as system overhead. An allocation of, for example, 0.25% could generate over a billion dollars for monitoring and verification. Mathematically, at US$150 per ton of carbon (roughly US$40 per ton of CO_2), a global program that covers 4.3 billion tons of carbon emissions would yield a fund of US$645 billion with US$1.6 billion available for monitoring and verification. US$645 billion directed into global agriculture would be a major new source of revenues in the US$5 trillion global agriculture market. These funds

[5] Paustian, K., Lehmann, J., Ogle, S., Reay, D., Robertson, P., Smith, P. (2016). Climate smart soils, *Nature*, 532.

would flow from industrial and urban areas of high emissions to rural areas where carbon farming was practiced, and under international agreements, could flow from the industrial north to the developing south.

Before we discuss how Blockchain will support the carbon-deposit, MRV, and payment systems, it is worth for us to understand better how soil absorbs carbon from the perspective of agricultural science.

16.4 Agricultural Practices and Soil Health

Plants use carbon dioxide for food and absorb it through photosynthesis during which the CO_2 molecule is broken down and the oxygen atoms are released. The carbon is converted into carbohydrates, some of which are used for plant growth, although the rest are directed down through the roots into the soil where bacteria and mycorrhizal fungi convert the carbon into stable forms of soil, such as humus, where it remains effectively sequestered. The amounts of carbon in soil, along with the vitality of the soil microbiology, are key indicators of soil health and productivity.[6] Industrialized farming techniques have generally focused on productivity alone and not on soil health. In fact, these practices often leave soils depleted of their vitality. Farm practices that leave soil bare and exposed lead to losses of soil carbon through erosion and oxidation to the atmosphere. Other farm practices such as the heavy use of nitrogen fertilizers, herbicides, and pesticides can also have a deleterious effect on soil health by breaking up the delicate web of bacterial and fungal biology. Healthy soils are dark, moist, and rich, rather than pale, hard, and dry. Healthy soils absorb water more effectively than depleted soils, allowing them to weather floods and droughts better than hard packed soils in which water runs across the top of instead of soaking in.

The practice of restorative agriculture places soil health as a central concern alongside crop productivity. The primary practices consist of minimizing (or completely eliminating) tilling, while also using cover crops so that the soil is never left bare. Other practices such as the spreading of manure and compost add carbon to the soil, but, more importantly, stimulate the growth of bacteria and fungi. Some rotational and cover crops, such as legumes, are good at fixing nitrogen in the soil and allow farmers to minimize the use of synthetic nitrogen fertilizers.

Estimates of global soil sequestration capacity vary among soil scientists. Dr. Rattan Lal from the Ohio State University estimates the global sequestration capacity at around 3.8 Gt per year.[7] Other scientists such as Dr Johannes Lehmann from Cornell University take a more expansive view and estimate that global soils could absorb as much as 8 Gt per

[6] Lal, R. (2014). Societal value of soil carbon, *Journal of Soil and Water Conservation*, 69(6).

[7] Lal, R. (2010). Managing soils and ecosystems for mitigating anthropogenic carbon emissions and advancing global food security, *BioScience*, 60(9), 708–721.

year.[8] Experts such as Eric Toensmeier estimate that sequestration rates per acre can vary from ¼ to 10 tons per acre and more.[9]

Much of the discrepancy among scientists results from differences in how deep the soil carbon is measured. Soil science research often focuses on topsoil (i.e., the top 15–30 cm), but many plants have root systems that extend down 2 m or more. The deeper the root system, the farther down the carbon can be deposited. From a carbon farming perspective, deeper root systems are better because they expand the overall soil sequestration capacity. Sequestration rates on any given parcel of land will vary over time and can eventually saturate, it is a highly dynamic system. New deep-rooted crop varieties could be developed that are intended to sequester as much carbon as possible. For the purposes of measuring carbon sequestration and carbon farming, it would be important to conduct deep soil measurements.

The Rodale Institute, an organic farming research institute, has conducted decades of farm systems trials which compare conventional farming techniques with regenerative organic farming techniques side-by-side. In a 2014 paper,[10] Rodale Institute claimed that "on-farm soil carbon sequestration can potentially sequester all of our current annual global GHG emissions of roughly 52 $GtCO_2e$." In 2003, Rodale published experiment results that showed their techniques were consistently able to sequester 0.5 tCO_2 annually in the top 1 foot (30 cm) of soil per acre. In addition to increasing soil carbon from 15% to 28%, soil nitrogen was also increased from 8% to 15% resulting in increased soil microbial activity, and specifically mycorrhizal fungi, which plays a key role in beneficial soil carbon cycles[11] (Fig. 16.2).

16.5 Policy Proposals

As foreshadowed at the beginning of this chapter, there are two major climate policy proposals that were combined to provide the mathematical foundation for the carbon-deposit concept. The French Ministry of Agriculture was the first governmental organization to embrace regenerative agriculture as a climate change solution at the COP21 meetings in 2015. The CLC in the United States is a private group promoting a carbon tax, the group includes leading conservative politicians as well major fossil fuel firms like

8 Lehmann, J., & Kleber, M. (2015). The contentious nature of soil organic matter, *Nature*, 528(7580), 60–68.

9 Toensmeier, E. (2016). *The carbon farming solution*. White River Junction, VT: Chelsea Green Publishing.

10 Rodale Institute. (2014). *Regenerative organic agriculture and climate change: A down-to-earth solution to global warming*. Retrieved from https://rodaleinstitute.org/assets/WhitePaper.pdf.

11 Rodale Institute. (2003). *Organic agriculture yields new weapon against global warming: Groundbreaking study proves organic farming counters greenhouse gases*. Retrieved from http://www.strauscom.com/rodale-release/.

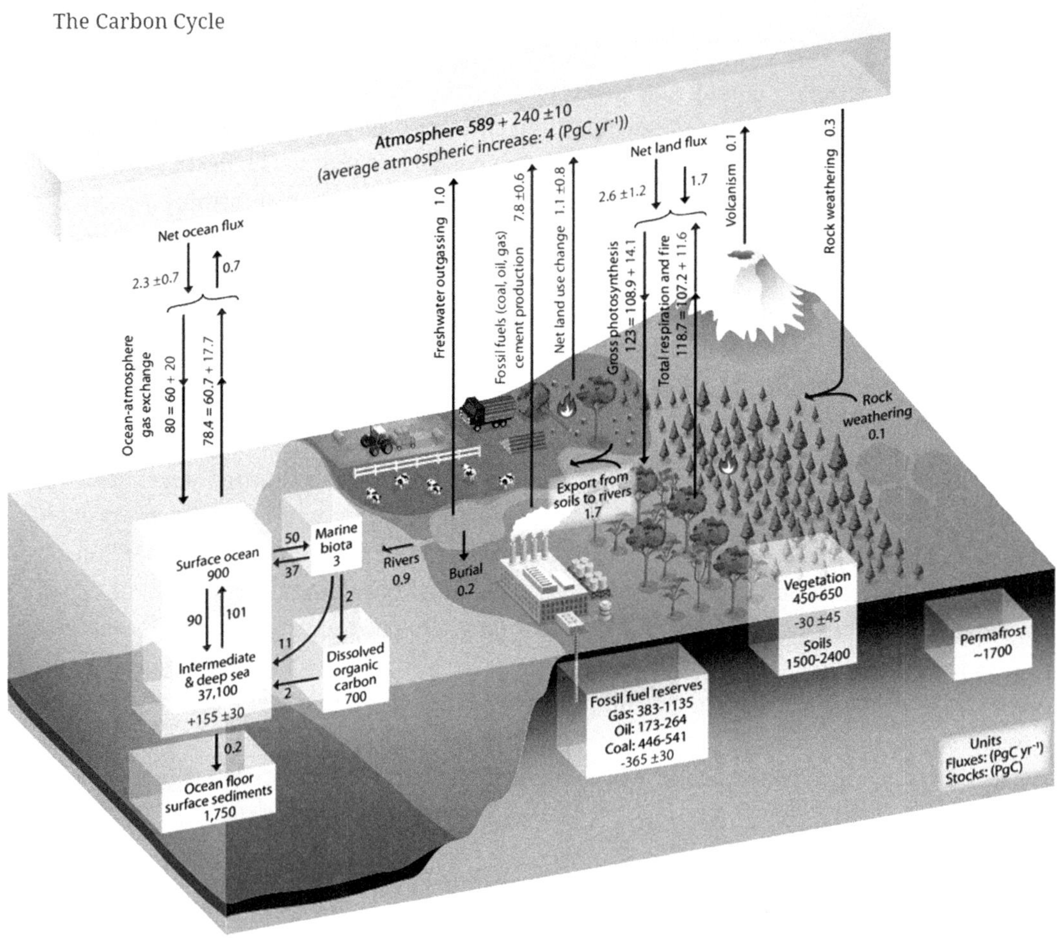

The numbers represent carbon reservoirs in Petagrams of Carbon (PgC: 10^{15}gC) and the annual exchanges in PgC/year. The black numbers and arrows show the pre-industrial reservoirs and fluxes. The red numbers and arrows show the additional fluxes caused by humans averaged over 2000-2009, which include emissions due to the burning of fossil fuels, cement production and land use change (in total about 9 PgC/year). Some of this additional anthropogenic carbon is taken up by the land and the ocean (about 5 PgC/year) while the remainder is left in the atmosphere (4 PgC/year), causing rising atmospheric concentrations of CO_2. The red numbers in the reservoirs show the cumulative changes in anthropogenic carbon from 1750-2011; a positive change indicates that the reservoir has gained carbon.

Figure 16.2
The global carbon cycle and the scale of planetary carbon sinks.

ExxonMobil who have historically opposed climate change policy, illustrating the shifting political winds and broad based support for the proposal.

The French Ministry of Agriculture's "4 per 1000" proposal is intended to promote the use of soil as a carbon sink in the fight against climate change. "4 per 1000" refers to increasing soil carbon content by 0.4% per year compounded annually. In this model, soil carbon measured down to a depth of 30 cm (approximately 1 ft) equals around 700 Gt. Increasing this topsoil content by 0.4% annually would equal 2.8 Gt carbon drawn down every year. Measuring soil carbon down to 2 m yields 2400 Gt. Increasing that count by 0.4% per year would be a drawdown of 9.6 Gt per year, which is more than enough to offset anthropogenic emissions completely (Fig. 16.3).

The CLC's proposal centers on a carbon tax of US$40 per ton combined with dividend payments to citizens, so that all the money that is raised by the carbon tax is paid back directly to the public. The purpose of the dividend payments is to both help citizens afford the higher energy costs, whereas also avoiding disputes over alternative uses for the funds. The proposal has met with wide acceptance across the political spectrum and is significant for including the endorsements of major fossil fuel firms, who have at times resisted taking action on carbon emissions and climate change. This plan is laudable for its intentions to be equitable and fair; however, it does not go far enough to completely solve the emissions problem.

The CLC's carbon tax is estimated to be able to reduce emissions by 25%–30%,[12] primarily from the electric power sector, where coal is readily replaced by natural gas, renewables, and nuclear power. Emissions from heavy-duty transportation (e.g., ships, planes, and trains) and energy-intensive industry remain steady because fuel switching options are limited and fuel price sensitivity is inelastic. This explains why soil sequestration is so important, it is the only option that can cost-effectively soak up carbon emissions on a global scale. Even as efficiency measures and electric vehicles used for light duty transportation work to eliminate some carbon emissions, other sectors of industry are already highly optimized and have little or no options to move away from hydrocarbon fuels. High temperature manufacturing, such as for concrete and steel, requires fuels, not electricity. Likewise, the superior energy density of liquid fuels over batteries ensures that some form of carbon-based fuels will continue to dominate military, aviation, maritime, mining, and other heavy-duty transport sectors.

Using the funds raised from the carbon tax to pay farmers and ranchers for carbon farming is logical as it offers favorable economics, environmental and emission reduction benefits, and can ultimately result in a global solution for net-zero emissions (Fig. 16.4).

[12] Climate Leadership Council. (2017). *A winning trade*. Retrieved from https://www.clcouncil.org/wp-content/uploads/2017/02/A_Winning_Trade.pdf.

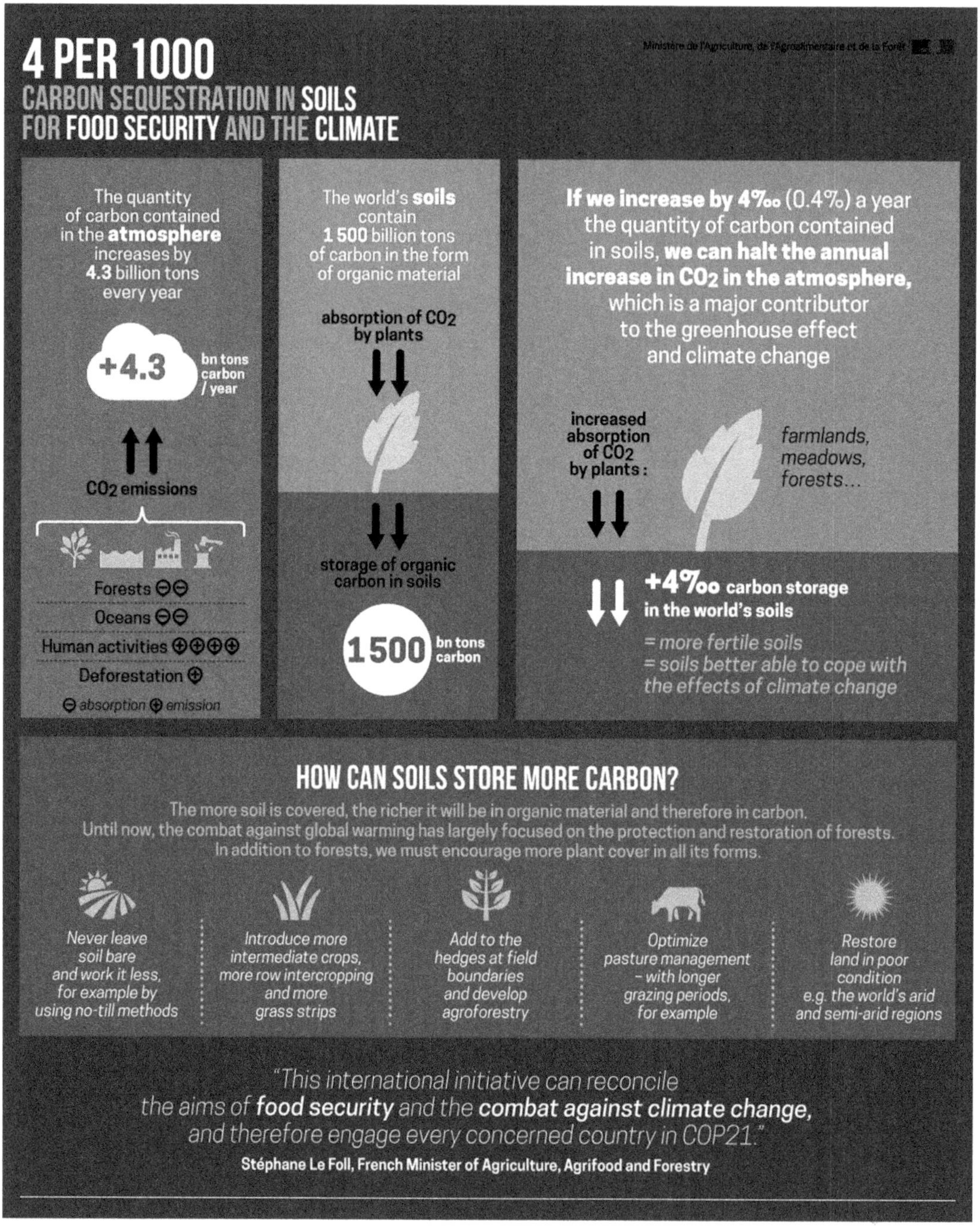

Figure 16.3
French Ministry of Agriculture, 4 per 1000 infographic.

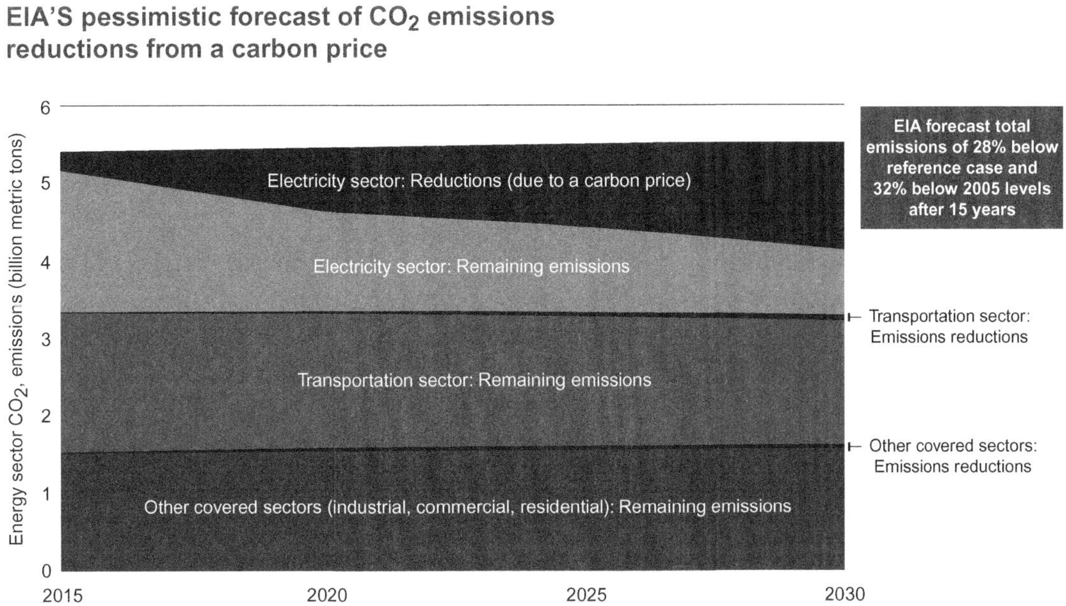

Notes: Data from EIA'S *Annual Energy Outlook* 2014. EIA Reference Case (dark + light blue) scenario is a forecast of the US energy system assuming only "on the books" policies as of late 2013, and its carbon pricing scenario (light-blue area only) scenario reflects a $25 per metric ton carbon price implemented in 2015 and increasing 5% per year thereafter.
Emissions from the transportation, industrial, commercial, and residential sectors exclude emissions from electricity generation.

Figure 16.4
The US Environmental Protection Agency forecasts emissions reductions resulting from a carbon tax by industrial sector. Electric power sector shows sharp reduction in carbon emissions because fuel switching options are readily available. Heavy-duty transportation and high temperature manufacturing sectors have few options for fuel switching and their emissions remain constant.

16.6 Soil Carbon Monitoring and Verification

Soil science is complex, and carbon is not static. There are many methods for measuring carbon both in the soil and in the atmosphere. Some methods are direct and involve the use of instruments to directly count carbon in a sample, whereas others are based on statistical models. Satellites, drones, and remote sensing technologies are improving and have a role to play in providing data. Large-scale measurements of soil carbon that can be relied upon to be accurate for the purpose of financial transactions are feasible but would be labor-intensive to conduct.

An example of a comprehensive tool used to measure soil carbon on farms is the COMET-Farm tool developed by the USDA (US Department of Agriculture) and Colorado State University. COMET-Farm is a whole-farm carbon accounting system that guides users through a process of describing their farms and farming practices. The tool makes extensive use of satellite imagery, where users can define their parcels of land on the maps and then document the practices on that land in order to quantify their GHG emission resulting from both land-use practices and fossil fuel consumption. By documenting their land and farm practices, farmers can evaluate alternative methods that may allow them to improve their carbon balances.

16.7 A Blockchainized Soil Carbon Accounting Platform

Blockchain technology provides an effective accounting platform for the carbon-deposit system. Blockchains can be designed to be open or closed systems. Bitcoin cryptocurrency is the most well-known example of an open Blockchain system that allows anonymous participants. One major down side of open architecture is that it requires a complex and incredibly energy-intensive system of crypto-mining to ensure system security. A closed Blockchain system allows only authorized users to write to the Blockchain, avoiding the need for costly crypto-mining. Hyperledger is an open-source Blockchain platform intended for business designed to be a closed system. Hyperledger is run by the Linux Foundation with the contributions of major corporations such as IBM. The closed system can still be open for reading by the public, making auditing transparent, but only authorized participants are allowed to write to the ledger.

Blockchain applications work well in business processes where there are multiple participants from different organizations who need to be able to agree on transactions and trust the record-keeping. The carbon-deposit system could benefit from the full suite of Blockchain attributes and smart contracts. The participants on the network would include carbon emitters, carbon farmers, verification authorities for both emissions and sequestration, and system auditors. Smart contracts are protocols or rule sets embedded within the Blockchain that are largely self-executing and enforce a contract condition. Terms are specified within the contract and executed when specific conditions are met. Smart contracts ensure that transactions are carried out instantaneously and consistent with the terms of the predefined contract.

In practice, the carbon-deposit system would require that emissions be properly accounted for in the same manner that emissions would be assessed for a carbon tax. These records would be entered onto the Blockchain network and checked off by a certifier who would also be identified on the Blockchain. A single Blockchain would be dedicated to a single unit of carbon, i.e., 1 ton of carbon. Carbon farmers would have their efforts monitored. When a certification authority signs off on a ton of carbon being successfully sequestered, then that data would be entered into the Blockchain. Smart contract features in the Blockchain would enable participants to be automatically paid when conditions are met. Verification authorities would receive their fees and the carbon farmer would get paid. The Blockchains can be made viewable to the public so that third parties can audit the transactions to guard against fraud.

The challenge in the whole system, indeed, lies in accurately accounting for carbon on both the emissions and sequestration sides of the equation. Soil carbon content is in a constant state of flux and depending on processes occurring on any given parcel of land. Soil scientists will need to develop tools and statistical models that are accurate enough to be considered fair and reliable in the marketplace. These challenges are resolvable with modern technology, but the efforts will likely be time-consuming and require a lot of hands-on effort in the fields as well as the introduction of new technology.

16.8 Carbon-Deposit System—An Old Concept Run in a New Way

An illustrative historical example of intentional soil improvement can be found in the 1928 guidebook for the USDA's Arlington Experiment Farm located on the grounds of what is now Arlington Cemetery near Washington, DC. The Arlington Experiment Farm was located at the site from 1900 until World War II, when it was removed to make room for construction of the Pentagon. The farm thrived during those years though the land was originally in very bad shape. The guidebook describes the efforts taken to restore the soil to health.[13]

> The land was in poor condition for agriculture when the department acquired the farm in 1900. Not only had the cultivation of the land been neglected since 1861, but much of the top soil had been removed for lawn-making in the Arlington Cemetery. It was therefore necessary to devote much attention to the improvement of the soil in order to bring it into suitable condition for the purpose for which it was intended. This has been accomplished largely through the use of cover crops and stable manure. **Although it has been a tedious, expensive process, it has afforded an interesting demonstration in soil improvement**.

The successful efforts to restore the soil at the Arlington Experiment Farm took place over a century ago, but the practices described are the same as what is needed today; except today, it is needed across the entire Earth. Cover crops, manure, and the farmer's intention of rebuilding the soil are the same tools we use today, and the insights from Arlington that the process is tedious and expensive also still holds true. Among the challenges is that farmers do not typically get paid to practice soil restoration; they only receive income from crops they are able to harvest and bring to market. Farmers who practice soil restoration today only do so based on their own private calculations of improved yields and reduced inputs, and perhaps a personal moral calculation that soil restoration is better than soil depletion, but not because of any overt market rewards.

[13] United States Department of Agriculture. (1928). *Arlington experiment farm—A handbook of information for visitors*. Washington, D.C.: USDA.

The carbon-deposit system would provide the rewards for soil restoration and inject hundreds of billions of dollars into farms globally. This additional income could transform rural economics, particularly in the developing world where incomes are a fraction of those in America and Europe. The carbon-deposit system leverages human nature by putting money on the table that will drive people's behavior. If you pay farmers fairly to put carbon in the soil, they will do it. And the carbon-deposit system, by directly tying emissions to sequestration on a ton-for-ton basis, is logical, fair, and offers an incentivizing mechanism to achieve net-zero emissions. Ultimately, the embrace of carbon farming globally offers a turning point in humanity's battle with climate change.

CHAPTER 17

Blockchain Ecosystem for Carbon Markets, Environmental Assets, Rights, and Liabilities: Concept Design and Implementation

Anton Galenovich[1], **Sergey Lonshakov**[1,2] **and Alexey Shadrin**[3]

[1]DAO Integral Platform for Climate Initiatives, Moscow, Russian Federation [2]Airalab[1], Togliatti, Russian Federation [3]Russian Carbon Fund[2], Moscow, Russian Federation

17.1 Background and Rationales

The problem of social costs, negative externalities and collateral damage (especially related to environmental issues) has remained in the focus of economics for over a century. These issues have become critically important for economic governance in the context of climate change. Basic questions which need answers in order to resolve the issue are: how to measure negative impact; who is liable; and how to settle. Ronald Coase has proposed the most efficient market-based approach to the problem of social costs. This approach has introduced the concept of limited right to perform activities harmful to the third party and provided basis for a peer-to-peer (P2P) settlement of reciprocal damage. Nevertheless, the most common model still implies more or less arbitrary assignment of the answers to these questions by governments. Social cost of carbon is a prominent example of applying this approach to resolve environmental issue of global scale, and Blockchain is so far the most rational and logical technology to actually implement it.

According to the famous "Coase Theorem," any allocation of limited resource is equally effective subject to free market trade and zero transaction costs. It means that transaction costs play a crucial role. Zero transaction costs are only a hypothetical case for "spherical resource in a vacuum." In this sense, the government should, in the first place, take care of maximizing efficiency of institutions, minimizing transaction costs, which ultimately means

[1] https://github.com/airalab

[2] https://russiancarbon.org/

Transforming Climate Finance and Green Investment with Blockchains.
DOI: https://doi.org/10.1016/B978-0-12-814447-3.00017-3

minimization of centralized interventions. At the same time, any allocation of the resource is acceptable, provided it remains limited.

The objective of this chapter is to demonstrate how existing carbon markets can be provided with a "Blockchain option" or "Blockchain representation." In the medium-to-longer term, it is possible to create a way for all environmental markets to evolve toward truly decentralized and free personal market choice model.

Furthermore, as costs and underlying values of externalities can be subjective, P2P evaluation is important for accounting for negative impact costs and mitigation benefits. Advancement of technologies such as Blockchain crucially lowers transaction costs for P2P interactions and thus provides for further development of Coase paradigm. Application of Blockchain technology to mitigat collateral socio-environmental damage of economic activities relies on a concept of decentralized P2P and public evaluation of negative impact, distribution of liability, and settlement in the form of mitigation.

The climate change challenge is universal with multiple and extremely diverse sets of stakeholders. Market instruments to mitigate climate change risks and damages are developing globally. They include the Paris Agreement concept of "internationally transferred mitigation outcomes." Although every carbon credit basically represents the right to emit one tCO_2e, regulatory, legal, commercial and trade, transactional, inter-jurisdictional border barriers prevent their fungibility, which gravely undermines their economic and environmental efficiency. There has been no common space fabric, financial instrument, and ecosystem, which would be universal, reliable, easy-to-use, and transparent. There needs to be an ecosystem that would allow diverse stakeholders, governments, civil society, businesses, and individuals to participate in mitigation activities, register quantified emissions reduction commitments, invest in environmental damage mitigation projects, offset carbon footprint, and acquire and trade mitigation outcomes. In a way, the Paris Agreement provides a global policy framework to address this issue.

17.2 Decentralized Autonomous Organization Integral Platform for Climate Initiatives (DAO IPCI[3])

At the DAO IPCI[1], Blockchain technology and smart contracts we have designed for carbon market instruments involve the introduction of protocols for carbon markets' interactions to balance environmental damage with environmental solutions.

The key lies in designing, developing, and actually launching Blockchain-based business processes for carbon markets in the form of smart-contracts. There are existing Blockchain platforms that can be leveraged to perform these functions under independently managed rules, allowing for diverse climate mitigation programs to interact to the extent appropriate.

[3] http://ipci.io/

Users have tested viability of decentralized application to register assets, rights, and liabilities to issue transfer and trade asset-based environmental units using a common set of adjustable, ready-to-use modules, and smart-contracts.

Further plans of research and development are extensive. They include a range of practical solutions to address scalability issue and different Blockchains' communication issue, if they become actually challenging.

17.2.1 Advancing the Current Market Approach

One of the main barriers to the full introduction of environmental markets, especially decentralized P2P evaluation schemes, is the inertia of existing regulatory mechanisms, corporate interests, and the perception of "the evil you know is better than the evil you don't." "Existing systems [such as command and control regulation of pollution], with all of its imperfections, was at least understood and capable of being manipulated by learned and skilled industry and regulatory professionals."[4] "Fundamental rationale for emissions trading is that industry need not be told how to achieve inexpensive emissions reductions; industry only needs to be given the freedom to develop these reductions in a way that assures positive environmental outcomes."[5] Many employees of government and corporate environmental agencies may have faced the pressure of re-training or even the threat of losing their jobs, and their solidarity in opposition to the new model has been quite understandable.

Market-based approach is fundamentally quantity or resource-based. As long as the resource is quantifiable following recognized methodologies, self-executing algorithms can substitute many regulatory functions. It would seem feasible for the partisans of market (quantity-based) approach to focus on opportunities derived from the development of "quantified greenhouse gas (GHG) emission limitation and reduction commitments" concept-based model.

17.2.2 The Context for "Mitigation Tokens"

According to The Intergovernmental Panel on Climate Change (IPCC), limiting the warming caused by anthropogenic CO_2 emissions alone with a probability of greater than 33%, 50%, and 66% to less than 2°C since the period of 1861–1880, requires cumulative CO_2 emissions from all anthropogenic sources to stay between 0 and about 5760 $GtCO_2$, 0 and about 4,440 $GtCO_2$, and 0 and about 3,670 $GtCO_2$ since that period, respectively. An amount of 1,890 [1,630 to 2,150] $GtCO_2$ was already emitted before 2011. Therefore, to provide for acceptable level of climate risk mitigation, the volume of future emissions since 2011 should stay within approximately 750 $GtCO_2$.[6] Similar to other scarce natural

4 http://www.e5.org/downloads/ETBrussel210202/Palmisano_commentsDirectiveProposal.pdf
5 http://www.e5.org/downloads/ETBrussel210202/Palmisano_commentsDirectiveProposal.pdf
6 http://www.ipcc.ch/pdf/assessment-report/ar5/wg1/WG1AR5_SPM_FINAL.pdf

resources, the more we exploit our atmosphere, the more expensive it gets, and the more costs we bear. Thus, for 2017, conservative evaluation of anthropogenic GHG emissions budget left would be 600 $GtCO_2$. This is the number chosen to set the ultimate cap for emission of "Mitigation Tokens" in DAO IPCI[1] Blockchain (which will be introduced on p.[x]).

In the climate change political debate, there are countries presumed to bear most responsibilities for accumulated damage and countries presumed most vulnerable to the damage. Moreover, future generations are the party to suffer most damage. Blockchain could provide a technological solution for quantified emissions reduction targets to be assigned to countries which value them the most. Therefore, the market design should provide for the interests of forthcoming generations by long-term budgeting of the resource; and probably by developing specific long-term market instruments.

17.2.3 The Architecture of the Emissions Trading Market

Quantified emissions reduction targets and commitments-based long-term budgeting of emissions are fundamental for environmentally sound mitigation policy. Presently, it is difficult to compare the ambition of emissions reduction targets and track progress toward meeting it, because countries have expressed them in a wide variety of forms. Only the quantified GHG emission limitation and reduction commitments and quantified commitments-based compliance units are inherently fungible on global scale.

The property rights issue should be resolved to provide for compliance units to become a commodity before the creation of the market for carbon compliance units, representing legal rights to emit CO_2, is possible.

Theoretically, it would be just natural and logical for the "peers" to formulate fair method of allocation of limited resource. The starting point or the baseline could be natural rights of ownership of the resource. For emissions markets, the principle is known as "grandfathering." Allowances are distributed on the grounds of claims, though in limited quantity, and assignment of its share to particular owner is subject to efficient and careful exploitation. Thus, it is the company itself, who is interested to claim a certain portion of the resource, to claim specified amount of annual greenhouse emission rights and justify the claim by efficient exploitation of the resource.

17.2.4 Operationalizing the Paris Agreement with Blockchain

The Paris Agreement reached in December 2015 demonstrates the global commitment to put the world on track to avoid catastrophic climate change. It has established an official process in which countries regularly need to propose, review, and renew existing contributions—known as Nationally Determined Contributions (NDCs)—in order to achieve this target. What is unique about the Paris Agreement, as compared to the Kyoto Protocol, is its bottom-up nature. For example, it leaves considerable discretion to national governments on the form, scope, and timing of their NDCs which highlight countries' current vision of strategic priorities, choice of policy instruments, and expected outcomes in areas of mitigation and adaptation, in order to achieve an economy-wide low-carbon transition. While enabling broad participation, this creates several challenges for assessing, tracking, and comparing the individual and collective impact of the NDCs, as well as ensuring their adequate design and implementation.

Article 6 of the Paris Agreement describes two types of units, which may be tokenized: internationally transferred mitigation outcomes (ITMOs) and units issued under Sustainable Development Mechanism.

To "Blockchainize" the implementation of Article 6 of the Paris agreement, the following design has been developed with most of the modules and smart-contracts needed already in place:

Paris Agreement	DAO IPCI Blockchain Ecosystem	
Article 6	DAO Core/Modules	
Voluntary cooperation		
1. The Paris Agreement recognizes that Parties may choose to pursue voluntary cooperation in the implementation of their NDCs to allow for higher ambition in their mitigation and adaptation actions and promote sustainable development and environmental integrity.	*"Quantified Commitment" and "Quantified Impact" Module*[a] NDCs should be reflected as "Quantified Commitments" and "Quantified Impact."	*Creation of ITMO Market Module by UNFCCC Secretariat independent DAOs of the Parties* Parties that have chosen to pursue voluntary cooperation create independent Operators DAOs with common market coordinated by UNFCCC Secretariat for all Operators, which chose to cooperate.
2. If Parties choose to cooperate by exchanging ITMOs, they shall promote sustainable development and ensure environmental integrity and transparency, including in governance, and shall apply robust accounting to ensure, inter alia, the avoidance of double counting, consistent with guidance adopted by the Conference of the Parties serving as the meeting of the Parties to this Agreement (CMA).	*Issuance of environmental units via Security Reserve or Security Deposit Contracts* • Common market token is an ITMO. • Once issued, ITMOs are deducted from the host country's NDC, which arithmetically means increase of Quantified Impact and decrease of mitigation contribution (Quantified commitment) by the amount of tCO_2e. • Double-counting is avoided; transparency is ensured by inherent properties of public Blockchain. • Respective Party Operator and Independent Entity are responsible for compliance of ITMOs with sustainable development and environmental integrity principles.	

(*Continued*)

(Continued)

Paris Agreement	DAO IPCI Blockchain Ecosystem
Article 6	DAO Core/Modules
3. The use of ITMOs to achieve NDCs shall be authorized by participating Parties.	*Complier Contract* Authorization of non-Party actors use (transfer or trading) of ITMOs is provided and ensured by the protocols for Issuance. ITMOs used by non-Party Actors are not counted as contribution of any Party to the Agreement.Use of ITMOs may be used only via respective Party (Operator) Complier Contract(s). Only the units "burnt" (irrevocably retired) at specific Compliers' Contract(s) may be counted as additional contribution.
Sustainable Development Mechanism	
4. A mechanism to contribute to the mitigation of GHG emissions and support sustainable development is hereby established under the authority and guidance of the CMA for use by Parties on a voluntary basis. It shall be supervised by a body designated by the CMA, and shall aim:	*Creation of independent DAO (Operator), the List of Independent Entities and SDM Token Market by an* UNFCCC body designated by the CMA
a. To promote the mitigation of GHG emissions while fostering sustainable development;	
b. To incentivize and facilitate participation in the mitigation of GHG emissions by public and private entities authorized by a Party;	*Issuers addresses and contracts* The Operator shall approve the addresses (list as SDM Agents) and Issuers' contracts only for the Issuers' listed as Agents by respective Party Operator (see Point 2: *Issuance of ITMO via Security Reserve or Security Deposit Contracts* above)
c. To contribute to the reduction of emission levels in the host Party, which will benefit from mitigation activities resulting in emission reductions that can also be used by another Party to fulfill its NDC; and	*Complier Contracts* SDM tokens (emission reductions) shall be used by another Party to fulfill its NDC via respective Party (Operator) Compliers Contracts. Only the units "burnt" (irrevocably retired) at specific Compliers' Contract(s) may be accounted to fulfill NDC. (see Point 3 *Complier Contract* above)
d. To deliver an overall mitigation in global emissions.	*Quantified Impact module* Emission reductions represented by SDM tokens should actually reduce registered quantified impact or restrain quantified commitments

(Continued)

(Continued)

Paris Agreement	DAO IPCI Blockchain Ecosystem
Article 6	DAO Core/Modules
5. Emission reductions resulting from the mechanism referred to in the SDM shall not be used to demonstrate achievement of the host Party's NDC if used by another Party to demonstrate achievement of its NDC.	*Issuance of environmental units via Security Reserve or Security Deposit Contracts.* *Quantified commitments and Quantified impact module* Once issued, emission reductions (represented SDM tokens) are deducted from respective Party NDC, which arithmetically means increase of Quantified Impact and decrease of mitigation contribution (Quantified commitment) by the amount of tCO_2e reflected in SDM tokens.
6. The CMA shall ensure that a share of the proceeds from activities under the mechanism referred to in paragraph 4 of this Article is used to cover administrative expenses as well as to assist developing country Parties that are particularly vulnerable to the adverse effects of climate change to meet the costs of adaptation.	*Commission fees* Commission fees in DAO IPCI[1] are established by independent Operators.

[a]This module is under development at DAO IPCI[1].

17.3 Principles of Concept Design

The design objective is to provide common space, common fabric, financial instruments, and infrastructure that would be universal, reliable, transparent, easy-to-use, and allow both businesses and individuals to register and invest in climate mitigation projects to reduce emissions and acquire and trade mitigation outcomes (ITMOs).

This Blockchain ecosystem developed by DAO IPCI[1] is smart contracts-based digital environment to minimize transaction costs and enable issuance and transfer of carbon units, ITMOs. It is highly reliable, transparent, and manipulation-proof.

Balance of self-sufficiency and environmental integrity is critical. Attempts to create carbon emissions-related Blockchain systems and cryptocurrencies would fail if they lack high quality underlying to support them. DAO IPCI[1] is based on the assured environmental assets, including climate mitigation outcomes in the first instance, as the underlying.

Independent Entity assures the existence and quality of the underlying asset and preclusion of double spending. The latter is further supported by the internals of Blockchain technology. Operator and Independent Entity ensure fundamental link of underlying assets and units issued through proof-of-asset protocol,

17.3.1 Advantages of DAO IPCI's[1] Blockchain Ecosystem

DAO IPCI[1] is designed to be truly decentralized public Blockchain-ecosystem; it aims to be a common business space to attract financing from investors. Mitigation project financing is no longer limited by financial capacity, location, or legal status with minimization of transaction costs, increased reliability, and transparency of the whole process, which would be free of interference, interventions, manipulations, or falsifications. Advantages of a Blockchain-based ITMO trading system inherent from defining attributes of Blockchain include:

- The Blockchain system as a whole is relatively immune to political, administrative, regulatory interventions of the governments.
- The open-source code provides security as there is no need for anyone to trust anyone except the source code, the protocols, and the smart-contracts.
- Decentralization is ensured starting the level of different mitigation programs operating in the same digital environment. There are no technical restrictions as to who may register and launch mitigation program.

Furthermore, DAO IPCI[1] basic design provides for technical possibility of actual interaction of different asset-based systems on the Ethereum Blockchain. The modules of different DAO, e.g., energy assets-based, may be included into mitigation program DAO, and vice-versa, and reflected in the decentralized application.

The primary elements to provide for interaction of different mitigation programs in DAO IPCI[1] as described on Fig. 17.1 are Mitigation Token (MITO) and MITO Market. DAO IPCI[1] MITO is sought to provide for exchange of numerous and diverse asset-based tokens representing societal costs mitigation outcomes. Namely and in the first instance, GHG emissions limited rights and reductions (carbon emission quotas and credits) are the assets that become fungible via MITO and MITO market. Distinctive characteristics of the two types of DAO IPCI[1] tokens are that only one of them, MITO, is an exchange vehicle or digital currency, whereas the rest of tokens represent diverse mitigation units as underlying. Other than MITO, tokens, i.e., environmental units, are issued via coordinated actions of issuers, operators, and independent entities. Furthermore, mitigation compliance units like carbon emission quotas and credits, other emission or effluent credits, renewable energy certificates, and even quantified social benefits ("environmental units") are regulated by a wide variety of programs, and in principle, nothing can prevent jurisdictions, entities,

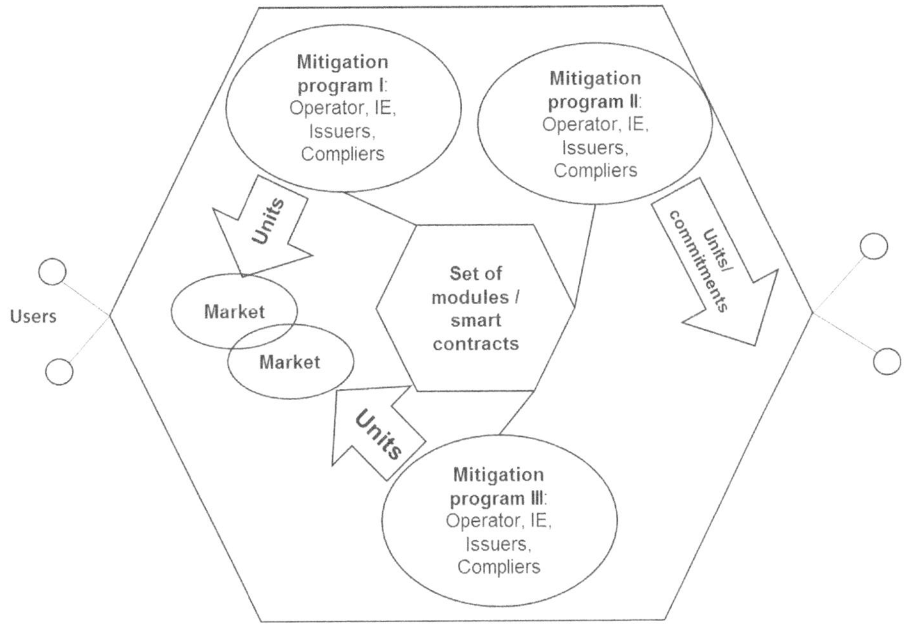

Figure 17.1
DAO IPCI[1] Blockchain basic design.

businesses, Non-Governmental Organizations (NGOs), or individuals from launching new and independent result-based mitigation programs in DAO IPCI[1]. Mitigation Token (or MITO) is the key element to provide for transactions' efficiency and integrity of DAO IPCI[1] ecosystem.

17.4 Environmental Units

Various environmental units ("units" on the Fig. 17.1) are issued in DAO IPCI[1] Blockchain under rules and requirements of specific environmental mitigation programs. These programs may include: mandatory, voluntary, and pilot environmental market programs, emission (or effluent) trading schemes (cap-and-trade programs in particular), offset credit, carbon tax credit-based and hybrid programs, renewable standards and renewable energy certificates-based programs, other environmental mitigation market-based programs, and generally social costs mitigation programs.

Rules and requirements of climate programs include standard elements: regulator (operator), verification by independent entities, limits, validity periods, tools to cover risks, market institutions (trading), etc. In DAO IPCI[1], these standard elements constitute system of smart contracts, adjustable for specific program requirements.

Environmental units are issued on DAO IPCI[1] Blockchain:

- Directly to the Issuer or Complier subject to program Operator's approval on the grounds of verification of mitigation outcomes or quantified commitments compliant with program rules, requirements, and criteria;
- On the grounds of the program Operator's decision to accept carbon units issued by alternative programs and accounting platforms in their original form or convert and exchange such external units for the program units subject to compliance with the program requirements and criteria and confirmed cancellation of corresponding registry entries and units turnover; and
- Under any climate change or environmental mitigation market-based programs subject to integrity of DAO IPCI[1], preclusion of double-spending, and verification of integrity of digital units with environmental underlying asset for the token.

Issuance of environmental units requires reservation of specific share of the units at the Security Reserve Contract for a specific period, established by the Operator on the grounds of independent assessment of related risks. The units would be burnt (retired) in case they are recognized as void, so that total amount of digital environmental units issued to the platform would in any event be equivalent to underlying. After the reservation period is over, the units are returned to the Issuer's account. Security Reserve Contract may be used to withhold environmental units in order to avoid or correct input issuance date mistakes.

Issuance of the units may be performed through Security Deposit Contract, which provides for issuance against collateral withdrawn in case the Issuer does perform verification commitments.

Collateral damage of production, consumption, transaction of goods and services has become crucially important competitive factor, and mitigating this damage is now customary for many market activities. For example, governments have to avoid deforestation as a result of palm-oil production. Offsetting carbon footprint scheme ensures irreversible burning of the units at the Compliers contract.

The trading of environmental units depends on the following initial core modules and operations being functional:

- Transfer of the units between Issuers and Users;
- Acquiring of the units to offset carbon footprint and burning of the units for the purpose of offsetting carbon footprint and other operation;
- Reservation of the units and withdrawal of the units reserved to the Issuer; and
- Placing and executing sell/buy orders.

17.5 Mitigation Token

Unified internal current token in DAO IPCI[1] ecosystem is "Mitigation Token," MITO. Mitigation Token is issued based on rigid limitation of the potential amount of emission, strict adherence to the interests of the token holders, and that of ecosystem stakeholders.

This unified internal current token is the key element to ensure transactions efficiency and economic integrity of the DAO IPCI's[1] ecosystem. Internal current token is an exchange unit derivative to environmental mitigation units' turnover, and represents internal capital and value of the integrated Blockchain ecosystem. MITO holders receive the right to exchange them for environmental units at MITO market. As MITO emission model limits their amount to the sum of environmental units accessible via environmental unit registries—"the MITO pool of DAOs," MITO serves initially as a digital carbon pricing instrument, a digital CO_2-equivalent of carbon cost.

17.5.1 MITO Operational Model

The following figure describes one of the possible scenarios of further emission which MITO covers.

The ultimate theoretical limit for emission represented by MITO has been capped at the level of 600 billion units—a conservative end of the global carbon budget.

Emission represented by MITO is designed to directly correlate to the sum of environmental units secured and accessible through the registries. To avoid misrepresentations (or fraud), the emission has to be approved by the pool of decentralized autonomous organizations, which have agreed to coordinate digital currency policy, and distribute newly issued tokens.

Every DAO IPCI[1] operator is duly authorized by the environmental program entering DAO IPCI[1]. An operator is eligible to issue MITO under basically the same terms and conditions in the amount directly corresponding to the sum of environmental units secured and accessible through the registries.

To secure the interests of MITO holders, avoid dilution of previously issued tokens' value, substantial part of the new MITO should be allocated to existing holders (Fig. 17.2).

17.5.2 Proceeds from MITO

MITO is inherently appropriate and designed for executing DAO IPCI[1] smart contracts, including placing and executing buying and selling of environmental units, with commission fees paid to support maintenance of DAO IPCI[1].

The proceeds from MITO are of nonprofit nature and targeted for research and development of DAO IPCI[1] digital ecosystem.

17.5.3 Contributions of MITO to Address Climate Change

With MITO, there can be long-term budgeting of annual which allows for precise planning of exploration, extraction, and supply of fossil fuel resources. Ultimately, MITO avoids

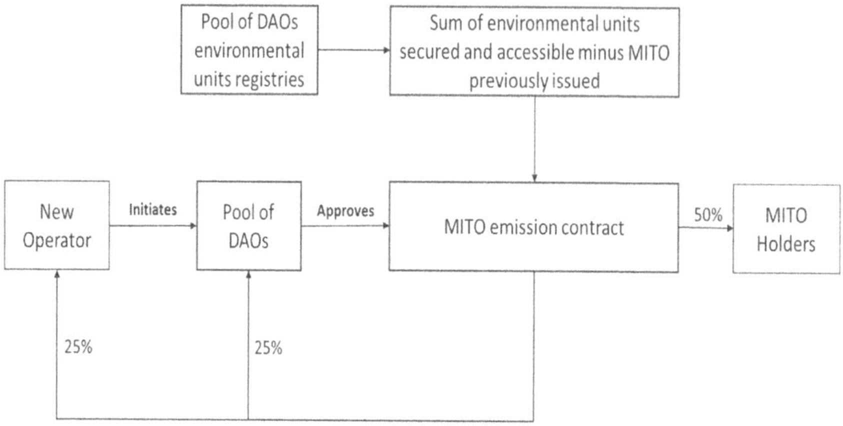

Figure 17.2
Emission of DAO IPCI[1] internal token —MITO.

multibillion losses for investors attributable to unsustainable exploitation of our natural resources.

17.6 The Way Forward for DAO IPCI[1] Blockchain Ecosystem

Long-term prospects of DAO IPCI[1] development are limited only by its functional capacity as the trends are evidently in favor of the expansion of the environmental asset market, specifically carbon markets. They will expand both in scale and number. Eventually, these markets will be linked and integrated as a common market space with fungible instruments. DAO IPCI[1] is exactly a prototype of such market space and units.

Although the initial objective is to provide existing markets with a "Blockchain option" or "Blockchain representation," as indicated at the beginning of this chapter, the goal is also to create a way for environmental markets to evolve toward truly decentralized and free personal market choice model.

Near-term prospects rely primarily on demand from major corporates and regional (subnational) climate programs (including global pilot market mechanism for international civil aviation), carbon footprint off-setting programs, and from consumers.

DAO IPCI[1] has the following ideas for further development:

- Detailed tracking of the environmental units origin (supply-chain);
- Introduction of secured-by-collateral quantified commitments-based environmental units;
- Mechanism for joint off-setting of carbon footprint by Supplier and Consumer applicable at retail level as well as up to the level of supporting carbon-neutral export programs;

- Linking DAO IPCI[1] with programs and systems based on physical measurement and Internet-of-Things (IoT)-based monitoring of anthropogenic climate impact in real-time mode;
- Mechanism to support performance under Green Bonds commitments;
- Mechanisms and fungible instruments to link different GHG emissions limitation and reduction/removal systems, schemes, programs, and standards;
- Development of over-the-counter transactions and links with environmental, carbon, securities and commodities exchanges;
- Development and introduction of environmental unit-based derivatives;
- Mechanism to support hedging volatility of prices for different schemes, systems, and programs carbon compliance units;
- Upon reaching adequate level of readiness and matureness of the ledger, development of virtual investment structure shall be considered.

Scalability of the DAO IPCI[1] Blockchain Ecosystem: a vision of Blockchainized mitigation project supply chain

The next step of our Blockchain ecosystem would provide for more accurate tracking of unit movement along a supply chain. The supply chain concludes with an approved and verified environmental unit, for example, an emission reduction credit. This supply chain may include:

1. A mitigation project concept;
2. Project concept developed in a standardized format;
3. The emerging concept being supported with engineering and financial documentation (validated by an independent entity);
4. The well-annotated concept submitted to regulator (operator of the program) for public comments and official approvals; and
5. Environmental units are registered for use (sale or compliance).

Blockchain allows users to track the time- and date-stamped ownership of electronic asset throughout the lifecycle on a supply-chain. On a decentralized ledger, an asset owner is allowed to hold assets and transfer or sell it to another peer; or add more information to the initial assets so that a documentation chain is credited.

Tracked data could include but might not be limited to:

- Name and contact information of person(s) and firms entering data;
- Details of the legal entity;
- Attributes associated with an environmental unit (credit)
  - Deforestation impacts
  - Water management impacts
  - Biodiversity impacts
  - Gender impacts

    - Health impacts
    - Number of GHG emission reduction projected by year
- Time, date, and location of data entry;
- Testing, measurement, and certification protocols used;
- Third-party attestations; and
- Details of auditor's insurer.

Each document is linked on an electronic chain so that the entire audit trail is visible to all stakeholders. As of today, every DAO IPCI[1] Blockchain transaction is supported by documents uploaded to "Inter-Planetary File System."

Blockchain can provide the auditability and liability assignment features industry demands nowadays.[7] The higher degree of auditability on a Blockchain-supported supply chain also facilitates the apportioning of liability because every claim can be traced back to the responsible party. Therefore, the end user of a carbon credit can be better informed of their reputational and commercial risks associated with an environmental unit purchase.

[7] Authors are grateful to John Palmisano for the contribution.

CHAPTER 18

How a Blockchain Network Can Ensure Compliance With Clean Development Mechanism Methodology and Reduce Uncertainty About Achieving Intended Nationally Determined Contributions

Steven Dunkel[1]

Rochester Institute of Technology, Rochester, NY, United States

Abbreviations

BTC	Bitcoin
CDM	Clean Development Mechanism
CDM EB	Clean Development Executive Board
CER	Certified Emission Reduction Credits
DOE	Designated Operational Entity
GHG	Greenhouse Gas
INDC	Intended Nationally Determined Contributions
MAP 2017	A two-year business plan from 2016–2017, and a one-year management plan
NMM	New Market Mechanism
SDG	Sustainable Development Goals
UI	User Interface
UNFCCC	United Nations Climate Change Conference
UN BOA	United Nations Board of Auditors

18.1 Introduction

Policies and actions to achieve all 17 United Nations' Sustainable Development Goals (SDGs) by 2030 are bolstered by ingrained features of Blockchain technology. A global Blockchain would distribute a data stream to measure the cumulative impact of small- and large-scale Clean Development Mechanism (CDM) projects toward reaching SDGs. The CDM is a mechanism created under the Kyoto Protocol to increase inclusion of developing

[1] http://blockchained.blogspot.co.uk/.

Transforming Climate Finance and Green Investment with Blockchains.
DOI: https://doi.org/10.1016/B978-0-12-814447-3.00018-5

countries in the global effort to reduce or limit emissions (CDM accreditation procedure, n.d.). The Kyoto Protocol emphasizes the importance of cooperation for the success of CDM projects. Comparability, transparency, and effectiveness are specifically highlighted as success metrics toward CDM implementation. The development of projects under the CDM are funded from resources in part provided by developed countries and multilateral development funds. In 2015, the United Nations Board of Auditors (UN BOA) reviewed the operations of the CDM and noted the huge potential of Blockchain technology in streamlining the CDM (Report of the UN BOA on the financial statements of the United Nations Framework Convention on Climate Change (UNFCCC) for the year ended December 31, 2015 (2016)). The Review indicates that a low demand for Certified Emission Reduction (CER) Credits is preventing the implementation of many new CDM projects that remain unimplemented, and leaves other projects stranded or operating at reduced capacity.

Of the 7,776 CDM Project activities that were registered as of July 2017, only 91 projects were registered in 2015 which is a decrease from 297 projects in 2013 (an overall decrease of 69%) (Meeting report – CDM Executive Board ninety-fifth meeting, n.d.). The number of credits issued also followed this trend sharply declining from 1510 in 2013 to only 574 in 2015 (an overall decrease of 62%). This trajectory continues in the number of applications received during the same period from 2030 submissions in 2013 to 474 in 2015 (an overall decrease of 77%). This decline was noted in an assessment as stemming from restrictions on CERs depending on the country and project type (FCCC/KP/CMP/2016/4, p. 10). These trends are problematic for the CDM as a share of the proceeds from fees collected during the issuance of certificates pays for administrative costs of the mechanism itself and to the external financing vehicles like the Adaption Fund (Adaptation Fund, n.d.). New tools that create demand for CERs can be provided to all project participants, from the bottom-up, by sharing the same information simultaneously using a Blockchain network (Rules of Implementation for The Joint Crediting Mechanism JCM, n.d.).

Increasing the demand for CERs, and boosting participation in CDM projects and activities should be a priority of any stakeholder looking to achieve the SDGs by 2030. The Paris Agreement is the current ideology for transitioning from ideological sustainability goals into directly impacting carbon emissions through actions. Changing the infrastructure of work streams within the CDM will take time that is already scarce as we approach 2030. Resources are limited and the CDM is operating with a US$43 million deficit (UN BOA Review of CDM, p. 6). These limited resources may impact the adoption rate of recommendations in this chapter but discussions need to occur today to be more pragmatic about achieving the SDGs by 2030. Using a Blockchain for authentication and issuance of CERs lowers operational costs of the mechanism by creating a more reliable instrument to collect administrative fees vital to the longevity of the CDM.

18.2 Blockchain Enhancement of the Function of the CER Marketplace

The CDM provides a way for developing countries to earn revenue for CER projects measured by 1 mt of carbon dioxide (CDM accreditation procedure Report of the United Nations Board of Auditors on the financial statements of the UNFCCC for the year ended December 31, 2015 (2016)). CDM guidelines dictate that a standardized electronic database must be established to provide accurate records of issuing, holding, and acquiring CERs (FCCC/KP/CMP/2005/8/Add.1, Appendix D, 2006). A Blockchain network would meet and support this requirement. The CDM registry is currently the platform that CDM projects funnel information through and the registry is open to public access. This registry is where a Blockchain can recall information from in order to validate or verify project data and vice versa. The Blockchain would host information on an immutable digital ledger for CDM projects, whereas simultaneously facilitating the transfer or issuance of credits, monitor activities in a supply chain, or collect fees at different touch points in the project cycle all within a matter of minutes.

CERs can be traded or sold by developed countries to meet their emission goals under the Kyoto Protocol. Data recorded on a Blockchain increases the attractiveness of carbon-reducing mechanisms as a source of funds by improving the liquidity and fungibility of issued credits (Report of the UN BOA on the financial statements of the UNFCCC for the year ended December 31, 2015 (2016)). The issuance and exchange of carbon credits is in a sharp decline after the first commitment period of the Kyoto Protocol (Report of the UN BOA, p. 10). During the first commitment period, there were 1,475,793,701 CERs issued, whereas the second commitment period has seen only 371,352,202 CERs issued (a 75% overall decrease) (CDM Issuance of CERs (n.d.)). CERs issued on the Blockchain which can be exchanged with other Blockchain tokens creating a bridge for moving financing through one mechanism to other financing vehicles under the CDM or providing additional liquidity. This includes Voluntary Carbon Markets that currently have no means of exchanging between one another. This exchangeability is critical for increasing funding to projects that aim to reduce global GHG emissions and empower developing communities with no access to traditional financing channels.

Any Blockchain used to meet SDGs or Nationally Determined Contributions (NDCs) will be eventually adaptable (chain-agnostic) and work in conjunction with legacy technologies. Legacy technologies encompass database solutions, cloud data storage, and physical recording. An ideal environment to reduce uncertainty and risk of Blockchain applications would be one or more distributed networks that enables multiple types of transactions and/or carbon credit schemes. This redundancy reduces uncertainty and provides additional checks and balances to the mechanism as a whole across the project lifecycle. These Blockchain networks ideally interact with multiple actors and provide identifiable information upon entry into the network for regulators and enforcement agencies.

Currently, the CDM manages the exchange of credits and the voluntary cancellation of certificates if the project participant wishes to exchange the CER in a voluntary market. If the CDM wishes to expedite these requests to free additional funding sources for project participants, a voluntary cancellation of a CER should be issued on the Blockchain to allow credits to be exchanged from the CDM to a voluntary market. Currently, a certificate is issued with a serial number to prove the cancellation of credits on the CDM registry. A bitcoin transaction could be recorded in tandem or as a replacement to that serial number on the same document. Any actor can reference the recording transaction on the Blockchain and verify the certificate's authenticity.

The fees tied to CER issuance can remain the same regardless of whether the CER is issued through a Blockchain or legacy technologies. The added benefit of using a Blockchain manifests by increasing transparency for all actors throughout each project cycle or touch point. Assurances of capturing fees will cover additional administrative costs to implement Blockchain integrations within the CDM Registry while being immutable and, therefore, lowering the risk of any fraud. Exchanging data on a Blockchain are revolutionary because they are a trustless ledger. There are strict controls of how data are recognized and how payment is issued. Mismanagement of funds, excessive or egregious spending, and other malicious actions will be quelled due to the transparent nature of Blockchain data exchange. Fees can be set and collected from the initial creation of a certificate and dispersed through a Blockchain contract when registering the project. That contract will stay valid regardless of the time it takes to complete each project cycle and the controls are autonomous requiring small amounts of administrative oversight, which is in a stark difference to the current structure of the CDM. The CDM uses the following fees when issuing CERs (UN BOA Review of CDM, p. 11, 2016).

1. Approximately \$0.10 per CER issued for the first 15,000 t of CO_2 equivalent for which issuance is requested in a given year;
2. Approximately \$0.20 per CER for any amount in excess of 15,000 t of CO_2 equivalent for which issuance is requested in a given year.

Prices of CER's currently fluctuate around \$0.50 which increases the significance of these discussions for the long-term sustainability of the mechanism. The Blockchain provides a transparent compliance tool for charging fees and reduces the backlog of issuing CERs to project participants. This is because there are discrepancies with the delivery of CERs to participants in part because fees are difficult to collect and CERs are only charged when issued. A large backlog of CERs is approved but never moved into the CDM Registry to issue to CDM participants which leaves these fees uncollected (Report of the UN BOA on the financial statements of the UNFCCC for the year ended December 31, 2015 (2016)). Blockchain smart contracts create a workflow where issuance and fee collection can happen at the time of validation which removes the congestion of the current methods. Smart contracts can deploy funds (or voluntary credit schemes if designed) after a predetermined set of data has been recognized. Further information regarding smart contracts can be found in supplemental chapters of this book.

18.3 Use of Validation CDM Data on the Blockchain

Blockchains offer a low-cost technology for developing countries and project developers to maintain their emission reduction records. Blockchain services that offer data recording tools for assets, documentation, and supply chain management can also enhance the functionality of CDM projects (Dunkel, 2015). Using the Blockchain, in tandem within the current validation process, can reduce CDM project verification times by removing unnecessary validation periods. There are seven steps in a CDM project lifecycle, and six primary actors perform accreditation services for different phases (Annual report of the Executive Board of the clean development mechanism to the Conference of the Parties serving as the meeting of the Parties to the Kyoto Protocol, 2016). Each step in the project lifecycle requires verification or validation of the steps completed before it. Confirmation requires assurances that the Project Design Documents were filed correctly, accurately, and no changes were made from the time of submission.

A CDM Project has a determined crediting period for credit issuance. These periods are set for up to 10 years, for a fixed crediting scheme, whereas a renewable crediting scheme can last up to 21 years (CDM-EB93-A05-STAN, p. 14). These credit issuance vehicles are designed to be long-term sources of financing and credit emission reduction. To achieve the designed function of these issued credits, the information recorded and exchanged must be accurate and transparent over a potential 10 or 21-year commitment. Relevant parties may change and the integrity of past validations and verifications would be preserved if a Blockchain was used to facilitate the exchanging of data through each commitment period. These crediting periods come with rules for disbursement and are approved by the Designated Operational Entities (DOE) (Procedure Clean development mechanism project cycle procedure; Standard CDM accreditation standard). If a country produces negative emission reductions, a portion of credits is deduced during the issuance of certificates. The immutability of the Blockchain ensures accurate monitoring reports are generated and can be used for modeling, predictions, evaluations, market instruments, or when discussing new policies relating to the CDM.

The monitoring report includes performance results and reflects how many CERs would be requested which requires a signature by the DOE, the Secretariat, and the CDM EB (to allow a veto vote). The signatures would dictate that each validation or verification was performed in accordance to procedure and the CDM registry would facilitate this multi-signature transaction (Standard CDM accreditation standard (n.d.); Standard CDM validation and verification standard for project activities). If a project has not filed documentation properly that step is designated incomplete if the appropriate actor had not signed their portion of the transaction by the deadline dictated in CDM procedures. This creates a transparent and defined administrative process that can be understood by any participant regardless of their knowledge of each process in a CDM project cycle.

The secretariat of the CDM performs evaluations of DOEs, and issues performance benchmarks referenced in the procedure are measured by the weight of the requests to the number of requests that have been completed (Procedure: Performance monitoring of designated operational entities (n.d.)). The weights of certain auditor requests dictate the magnitude of the service the DOE is providing. The highest weight classifications are failures to identify problems with implementation, procedural compliance, and accuracy of documentation (CDM-EB58-A01-PROC, Appendices 1–3). An auditor performing the same functions referencing Blockchain documentation saves the current state of the document for the lifetime of that network which shortens future validation times of the highest weighted requests of DOEs.

Goal 9 of the SDGs seeks to create resilient infrastructure while promoting inclusive and sustainable industrialization and there is a call to foster further innovation (Transforming our world: the 2030 Agenda for Sustainable Development, 2015). Blockchain technology is the epicenter of increasingly innovative ideas which are disrupting financial and administrative services. Adopting a Blockchain data registration service within the CDM provides an inclusion tool for developing countries that do not have efficient ways of ensuring document preservation or validation. Preservation of data is one of the key features of Blockchain technology.

In addition to data preservation, the Blockchain provides an exchange platform for a wide array of funding instruments that support the efforts of the CDM. This platform provides a means to exchange tokens representing a CER or by using the Blockchain to record changes in ownership. Information pertaining to the funding source, how those funds are required to be used, and where these funds are spent are critical accounting functions. These accounting standards are required to receive funding from the Adaptation Fund which provides direct access to climate financing for developing countries with less than 10 CDM projects (Direct Access - Adaption Fund (n.d.)). It is important to observe the fiduciary standards for Adaptation Fund recipients as the institution draws funding from 2% of CERs issued by the CDM (Adaptation Fund (n.d.)).

Fiduciary standards of the Adaptation Fund stress transparency, financial integrity, and the capacity to undertake monitoring and evaluation. The Blockchain provides a platform to perform these functions without additional infrastructure or resources. Risk management is a characteristic of the Blockchain given its transparency and the assurance that only the private key holder can initiate and sign transactions. This is an improvement over current internet-based databases that are subject to tampering or mismanagement from human actors that are organizing information. Information is exchanged on the Blockchain in a structured manner at a predictable rate (Procedure Clean development mechanism project cycle procedure (n.d.); Dunkel, 2015). Another requirement for funding is the use of national, regional, or multilateral implementing entities; it provides an additional layer of compliance that can be touched upon further with features of the Blockchain.

18.4 How Blockchain Technology May Benefit the CDM

Using a Blockchain creates deviations from current CDM Registry account creation procedures by using private and public keys in lieu of a password. On the Blockchain, the DOE will use a private key as a digital representation of their signature during validation and verification. Using a private key signals that the user agrees to the current state of the transaction and all outputs are valid. Keys are not synonymous with passwords and should be handled with a higher level of scrutiny. If a private key is used for verification and validation purposes, responsible parties need to know that only an authorized DOE is signing the appropriate transaction. Various key distribution methods are available to wallets or devices to manage Blockchain keys but the use case of the Blockchain dictates how applicable each method is. The devices and instruments performing data collection and monitoring services must also be compliant with the methodology of the project and provide data in a useful way.

A wallet can be created by an individual developer or assigned by a wallet service which creates an onramp for capturing information about actors participating in the network. If the CDM registry acts as the on-boarding platform for issuing wallets, the Blockchain records each transaction that wallet ever signs. This creates a platform for the CDM that provides transparency and ensures financial integrity for monitoring assessments and evaluations. An additional benefit to this architecture is the creation of another avenue to collect administrative fees by requiring payment prior to being issued a wallet or after being registered as a DOE. This is a critical point in the workflow design using a Blockchain network.

Both public and private keys, including issuing wallet addresses, comprise the basic administrative structure of account creation on any Blockchain. Each wallet signature composing a transaction is recorded with that transaction on the Blockchain. Transaction information in the Blockchain is easily identifiable. This information includes who is signing the transaction, what that transaction includes, and where the data go once confirmed. An entire transparent ecosystem can be imagined exchanging information between internet applications, Blockchain services, and project registries using applications drawing from this information.

Recording and monitoring devices that can communicate with a Blockchain are hardware wallets that physically have key management software installed within the components of the device. Variations and forms of this wallet type have already been manufactured and commercialized that communicate with the Blockchain remotely using cellular communication in lieu of a computer (Report of the Conference of the Parties serving as the meeting of the Parties to the Kyoto Protocol on its first session (n.d.)). This would be ideal in a mechanism that relies on accurate recordings to measure emission reductions. To ensure validity of independent auditor signatures, CDM DOEs ideally will have a single hardware wallet device assigned to their person for signing and verification. Passwords or mnemonic phrases can provide additional assurances that only the authorized DOE is using a private key, but biometrics can also be used with appropriate hardware. How these keys are embedded and how those devices are manufactured are the queries outside the scope of this chapter.

Blockchain networks can assist CDM projects by providing a new issuance platform for accounting activities while maintaining compliance with current methodology and regulations. Nodes, in a Blockchain, refer to computers connected to the network using a software client that performs tasks to validate and relay transactions. The same equipment that is currently used to manage the CDM registry would act as a node for exchanging information on a Blockchain network without increasing operational costs or requiring additional staff.

Intermediaries within the CDM are necessary to ensure funds are transferred in a fair and equitable way. Complexity and decreased transparency occurs while exchanging data with communities that do not have equal regulations or banking infrastructure. Blockchains transparently show the data exchanging between each participating network and validate the ownership of the data harmoniously among all parties. The data collection methodology of CDM Projects also differs between industries reducing "robust accounting practices" (Report of the Conference of the Parties on its twenty-first session. Paris Climate Agreement, 2015). The Blockchain provides transparency, financial integrity, and offers accurate data to undertake monitoring and evaluation.

Assurances in the validity of data records are costly to provide in legacy mechanisms and regional financial uncertainty leads to increasingly risk adverse financing or investment decisions. Creating a bottom-up approach using a Blockchain network will comply with market provisions while maintaining the integrity of that data across databases (Report of the Conference of the Parties on its twenty-first session. Paris Climate Agreement, 2015). Integrated circuits of recording devices, outlined by the project manager, can have Blockchain private keys embed metadata into each record to prove the validity and uniformity of sensitive data. Embedding metadata, or supplementary information contained in each transaction compiled from the approved methodology of the project, creates an easy confirmation for conditional contributions. This is important as there are beyond 200 different approved methodologies in the CDM (CDM Methodology Booklet, 2016). How metadata are embedded into a transaction depends on the Blockchain, which is beyond the scope of this chapter. Commercial businesses offer this service today that can be used or mimicked by the CDM for future implementation (Kyoto Protocol to the UNFCCC, 1998).

A Blockchain network can enable CER transfers within the same platform and enables exchanging information with other platforms, including legacy technologies either physically or remotely. The cross-platform capabilities of a Blockchain allow broad participation from many actors, increasing the adaptability of the CDM, and other carbon emission reduction schemes.

18.5 Looking Ahead—Achieving the Sustainable Development Goals

CERs are the measure of overall impact of a project toward improving sustainable development and reducing greenhouse gas (GHG) emissions. As each country implements their own approach to meet their contributions toward the Paris Climate Agreement, there will be a need to establish efficient data streams. Each country is given flexibility in how they will use CERs to meet their Nationally Determined Contributions (NDC). NDCs form the path to meet each of the SDGs and each nation needs to create global partnerships to achieve SDGs goals by 2030. The Blockchain is a risk-adverse process to facilitate the success of achieving various SDGs. The SDGs encompass many avenues to improve the quality of life of each participant while complimenting the efforts of improving. Relevant goals that a Blockchain include improving economic growth and earning potential of regional communities (Goal 8) through technology adoption and facilitate growth. A Blockchain network also provides the groundwork to facilitate future technological adoption (Goal 9). By providing a data stream that is devoid of manipulation, communities can use Blockchain CDM projects to receive funding and adopt sustainable projects (Goal 11). Developing communities lack infrastructure or institutional support to comply with the requirements of some or many financing sources. A Blockchain network at baseline provides these necessities and attracts new partnerships (Goal 17) to exchange information and increase collaboration as each country advances toward achieving the SDGs. Data inputted on a Blockchain would ideally be viewed through the UNFCCC NDC Registry. Article 4 of the Paris Climate Agreement requires each country to publicly issue their NDC to the UNFCCC NDC Registry to monitor progress, and the responsibility to meet these NDCs falls on each independent country (Report of the Conference of the Parties on its twenty-first session. Paris Climate Agreement, 2015; NDC Registry, n.d.). There is no punishment for failing to meet these goals, nor is there a distributed reward for achieving specific metrics so further incentives need to be considered. It should be in each participant's best interest to pursue innovations that improve the outcomes of our actions toward bettering our planet.

18.6 Conclusion

Achieving all 17 SDGs by 2030 is ambitious, and the efforts to achieve these goals should be equally as ambitious in order to empower communities that facilitate emission reduction projects. Blockchain networks provide a channel to exchange emission reduction project data and enhancing the functioning of carbon markets, such as the CDM. Changes to the digital infrastructure of the CDM can be done in stages, and not all suggestions need to be implemented at once. It is recommended that the CDM strives to shorten validation times using a Blockchain to submit project forms, which with greater transparency and accessibility, can be achieved using methods and procedures currently in place.

Changes to the digital infrastructure of the CDM would require a developer with knowledge of creating Blockchain applications and a computer to act as a node to post transactions to the Blockchain which is transcribed on the CDM registry. A User Interface (UI) can be built on top of this data stream to simplify the CDM project cycle for all project participants. Projects that rely on certificates as a source of funding should have additional measures to exchange CERs interchangeably. Blockchain metadata aid in the exchange process by creating identifying markers to use for financial disclosure and increase transparency. Transactions on the bitcoin Blockchain can be recorded with enough space to contain 40 bytes of metadata. This additional information can direct to locations of applicable documents or provide additional administrative controls outside of exchanging bitcoin (Nakamoto, 2008). CER issuance on a Blockchain would also expedite required elements of the filing process. Monitoring reports could be quantified and tracked to measure the impact of individual projects toward their nation's NDCs and the SDGs. Automating the collection of administrative fees and the share of proceeds removes the burden from the project participant by collecting the fee at the same time as the issuance of a credit. The reporting structure of project workflows could change first to decrease project approval timelines and increase the efficiency of registration and authorization actions.

Using the innovative ideas contained in this book will increase the relevancy of carbon markets and draw new innovations to revitalize interest in emission reduction projects, such as the CDM, by simplifying processes and increase the issuance of credits. If carbon markets are going to continue to stay relevant beyond 2020 and reach the SDGs by 2030, a Blockchain solution should be considered. The SDGs are achievable with strong global partnerships. Creating incentives to use transparent, immutable, and adaptable platforms will assist in strengthening these partnerships and elevate developing countries by using the Blockchain to facilitate borderless transactions.

References

Annual report of the Executive Board of the clean development mechanism to the Conference of the Parties serving as the meeting of the Parties to the Kyoto Protocol (2016). *Rep. No. FCCC/KP/CMP/2016/4) United Nations Framework on Climate Change.* <http://unfccc.int/resource/docs/2016/cmp12/eng/04.pdf>. Accessed on 10.9.17.

Adaptation Fund *(n.d.) United Nations.* <http://unfccc.int/cooperation_and_support/financial_mechanism/adaptation_fund/items/3659.php>. Accessed on 11.6.17.

CDM accreditation procedure *(n.d., Rep. No. CDM-EB05-A02-PROC, Version 12.0) Clean development mechanism.* <https://cdm.unfccc.int/filestorage/e/x/t/extfile-20160928100413048-accr_proc01v12.pdf/accr_proc01v12.pdf?t = eGF8b3RwZWhsfDCNwqvpT4ift5Kd5p0OmHiW>. Accessed on 12.6.17.

CDM Issuance of CERs *(n.d.) Clean development mechanism.* <https://cdm.unfccc.int/Issuance/cers_iss.html>. Accessed on 12.7.17.

CDM Methodology Booklet (2016). *Clean development mechanism.* <http://cdm.unfccc.int/methodologies/documentation/1611/CDM-Methodology-Booklet_fullversion.pdf>. Accessed on 20.5.17.

Direct Access - Adaption Fund *(n.d.)* <https://www.adaptation-fund.org/about/direct-access/>. Accessed on 12.7.17.

Dunkel, S., (2015). *White paper: Securing digital assets on the Blockchain, Case Wallet Inc.* <https://choosecase.wordpress.com/2015/09/09/white-paper-securing-digital-assets-on-the-Blockchain/>. Accessed on 2.7.17.

Kyoto Protocol to the United Nations Framework Convention on Climate Change (1998), *Article 6, 12, 17. United Nations.* <https://unfccc.int/resource/docs/convkp/kpeng.pdf>. Accessed on 9.7.17.

Meeting report - CDM Executive Board ninety-fifth meeting *(Version 01.0, pp. 4-9, Rep. No. CDM-EB95). (n.d.). CDM Executive Board.* <https://cdm.unfccc.int/UserManagement/FileStorage/4TI0GBJOY6CL8QRAE3U72WM5HFK1XD>. Accessed on 3.6.17.

Nakamoto, S.: (2008) *Bitcoin: A peer-to-peer electronic cash system.* <http://bitcoin.org/bitcoin.pdf>. Accessed on 14.6.17.

NDC Registry. *(n.d.) United Nations Framework Convention on Climate Change.* <http://www4.unfccc.int/ndcregistry/Pages/Home.aspx>. Accessed on 12.4.17.

Procedure Clean development mechanism project cycle procedure *(pp. 7-20, Rep. No. CDM-EB65-A32-PROC). (n.d.). Clean development mechanism. Version 03.1* <http://cdm.unfccc.int/Reference/Procedures/pc_proc01.pdf>. Accessed on 11.6.17.

Procedure: Performance monitoring of designated operational entities *(n.d. Rep. No. CDM-EB58-A01-PROC. Version 03.1) Clean development mechanism.* <https://cdm.unfccc.int/sunsetcms/storage/contents/stored-file-20150807164219209/Accr_proc02.pdf>. Accessed on 17.7.17.

Report of the Conference of the Parties on its twenty-first session. Paris Climate Agreement (2015) *Rep. No. FCCC/CP/2015/10/Add.1. United Nations Framework Convention on Climate Change. pp. 2–4* <http://unfccc.int/files/home/application/pdf/paris_agreement.pdf>. Accessed on 10.7.17.

Report of the Conference of the Parties serving as the meeting of the Parties to the Kyoto Protocol on its first session, *held at Montreal from 28 November to 10 December 2005, Rep. No. FCCC/KP/CMP/2005/8/Add.1. (n.d.). United Nations Framework Convention on Climate Change, pp. 7–20.* <https://cdm.unfccc.int/Reference/COPMOP/08a01.pdf#page = 7>. Accessed on 13.7.17.

Report of the United Nations Board of Auditors on the financial statements of the United Nations Framework Convention on Climate Change for the year ended 31 December 2015 (2016) *United Nations Board of Auditors.* <http://www.un.org/en/auditors/board/pdfs/2016-08-26%20UNFCCC.pdf>. Accessed on 7.5.17.

Rules of Implementation for The Joint Crediting Mechanism (JCM) *(n.d. Rep. No. JCM_MN_RoI_ver01.0). Ministry of Economy, Trade, and Industry.* <http://www.meti.go.jp/policy/energy_environment/global_warming/pdf/JCM_MN_RoI_ver01_0.pdf>. Accessed on 10.5.17.

Standard CDM accreditation standard *(n.d. Rep. No. CDM-EB46-A02-STAN, Version 06.0) Clean development mechanism.* <https://cdm.unfccc.int/sunsetcms/storage/contents/stored-file-20140721152731014/accr_stan01.pdf>. Accessed on 1.7.17.

Standard CDM validation and verification standard for project activities *(n.d. Rep. No. CDM-EB93-A05-STAN) Clean development mechanism.* <https://cdm.unfccc.int/sunsetcms/storage/contents/stored-file-20170502114945162/reg_stan06.pdf>. Accessed on 1.7.17.

Transforming our world: the 2030 Agenda for Sustainable Development (2015) *Rep. A/RES/70/1) United Nations General Assembly Resolution adopted on 25 September 2015.* <http://www.un.org/ga/search/view_doc.asp?symbol = A/RES/70/1&Lang = E>. Accessed on 1.6.17.

Further Reading

Status of CDM MAP 2017 implementation *(n.d. Rep. No. CDM-EB95-AA-A02) Clean development mechanism.* <https://cdm.unfccc.int/UserManagement/FileStorage/5ICJP6T9NSQ14OMDV2XFAKELZ87HWG>. Accessed on 20.7.17.

CHAPTER 19

Networked Carbon Markets: Permissionless Innovation With Distributed Ledgers?

Adrian Jackson[1], Ashley Lloyd[2], Justin Macinante[3] and Markus Hüwener[4]

[1]Edinburgh Parallel Computing Centre, University of Edinburgh[1], Edinburgh, United Kingdom [2]Edinburgh Business School, University of Edinburgh[1], Edinburgh, United Kingdom [3]Law School, University of Edinburgh[1], Edinburgh, United Kingdom [4]Independent Scholar, Frankfurt, Germany

The purpose of this chapter is to outline the most important questions identified in relation to the connecting of carbon markets through the application of distributed ledger (DL) technologies, and share the authors' current thoughts on those questions.

19.1 Connecting Carbon Markets

Carbon markets[2] are highly regulated and vary in scope, design, implementation, and rules and standards across the globe. Therefore, any approach to link them should be able to connect diverse and heterogeneous carbon markets, not by equalizing them, but rather by recognizing the differences between them and placing a value on those differences. This is important for enabling a sustainable carbon price across various carbon markets; incentivizing investment in carbon mitigation projects and carbon-efficient manufacturing; and encouraging more transparent auditing and accountability for carbon emission mitigation in global carbon markets, thereby meeting the ambitions of the Paris Agreement.

Regulators in each carbon market control supply of tradable emission units and set requirements for carbon accounting. Therefore, any system to link such markets should be able to address these variations in policy and approaches while still enable trading between participants.

---

[1] https://www.epcc.ed.ac.uk/blog/2017/08/10/distributed-ledgers-carbon-markets

[2] In this chapter, the authors are focusing in the first instance on connecting government-instigated emission trading schemes (ETSs), as a first step toward the ultimate objective of being able to connect more generally a broader range of mitigation actions/carbon pricing measures (e.g., including carbon taxes, renewable energy certificate schemes and so on). Also, note that the expressions "market," "carbon market," and "ETS" or "ETS market" are used interchangeably throughout the chapter to mean the same.

Transforming Climate Finance and Green Investment with Blockchains.
DOI: https://doi.org/10.1016/B978-0-12-814447-3.00019-7

Furthermore, given that carbon emissions and climate change are global issues, we need a global solution accessible to developing and developed nations without barriers to adoption such as heavy investment in bespoke software or hardware for nations with limited resources, but still allowing participation on an equal footing for all.

19.2 What Types/Levels of Market Should We Support?

Although the overall goal is to connect numerous individual carbon markets to enable international transfers of mitigation outcomes, it is necessary to recognize the multiple levels where carbon trading or carbon accounting and tracking may occur.

19.2.1 Corporate Level

Firstly, individual corporations may need systems to allow them to undertake carbon accounting across various jurisdictions, locations, and suborganizations in which they have operations. Many businesses currently do this in some form to track data, be it with spreadsheets that are e-mailed among participants; or bespoke applications/data infrastructures. Different jurisdictions may require different accounting/tracking procedures, meaning a single spreadsheet or application may not be fit for purpose across all those jurisdictions. There may also be oversight and auditing requirements within a corporation.

19.2.2 Single Emissions Trading System (ETS) Level

Secondly, individual carbon markets (or ETSs), be they subnational, national, or regional, require credible systems to organize and administer their operations. These operations include assigning and distributing emission allowances to companies; allowing for trading of emission units among organizations (subject to the rules of the ETS); and allowing companies to surrender/report emission units for obligations when required (recorded through ETS) administrator-maintained registries A carbon market should be an open level-playing field in which various organizations should be able to participate. Traditionally, this right to participate has been mainly the remit of large, multinational corporations. But for climate mitigation movement to be a real success, the market should be able to account for organizations of all sizes,[3] including those without the funds to purchase and operate costly computing infrastructure.

19.2.3 Connecting Carbon Markets

Finally, the aim is to connect these individual ETSs to enable trading among different jurisdictions' markets. Individual ETSs are differentiated by varying timescales, rules of

[3] Or even for individuals to trade, if such is approved by the relevant jurisdiction.

auditing and accountability, as well as environmental performance standards. Given the diversified attributes of existing ETSs, at present, direct, unrestricted, trading of carbon credits among all markets is not possible.

An international agreement for the ETSs in all jurisdictions to be linked is unlikely to be feasible. Thus, we need a mechanism which recognizes the individuality of carbon markets provides for their interactions. The system will need to support different approaches of interactions among markets by governing how/when/whether they interact with each other. Essential for this mechanism is flexible membership. Individual jurisdictions should be allowed to connect or disconnect, as they deem appropriate. Voluntary participation is crucial in fostering ever-growing emissions trading systems and achieving the intent of the Paris Agreement of engaging the Parties in cooperative approaches voluntarily.

Ideally, the Parties can agree to create a single system, or at least a single set of tools (a *software infrastructure*) that can accommodate the demands at all three levels described above. These range from providing the functionality of carbon assets/exposure management for a single organization to availing a platform that can network multiple carbon markets. Although there is no necessity to have a single software infrastructure for cross-level use, uniform emissions trading infrastructure would provide the potential for multi-jurisdictional benefits, from enabling simple adoption and sharing of data to full-system carbon auditing and tracking, which is to ensure transaction parties' accountability and prevent fraud in the system. Further, as will be discussed in greater details, the technical issues that need to be addressed for such a single system inside a major/partitioned organization are applicable at the networked market level and vice versa. Therefore, a feasible technical solution must have the capacity to support multilateral participation in a network of carbon markets comprising various organization users.

19.2.3.1 A Single Software Infrastructure

Despite this discussion being about a single software *infrastructure*, it does not necessitate a single software *solution*. What is proposed is a data infrastructure that enables all of the functionalities outlined above; and creates a reference implementation that is usable without precluding others from implementing their own solutions and participating in the networked markets.

Some may question if it would be possible to create a solution which satisfies the requirements outlined without using distributed ledger technologies (DLTs). Yes, it would. Nonetheless, the core benefit that DLTs offer is the unique functionality to ensure that access to the data in the network is distributed and available to all participants (subject to any access restrictions deemed necessary in carbon markets/carbon trading[4]) and, hence,

[4] As we will discuss later in the chapter, we are not necessarily advocating fully visible, fully unpermissioned ledgers. There may be good reasons to restrict the access or permissions to changing some data in the system.

equal participation in the market. Other functionalities such as auditability and visibility of transactions can easily be enabled and extended/augmented as required, ensuring that carbon market(s) can be made as transparent, trusted, and liquid as possible.

19.3 What Is Distributed Ledger Technology (DLT)?

DLT covers a wide range of potential functionalities, from completely decentralized, permissionless systems in which anyone can participate, to permissioned and controlled systems which are very similar to current data networks controlled by a single or small number of entities. As such, it is necessary to understand different types of DLT for available options with which to design a DLT-based system to connect carbon markets; and the impacts these options would have on the functionality of carbon market linkages.

The following features (Table 19.1) are recognized as key for any system to be considered a DLT:

The following features (Table 19.2) are recognized as configurable[5] for a DLT depending on the application it is being used for and the environment in which it exists:

19.3.1 Designing for Use Cases

The distinction between functionalities that define a DLT versus those features that are configurable provides scope for designing DLTs which are quite specific to targeted Use Cases and User Groups.

Permissionless systems, for example, are open to all participants. Anyone may become a ledger node (i.e., hold a copy of the ledger) and add valid entries. The converse of this is a **permissioned system**, where only authorized entities can hold a copy of the ledger or participate in transactions. However, it may also be possible for a range of permissions to be applied to the system, i.e., permission may be required to become a ledger node but individuals may interact with ledger nodes without permissions; or anyone may be able to become a ledger node but may require permissions to add entries to the ledger; or there may be different permissions for adding entries compared to viewing entries; and so on.

To give this more structure from a design perspective, there are two different types of permissions that determine a participant's scope of interactions with the ledger:

- Hosting the DLT (i.e., having a copy of the ledger, being a node in the system)
- Interacting with the ledger (i.e., viewing, or adding to, the ledger)

[5] Configurable: Potentially present or not in the system; also, potentially different levels or types of this functionality present in the system.

Table 19.1: Design features defining a distributed ledger.

Feature	Description
Distributed infrastructure	The system must be composed of multiple entities, or nodes, that each have a full copy of the ledger. This implies that nodes can be added and removed from the system without affecting the overall operation of the system or the integrity of the data stored in the ledger.
Encrypted transactions	The ability of participants to use public/private key encryption to authorize/assign/ interact with transactions in the system.
Global consensus	The nodes in the system must have a mechanism for coming to agreement about what is a valid addition to the ledger and what is the current state of the ledger. This means that entries to the ledger can be added by different nodes and the ledgers held by the nodes in the system will be updated to reflect the changes made at individual nodes. This also implies that there is a system for deciding whether an entry, or set of entries, that has been added to one or more of the copies of the ledger is valid and reject entries that are not valid.
Immutability	The ledger is accumulative: once entries have been accepted into the ledger it should not be possible to go back and amend them. This is achievable by including some type of *hashing*[a] in the ledger, generating a key from the data in an entry in the ledger, and incorporating that key in the subsequent entries. This process makes it extremely hard to alter entries in the ledger without that alteration being noticed as any change to the data in the ledger will change the key (or hash) generated from that data. An audit of the ledger (passing back through the ledger, calculating the hashes for each entry and comparing them with the hash stored in the subsequent entries) will quickly highlight if entries have been changed.
	Immutability becomes more problematic in the distributed infrastructure as it is possible for different versions of the ledger to exist (remembering that each node has a version of the ledger and they may be in different states of being up-to-date). This is overcome by the global consensus model of the system, but this does need careful consideration to ensure that immutability is not compromised. There are mechanisms for addressing these issues which must be considered when designing the immutability and consensus models of the DLT.

[a]A "hash" is a one-way mathematical function that summarizes a piece of data as a piece of unique, fixed-size, short data. The hash function turns data into a key of random characters called a "hash."

Although it may seem strange to separate out these two types of permissions (after all does it make sense to be a node in the system if one cannot add or view transactions?), there may be use cases when entries can only be added to a ledger following external approval (e.g., in a regulated environment); or where visibility of entries in the ledger is restricted (e.g., if commercially sensitive data are present).

19.3.1.1 Using a Single Ledger for Carbon Markets?

One option would be to create a single distributed ledger that contains the emission units to be assigned, traded, and surrendered in all the ETS markets that will be involved in the network. This yields the benefit of providing a single ledger for all carbon trades, meaning

Table 19.2: Design features of DLTs that are configurable.

Feature	Description
Permissioning	DLT systems have traditionally been viewed as permissionless systems, in which anyone can participate and be part of the system. No one entity, organization, or system participant is in charge or has any more authority than any other participant. However, this is not necessarily a fundamental property of distributed ledgers.
Proof of work	One of the key features of permissionless cryptocurrency systems, such as Bitcoin, is the ability to decide when a transaction (addition to the ledger) is valid or not. With no permissions in the system there is no authority to consult on validity; therefore, a computational system for ensuring validity is required. This is tied up with the global consensus feature, which *proof of work* (also known as *mining*) schemes use to both control the availability of entries in the ledger (coins in the cryptocurrency) and ensure that the selection of which node can add an entry next is random in the system.
Smart contracts	When entries are added to the ledger it is also possible to automatically perform some actions associated with that entry or ledger. A *smart contract* is simply a set of operations to be performed automatically once the contract has been fulfiled.[a] In the context of DLTs, fulfiling the contract is likely to be the same as adding an entry into the ledger, although smart contracts could rely upon multiple ledger entries to be fulfiled.
Settlement, exchanges, or payment systems	Although DLTs can be used as cryptocurrencies and can also support many business processes, the actual transferral of money between parties, or settlement for physical assets, need not be part of the DLT. Indeed, there are numerous Bitcoin exchanges that operate to allow transfer between Bitcoin and Fiat currencies.

[a]Or in a more legalistic sense, it might be seen as the transactional terms and conditions embedded in computer code which allow automatic execution of the relevant transaction once precise conformity with those terms and conditions has been established.

that emission units traded between markets can easily be tracked and audited. However, the requirements to support different operational modes and timescales for individual ETS markets (and their respective emission units) make such a system unwieldy.

A single-level system would require multiple regulator (or administrator) nodes in the system, with all being able to issue and accept surrender of emission units. They would also potentially be entitled to authorize or block transactions of emission units issued by them or traded with organizations being regulated by them. Such entitlement would either potentially significantly reduce the responsiveness of such a DLT/market, or make the system very complex.

More critically, how different administrator nodes allocate emission units to entities, account for the differences in the respective markets; and the fact that each jurisdiction would need to continue maintaining and operating its own domestic registry (i.e., its part of the ledger), make a single market solution very hard to conceive. The translation of value between emission units issued by different administrator nodes on the same ledger would require sophisticated functionality in the systems surrounding the DLTs.

Finally, having a single-level DLT would make it difficult for carbon markets to connect or disconnect. It would require full migration of operations into the DLT for a market prior to joining, and full withdrawal of operations from the DLT into something different if a jurisdiction wished to disconnect.

Therefore, while a single-level DLT provides the requisite functionality for a market to operate, it is not compatible with the foreseen evolution of that market and its linkages with other markets. Hence, a broader view of the relationship between design and user adoption is required.

19.3.2 Designing for Diffusion

To have meaningful impact on emissions trading practices, any new solution has to be adopted by user groups that extract value, share knowledge about system use, invest in support, and thereby help to accelerate adoption by others. This enables the solution to diffuse across its intended markets, grow in scale and add value for users. Having designed a DLT for a specific set of Use Cases, it should be recognized that new users will bring new contexts and modalities that might not be predicted, but do need to be anticipated if the system is to achieve its design objectives.

The study of the diffusion of technology is a well-established field and relevant to the design of a DLT-based network infrastructure. The configurable design criteria discussed above offer justification for rejecting a single-level DLT design, including: complexity, responsiveness, and flexibility. These qualities are known to affect the rate of adoption and diffusion of any new technology. They were among the five characteristics for speedy adoption, first suggested by Everett (Rogers 1962), including: *relative advantage* over existing technologies; *observability* of those benefits; *compatibility* with existing needs and use patterns; relative *complexity* of the innovation and the degree to which it has *trialability*, where availability for experimentation is positively correlated with adoption speed.

Considering the configurable design criteria of *Permission*, it is reasonable to posit that each of the qualities associated with adoption and diffusion above may be somewhat dependent on the degree of permission (Table 19.3).

The associations above highlight the importance of the link between design decisions and the potential for an innovative technology to diffuse via adoption amongst its intended users. These associations also suggest that we can learn from earlier studies of the diffusion of information and communications technologies, including: (i) Open Systems development

Table 19.3: Potential relationships between DLT design and diffusion: Focus on permission.

Diffusion Characteristics	Potential Relationship with Degree of Permission
Relative advantage	Unpermissioned DLTs have applications, such as cryptocurrencies that have no fiat currency analog, and hence a significant relative advantage in that application area. Fully permissioned DLTs may operate with a central authority and offer an alternative to traditional systems where the locus of relative advantage is more constrained. *Relative advantage might therefore be seen as inversely correlated with degree of permission, but highly market specific.*
Observability and trialability	Unpermissioned DLTs may be used by anyone without restriction while fully permissioned DLTs require user identification and authentication that defines the type of access that is available to the user. To this extent, a potential user may experiment with the unpermissioned DLT and hence directly observe the purported advantages. Fully permissioned DLTs, however, require user engagement before full experimentation can occur. *Observability and trialabilty might therefore be seen as inversely correlated with degree of permission.*
Complexity	Complexity: the complexity that is "observable and trialable" by the user can be constrained by the degree of permission: strongly permissioned systems define the user access rights and, hence, constrain the range of features available to a user. Unpermissioned systems in contrast, not only require technical expertise, but also significant computing resources to engage with fully, representing the difference between using a cryptocurrency versus cryptocurrency ("bitcoin") mining. *Complexity might therefore be seen as inversely correlated with degree of permission.*
Compatibility	Compatibility: in markets where applications for DLTs are seen as improvements over legacy applications, the match between existing user needs and DLTs is easier to establish for fully permissioned systems.
	In terms of impact on diffusion, compatibility will be strongly market specific—while it is possible that an unpermissioned DLT will diffuse quickly across a global user base, so too will a fully permissioned system that is a clear upgrade path from existing systems for an established user base. *Compatibility might therefore be seen as positively correlated with degree of permission but highly market specific.*

and Open Innovation where, for example, as we adjust the degree of Permission we make the solution more "open" or "closed," impacting the potential for others to innovate at the "edge" of these technologies and generate new markets that accelerate diffusion. This is why diffusion is generally modeled as a nonlinear process, with the potential for exponential market growth. If it is assumed that an initial market demand exists then "Diffusion Potential" is highly sensitive to changes in "Degree of Permission" and must be carefully considered at the earliest design stage. This is illustrated in Fig. 19.1, along with the relationship between Degree of Permission and Openness discussed above.

Although it is clear from the design considerations applying to a single DLT for carbon markets that design needs to support the evolution of a market infrastructure, "designing for diffusion" illustrates just how nonlinear the diffusion process is, and hence the importance

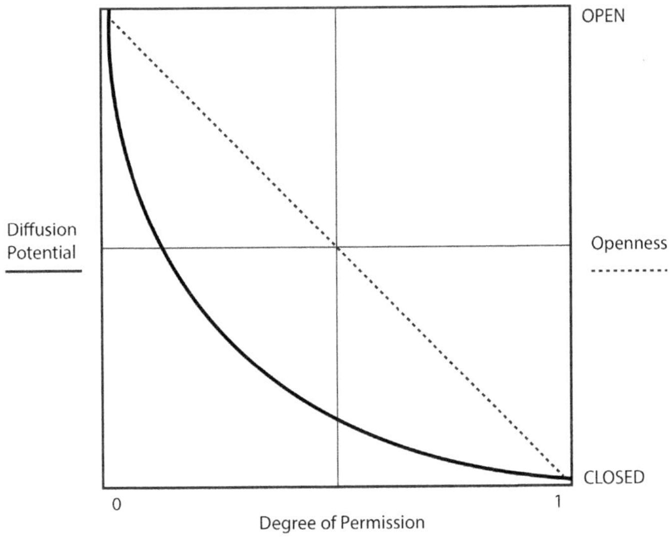

Figure 19.1
Diffusion potential as a function of Degree of Permission (see Table 19.3).

of incorporating this rationale into the design of any potential solution that will reduce emissions in accordance with the timetable defined in the Paris Agreement.

To better understand the barriers to diffusion, a more granular examination of the market structure and, in particular, User and Stakeholder groups is required.

19.4 DLT/Carbon-Matrix

The DLT/Carbon Matrix (Table 19.4) incorporates the linear and nonlinear elements of design. Applications at the level of corporate accounting and internal transactions over different ETSs will form the basic use case for any implementation scaling at the next level. The first level already carries the essential components for the following levels and diffusion requirement. If the functional requirements of each use case are not met then there is effectively no solution to diffuse; but as design decisions move from the left to the right, the network effects and diffusion impacts become stronger. These decisions are explored in the following text.

Table 19.4: DLT/Carbon matrix.

DLT/Carbon-Matrix	Ledger Use Case	Transaction Mechanism	Permissioning
Corporate level	1.	2.	3.
Single ETS	1.	2.	3.
NCM	1.	2.	3.

19.5 Ledgers (Matrix 1.)

19.5.1 What System Can We Have if We Don't Have a Single Ledger?

The alternative to a single-level structure is to have **multiple different ledgers for different carbon markets**. This would enable jurisdictions to run their own ledgers, manage their internal carbon market operations as required by local laws and regulations, but still connect to other markets for trades.

The challenge with multiple different ledgers is to create a distributed ledger: (i) within a single organization to manage its own carbon trading data; (ii) operate a single carbon market; and (iii) create a network among carbon markets. The same technology is used to operate different markets with different requirements, but can interoperate and allow movement of data between different levels.

Connection of ledgers is not a new concept. Indeed, mechanisms for connecting ledgers together such as, side chains,[6] inter-ledger protocol, as well as connection technology, such as that used by Ripple,[7] are actively currently being developed, tested, and used. We are proposing employing an approach broadly similar to the *side-chain* model. A side-chain model for linked carbon markets would allow different local carbon markets to connect to the global carbon market, without jeopardizing the core code of the global carbon market and putting potentially billions of dollars of emissions units at risk. Side chains, also known as intermediate ledgers, can join the federation of our primary market ledgers, enabling many primary ledgers (or *parent chains*) to trade/transfer resources.

A Carbon Market Ledger is the connection (*side chain*) between Single Market Ledgers for trading among these separate markets. Likewise, at the next level, the Single Organization Ledger is the connection (*side chain*) between Single Market Ledgers for accounting and tracking of assets across the markets an organization is involved in. However, we need to enable such functionality without lengthy waiting periods or delays between operations.

These intermediate ledgers provide a place to both undertake conversion of emission units between ETS markets, and ensure resources can be securely traded with both parties having visibilities of the resources to be traded and their availability. This can be achieved using two-way-peg-like mechanisms through the intermediate ledgers; although research is required in the safest way to integrate such mechanisms into DLTs without significantly

[6] This chapter has a good explanation of side chains and the two-way peg mechanism: https://blockstream.com/sidechains.pdf

[7] https://ripple.com/technology/

impacting ledger update rates while ensuring secure integration with ledger divergence reconciliation mechanisms.[8]

Also, our concept requires different types of participants in the overall ledgers, with administrator nodes able to interoperate with ledgers at levels higher or lower in the hierarchy. In keeping with the general requirements, required is the capacity to have different types of participants in the ledger for different operations (e.g., the regulator or administrator who can create and allocate emission units).

The attractiveness of this multilevel approach is that the DLT functionality can be used independently of other operations; for instance, a company can adopt the solution to manage its carbon assets even if the local market is not using such a solution. A local market can be created using this solution without requiring organizations participating in the market to use the solution for local accounting—and can use this approach without networking with other markets, or even before other markets are available. As different levels are separated, organizations or markets can change their approach or technology without impacting other markets or other involvement in these systems, allowing the system to diffuse easily.

The goal is to create a set of software tools that are sufficiently configurable to create and participate in any of the levels outlined, but do not mandate any particular level for their use. However, as the figure is illustrative it does not fully represent the required distributed infrastructure. With an operational solution, each of the nodes holds a full copy of the ledger and communicates, in a peer-to-peer fashion, with other nodes in the system.

19.6 Transaction Mechanism (Matrix 2.)

19.6.1 Are Smart Contracts Fit for Purpose?

Any system designed for carbon market trading and administration must be flexible enough to enable a range of metadata to be stored with emission units, and enable transactions through smart contracts. This is necessary to ensure emission units can be identified as being from a particular ETS, as part of a particular market cycle, for instance, and then not used in future market cycles (unless for instance, the jurisdiction allows banking, in which case this would be built into the metadata), or it may be required to ensure emission unit conversion rates and prices can be decided and applied automatically in trades.

[8] Because of the distributed nature of the ledger, different nodes in the system can have different versions of the ledger, which need to be reconciled at some point using the consensus mechanism. The use of interconnected ledgers may complicate this reconciliation, some further work is required to ensure it can be achieved without rolling back resources committed to remote ledgers or received from remote ledgers.

Having fully functional smart contacts requires the code implementing those contracts to be correct and secure; so, there is no way to change the contract once it has been added to the ledger. Part of the work that needs to be done to link carbon markets with DLT will be to design a safe way of executing contracts on the ledger. Indeed, to ensure emission units can be transferred between ledgers, automatic locking, and releasing of emission units would be desirable; so, participants from other markets can have confidence that double spending of emission units is not occurring.

19.6.2 How to Enable Consensus without Mining?[9]

Technologies like Bitcoin or Ethereum use "mining" to ensure that no entity is in control of the network and subvert the ledger. However, it has a significant computational overhead and, hence, energy and cost requirements that make it unsuitable as a technology that needs to scale globally.

To address this, this chapter has proposed **a permissioned distributed ledger that does not require mining to add emission units to the ledger**. However, there would still need to be a mechanism for ensuring consensus across the copies of the distributed ledger. Mining is used to enforce some level of randomness over the choice of which node in the system (which copy of the distributed ledger) gets to add its update to the ledger at any given point in the operation of the ledger. Without mining, there needs to be a method to ensure that no one node can hijack the ledger and alter it.

19.7 Permissions (Matrix 3.)

19.7.1 How Can Permissions Be Distributed?

The authors advocate a **permissioned ledger**. Indeed, one of the strengths of the system design proposed is the ability to have different functionality, different roles, for individual nodes, or participants, in the system, not through different software, but **purely through different permission levels assigned in the ledger**. To implement a permissioned ledger system and assign different roles to different participants, this chapter is proposing **the combination of two ledgers in our infrastructure at each level: one ledger for recording emission unit transactions; whereas, the other ledger for storing permission**

[9] Mining: The process of undertaking arbitrary work to decide which network participants can add an entry at a given point in time. Bitcoin uses proof of work, where every node wanting to add an entry to the ledger undertakes to solve a mathematical problem, the node that solves it first gets to add their entry to the ledger next. The random and variable nature of the mathematic problem randomizes which nodes can add entries to the ledger at any one point, distributing update rights and aiming to avoid the ledger being hijacked by any given node.

transactions, using an Authorization Unit (AU) to record permissions of organizations or individuals in the system.

Each node participating in the DLT will have a full copy of both ledgers, but what they can do with the ledgers, and how much data they can see, will depend on their entries in the permission ledger, their AU. Without an AU, they will only have basic access to the ledger. This may be "read only" access to all the emission unit transactions, or may be simply being a passive node on the network, depending on how the DLT is configured.

19.8 Conclusion

This chapter has considered some technological options for connecting carbon markets and argued that, although there are many options, the capabilities of distributed ledger technologies are well suited for doing so in a way that will help deliver the emissions reduction targets of the Paris Agreement.

Exploring use cases and DLT design flexibility indicates that design not only needs to be functionally capable to address the requirements of a viable market infrastructure, but also needs to address a complex and dynamic range of stakeholder requirements. By so doing, the solutions can diffuse quickly to build the required scale on a timescale that is compatible with the deadlines for the Paris Agreement targets.

To achieve this, the authors propose a solution based on a concept of **multiple levels of Distributed Ledgers** (what we describe as "**federated distributed ledgers**"), but with dual ledgers at each level (encompassing both permissioning data and market data) that can be deployed to enable markets. Hence, at each level, there is a set of ledgers encompassing permissioning, then a second set of ledgers that facilitate inter-market trading, or connecting of the markets. This approach offers a number of advantages including reduced transaction costs through: (a) making the transactions faster; and (b) eliminating intermediaries that both slow the transaction and charge fees.

In addition, the intended design emphasizes ease of adoption by participants, with the objective of making the software (and, if applicable, hardware) as accessible as possible, leveling the playing field and, hence, helping the solution diffuse quickly. Other types of engagement are also anticipated and need to be supported, with participants other than compliance entities likely to engage with the market also if there are opportunities for delivery of value-added services or direct arbitrage.

Consequently, it follows that to meet the functionality anticipated by the Paris Agreement, the technology platform needs to be designed to accommodate carbon markets that are different in their design, implementation, and ambition. Experience with DLT suggests that the technology is mature enough to deliver a robust functioning system, but specific design options to address potential barriers to adoption and diffusion require additional research.

Reference

Rogers, E. M. (1962). *Diffusion of innovations* (1st ed.). New York: The Free Press of Glencoe.

Further Reading

<https://blockstream.com/sidechains.pdf>
<https://ripple.com/technology/>

INTERLUDE V

SECTION 5

Blockchain for Better Green Finance Law Enforcement

Legal and regulatory frameworks are often the fulcrum of leveraging private capitals for green investment. And so, the enforceability of (green) finance regulations in every jurisdiction determines the level of trust investors put and the amount of green finance markets get. Despite the tremendous growth of green finance, especially the issuance of green bonds which surpassed the magic US$ 100 billion benchmark in 2017, there are still a number of factors undermining the enforceability of (green) finance regulations. They include the lack of common definition of "green," symmetric market information and fully transparent yet efficient processes to verify claims of corporations around green investments.

Green bonds, for instance, may expect to benefit immensely from the application of Blockchain technology. Blockchain enables low-cost yet trustworthy validation of the impact of green bonds for both issuers and investors. A pioneer working in this space is Stockholm Green Digital Finance. Regulations monitoring or scrutinizing the use of proceeds from the issuance of green bonds could be much more effective as Blockchain is capable of verifying claims linked to the "greenness" of the investment. Such "greenness" refers to the impact of a particular green bond on sustainability: outcomes of emissions reductions, fulfillment of adaptation needs, supply chains, fair trade, or even improvement of living standards for people in developing countries. In Chapter 20, How to Trust Green Bonds: Blockchain, Climate, and the Institutional Bond Markets, Owen Sanderson will elaborate on how Blockchain can create trust among investors and consolidate the credibility of the green finance market. In Chapter 21, Utilizing Blockchain for Better Enforcement of Green Finance Law and Regulations, Xiaochen Zhang, Professor Duoqi Xi, and her team at Shanghai Jiao Tong University, will provide an overview of the regulatory frameworks for green investment in China, as a case study, and the extent to which Blockchain innovations in the country's finance sector is enhancing the enforceability of relevant financial regulations.

> The government is like a battleship. It's very hard to turn – especially with technology.
>
> ***Shira Scheindlin, US District Judge, Southern District of New York at LegalTech, 2016.***

It is said the use of Blockchain technology for scrutinizing green investment would arise other issues. For one thing, the disruptive feature of Blockchain technology, especially "smart contract," which automates compliance in many sectors, can make traditional provisions of current regulatory frameworks irrelevant within the next decade or so. It necessitates a new paradigm for international legal cooperation against climate change in the Blockchain era. Indisputably promising is the huge potential of smart contracts to facilitate climate change actions by linking renewable energy and carbon accounting, reporting, and tracking on a micro- and macro-economic level. The databases of high interoperability can be connected to funding initiatives and trade settlements. For example, renewable energy trading on a Blockchain network can lock in value for taxes and penalties without the friction introduced by traditional intermediaries. To ease the authorities' turning toward this new direction, the international community needs to ponder how the Blockchain-based legal regime should be designed and unrolled for climate mitigation and adaptation finance on a global scale.

In Chapter 22, Blockchain and Smart Contracts: Complementing Climate Finance, Legislative Frameworks, and Renewable Energy Projects, the last chapter of this book, James Duchenne will discuss the paradigm shift that smart contract engenders across a range of fields. He will share his insights on promoting the interoperability of databases to improve the efficiency and transparency along the investment value chain—issuance, financing, trading, and reporting, which can complement current efforts in the industry.

CHAPTER 20

How to Trust Green Bonds: Blockchain, Climate, and the Institutional Bond Markets

Owen Sanderson
GlobalCapital, London, United Kingdom

At first glance, branding a bond "green" seems too trivial to have a purpose. Once a bond issuer says a bond is a green bond, then it is a green bond, whatever its other attributes. Would-be market guardians, from the Chinese state to the International Capital Market Association, have attempted to define the green bond market, but there are few international rules about what counts as "green" (International Capital Market Association, 2017; People's Bank of China, 2015).

This relative freedom has helped the market to explode in size in a short time and go global, financing everything from carbon capture in China to hybrid cars in California. The volumes of funding raised are large compared to other green initiatives—USD$103.9 billion so far in 2017, according to the Climate Bonds Initiative, and USD$81 billion in 2016—though still tiny compared to Dealogic's figure for total bond market issuance of USD$6.69 trillion in 2016 (Dealogic, 2016; Climate Bonds Association, 2017).

The rapidfire and ramshackle development of green bonds, though, opens the market to charges that it is merely a marketing tool, rather than a sincere means to channel the power of the institutional bond markets into climate finance.

Green bonds need to grow, but they need credibility, verification, and transparency. The market needs to create (or rebuild) trust, and this is what distributed ledger technology, or Blockchain, can deliver.

Today's green bonds, are little more than a promise of environmental virtue added alongside existing, firmer financial contracts; tomorrow, flows of carbon may start to be encoded alongside flows of cash using Blockchain. To get to this point, though, the market will have to confront formidable problems of cooperation and coordination.

Transforming Climate Finance and Green Investment with Blockchains.
DOI: https://doi.org/10.1016/B978-0-12-814447-3.00020-3

In any case, the green bond market is still finding its feet. As a capital markets instrument, green bonds are only 11 years old, and have only hit their stride in the last six. The World Bank issued the first labeled green bond in 2008, which followed an equity-linked issue by the European Investment Bank (EIB) in 2007 (World Bank Group, 2017). Since then, there have been plenty more "firsts." The first private sector bank green bond, the first sovereign green bond, the first corporate issuer of a green bond, the first North American issues, Chinese, Indian, and others.

Each corner of the bond market now comes with a green variant—there are green "high yield" issues (these used to be called junk bonds), green securitizations, green covered bonds, and green project finance financings (Climate Bonds Initiative, 2017).

This, in turn, has sprouted a subindustry proving the worth and value of green bonds. CICERO, a Norwegian nonprofit, was the first organization to start "verifying" green bonds, and has been joined by private-sector consultancies such as Sustainalytics and Oekom. More recently, credit rating agencies Moody's and Standard & Poor's, whose main businesses involve rating bonds based on traditional financial metrics, have branched out into offering assessments of "greenness" (Sanderson, 2016).

20.1 Bells and Whistles

Like any other bond, green bonds are financial securities in which the bond issuer—the organization which wants to borrow money—sells a contract promising repayment and a specified rate of interest.

Beyond that comes various optional extras. The promise to repay can be backed by assets. Issuers can add credibility by agreeing to limit their actions, or by giving buyers of the bonds extra rights. Bond contracts spell out these rights explicitly, giving investors different blends of security and income, as well as making arrangements for nonpayment, security over assets, changes in tax rules, and other features.

The "green" part is one of those other features. Typically, a borrower will voluntarily agree that any proceeds from selling the bond will be used for some environmentally friendly purpose. For a town or local authority, this could mean upgrading its fleet of buses, for a water utility, installing new water treatment works, and for an electricity company, perhaps funding a wind farm.

When banks issue green bonds, these are usually backed by "green" mortgages—lending secured by homes which meet certain environmental standards—or by loans for some of the purposes above. The first green bond issued by an insurance company, a deal from Australia's QBE Group in April 2017, was backed by a portfolio of other green bonds (Davies, 2017).

20.2 Seeking Purity

The benefits of this are debatable. In the case of a bank, lending to "green" homes means in practice lending secured by new buildings, which are constructed to high levels of environmental efficiency in the European Union—green bond issues by the Dutch banks ABN Amro and Rabobank, and the German lender Berlin Hyp fall into this category.

New homes are harder for banks to value, but in theory their more frugal energy use should give their occupants more disposable income, all other things being equal, rendering new buildings more useful as collateral (European Banking Federation, 2017).

However, banks will lend extensively against new homes wherever there are new homes to be built—whether this is branded green or not makes little difference to their activities, or to the relative environmental quality of the buildings. European rules on new buildings through the Energy Efficiency Directive have a much larger impact on greening the European housing stock than whether mortgages are financed by green bonds (European Commission, 2016).

Around the green bond market, there is also a broader market for "ESG" (environmental and social governance) bonds, with even woollier standards. Lloyds, a major UK bank, issued an "ESG" bond which was largely backed by lending in deprived areas of the UK, predominantly Scotland, business it would almost certainly have been involved in irrespective of the ESG bond (Lloyds Banking Group Investor Relations, 2014).

Banks are always going to be at one remove, because their business is lending. Corporations have a straighter path to greening their bond issues, though it is still morally complicated.

A bond issue by Repsol, the Spanish oil company, in May 2017 (the first green bond from an oil company), offers a good illustration of the problems (Turner, 2017).

The bond was used to finance "energy efficiency" projects—improving the quality of turbines, purchasing equipment to lessen the environment impact of oil refining, equipment to cut flaring and venting of natural gas, to reduce the likelihood of pollution, and so on (Repsol, 2017).

But for many observers, as well as investors, the fact that the company exists to extract, refine, and sell hydrocarbons means its bonds can never be green. A world sincerely targeting a limit of two degrees of warming needs to stop burning hydrocarbons almost immediately, rather than tinker around the edges to make the business of emitting carbon marginally more efficient.

The alternative argument is that engaging polluters is by far the most powerful form of green leverage. Repsol expects the projects financed by its deal can save 1.2 mt of carbon annually by 2020 (Repsol, 2017).

Contrast this to Dutch bank ABN Amro's green bond framework, which expects to save 2,480 t per year lower carbon compared to a conventional issue for every €100 million of bonds issued (ABN Amro, 2015).

For a bond of the same €500 million size as Repsol's debut deal, that works out at 12,400 t of carbon—around 1/1000 size of Repsol's promised carbon cuts for the same size of bond.

Such a simple calculation glosses over formidable methodological issues, but the disparity makes intuitive sense. Greater environmental improvements come from reducing the impacts of bigger emitters, rather than by optimizing the environmental credentials of activities which already emit little carbon.

But if the aim of green bonds is to allow bond buyers to claim they are purely financing green activities, rather than to make the largest carbon reduction possible, ABN Amro's issue might still prove more attractive to some. It's a philosophical tension at the heart of the market.

20.3 Costing Carbon

Although the issues from ABN Amro and Repsol illuminate some of the ambiguities involved in green bonds, "pure green" or (dark green, as the market prefers to style it) financings exist. But there's still controversy over whether green bonds can cut the costs of these projects relative to conventional funding.

When Dong Energy, the Danish power company, raises a loan to build an offshore wind farm in the North Sea, lending to the project will be a straightforward green financing which clearly cuts carbon emissions. Every turbine installed on the sea bed will help to displace coal power in the UK and European electricity grids.

It is likely that this loan will ultimately be funded by green bonds. Dong has sought and won lending commitments from public sector bodies like the EIB, which has a large and sophisticated green bond programe (European Investment Bank, 2017).

Nevertheless, although the loans will be clearly green, after a project can be demonstrated to achieve certain standards of environmental credibility and a carbon hurdle rate, more carbon reduction doesn't mean lower financing costs. EIB provides cheap lending to all kinds of projects—the greater environmental friendliness of Dong's wind farms does not reduce the cost of borrowing from the EIB any further.

This is the market convention both in green lending, that will ultimately back green bonds, and in green bonds which go to finance carbon reduction projects directly. In theory, green bonds can be placed to a wider audience than conventional issues—everyone that would ordinarily buy a particular bond, plus specialist green investors as well—but in practice, issuers rarely try to push the prices of their green bonds too far from conventional issues.

A study from the Bank of International Settlements recently found green bonds paid an interest rate 18bp (0.18 of a percentage point) lower than comparable conventional issues—but the study only looked at 21 issues across four years (Packer, 2017). Senior funding officials at the EIB, the largest green bond issuer, have estimated the difference at closer to 3bp (0.03%), although some green bond bankers deny there is a difference at all, while acknowledging that green bonds often perform better after they have been issued (McGlashan, 2017).

Some see pushing for green bonds to be issued at a lower interest rate as an aggressive move—while it might incentivize more issuers to join the market in the long term, it hurts investors with a green mandate in the short term. Rather than punish green investors, most issuers are inclined to help them, by making sure they receive more of the bonds they request than investors without an environmental mandate (McGlashan, 2017). Bankers selling green bonds frequently talk up the proportion of green or "impact" investors involved in buying new deals, but are wary of talking up any possible price advantages of issuing green bonds.

20.4 The Market that Might Have Been

The issue over whether green bonds should be better priced than conventional bonds is really, in part, a question about what kind of green bond market should exist.

Some imagine a market tightly tied to carbon reduction, in which investors compete not only for the most profitable investment portfolio, but also for the portfolio that achieves the greatest demonstrable benefit to the planet. Investors would prize most highly the bonds related to the best carbon-cutting outcomes—and compromise their financial return to do so. This would bring market forces to bear on companies through the bond market, allowing cleaner industries a lower cost of funding, and therefore higher profitability.

To make this dream a reality, carbon needs a high and tradeable price. Investors running socially responsible or environmental portfolios would need to be compensated for the interest payments they forego by receiving financial benefit from the carbon they help to save.

Because that carbon price does not yet exist, this utopian green bond market shows no sign of emerging. Instead, the existing green bond market is an ordinary bond market where investors compete on ordinary capital markets measures of financial creditworthiness and market pricing—but with a gate at the entrance, a greenness hurdle that issuers and investors must pass.

Once they clear the green hurdle, it is business as usual.

20.5 Prove Yourself

Whether considering the existing green hurdle-based market, or the more carbon-attuned market that might exist in the future, issuers still need a way to demonstrate their greenness.

An issuer can technically make their bond green simply by saying so—there is no official definition of a green bond across most of the world—but most choose to go further, in an attempt to show investors that they are acting in good faith.

The way this works today is through "approved verification" and "second opinions," though there are large variations around the world. North American issuers, for example, often go without any external verification (Sanderson, 2016). One of a few specialized firms reviews the documentation associated with a green bond issue, and perhaps does a deeper due diligence dive into a company's structures, controls, and business practices. There are a limited number of these verification firms, which are usually accredited by the Climate Bonds Initiative. The verification firm will then issue a second opinion, which is provided to investors considering buying the bond issue. There is no standard format for this second opinion, though it usually includes similar information, and it's optional in most of the world (Sanderson, 2016).

The "Green Bond Principles," another voluntary set of guidelines for the market developed under the auspices of the International Capital Market Association, sets out what investors can expect from a green issuer, and includes, in its latest version, a recommendation that issuers offer some kind of external review for their green bonds (International Capital Market Association, 2017). There is no penalty for failing to abide by the Principles—they are simply a statement of best practice and a way to codify what is "normal" in the market.

The striking feature of this system is how light-touch it is. All elements are voluntary, and the external review requirements in the Green Bond Principles assume the second opinion provider will be paid by the green bond issuer or the banks it has picked to work on the issue. This creates an obvious conflict of interest if any green bond issuer did want to gain green branding without doing to the hard work of carbon reduction.

After that, external oversight often vanishes entirely. Normal market practice in the green bond market includes little or no requirement to make sure the promises made in an issuers' green bond documentation are kept, and no sanctions whatsoever for an issuer which does not keep its promises or fails for other reasons to meet its carbon-cutting targets.

This is in stark contrast to financial information—offering securities to the public in any developed market requires regular disclosure of audited financial statements. Deliberate falsification of a company's finances can be criminal, and companies must release any material changes in financial conditions immediately, through a heavily regulated process.

On the other hand, usually there's no penalty at all in the documentation for failing to maintain a requisite level of greenness—investors might choose to sell the bonds a company has issued, if it uses the proceeds to back oil exploration not solar panels—but it is untested.

No issuer has yet failed to keep up their green commitments (not that they are obliged to report either way), so no investors have been forced to consider whether they are willing to sell a bond whose financial characteristics they like but whose environmental characteristics have deteriorated.

Although green bond market practice includes few legal sanctions to make sure green issues fulfill their environmental promise, issuers do have strong incentives to behave honestly.

Given that doing a green bond is voluntary and, as we've seen confers minimal financial benefit (any cut in interest rate paid will likely be eaten up by higher fees for structuring the issue), so there is no incentive for issuers to produce a fake green issue or push the boundaries of what investors will tolerate.

Issuing a green bond is a public statement of environmental commitment—one which would be utterly undermined by failing to keep green promises or deceiving external review providers or investors.

If an issuer is having trouble meeting stringent green requirements, far better to avoid green bonds altogether. There is no penalty for not doing green, but lots of potential embarrassment for doing green and failing.

20.6 The Culture of Green Bonds

But however close external supervision of green bonds becomes, this does not greatly change the basic benefits of the market much. However closely issuers track carbon and measure projects, and however keen-eyed their external reviewers are, the market still needs to deal with problems of addtionality and cost of funding.

However good the measurement of the carbon cut by a certain project or bond issue, if green bonds deliver financing at almost the same cost as ordinary bonds, to projects which would have occurred anyway, they do little besides making those involved feel good.

Inside this conclusion, though, is a kernel of hope, which is that the very act of creating and issuing green bonds; especially, those with rigorous impact reporting and verification, changes the culture inside corporates (Sanderson, 2016).

Instead of seeing care for the environment as an outgrowth of corporate social responsibility—and therefore politically weak, segregated from the core business, green bonds yoke the finance and treasury functions of a corporate, typically some of the most powerful actors in decision-making, to the project of carbon reduction.

Simply by encouraging the measurement and analysis of carbon emissions, and placing this power with the accountants and financiers in corporate treasury, green bonds lay the groundwork for a broader cultural transition, in which rigourous carbon impact assesssments and carbon reporting become the norms, even for projects falling outside the perimeter of the green bond programe.

Green bond issues typically involve board members and chief financial officers, as well as the treasury and funding teams who handle ordinary bond issues. They are strategic investor marketing exercises at the highest level. As they usually come with sponsorship from the most senior corporate executives, they are a powerful tool for promoting a carbon-cutting culture inside corporates.

This alternative explanation of the benefits of green bonds runs both ways—doing a green bond might help to embed sustainability in corporate culture, but at the same time, the corporates willing to do green bonds are typically those which already consider themselves as forward-thinking, environmentally astute organizations. Even for companies which already have strong environmental cultures, the simple act of starting and maintaining a green bond programe can push them to become more rigorous and disciplined about their carbon cutting activities.

20.7 The Market of the Future

The "Green Bond Principles," and other, voluntary standard-setting organizations like the Climate Bond Initiative, must strike a balance between encouraging issuers to join the market by keeping standards manageable, and maintaining the market's credibility. What use is a mark of quality that is too easy to attain?

Governments are also getting involved. The Chinese state, through the People's Bank of China and the National Development and Reform Commission, mandates how green Chinese green bonds must be, although to a different standard from other geographies, because "clean coal" projects are included (People's Bank of China, 2015) The Chinese green bond market is, like the broader Chinese bond market, mostly closed to international investors, and often involves transactions between state-owned companies, state-owned banks, and state-backed local governments. In India, which can be similarly difficult for international investors to access, the state also sets standards for green bonds through the Securities and Exchange Board of India, the local securities regulator, whereas Singapore has offered subsidies for issuers to obtain external reviews of their green issues (Wong, 2017).

Europe, which hosts a more open green bond market, is making moves in the same direction.

The European Commission has a "high level expert group" on green finance, which recently published a wishlist of recommendations for green financing, including a call for a European green bond standard (European Commision HLEG on Green Finance, 2017). Although the recommendations are a long way from becoming law, the report has heavyweight sponsorship from within the Commission, including the Commissioner for Financial Services, Valdis Dombrovskis.

In the shorter term, the Luxembourg Stock Exchange—listing venue for many of the international bond issues offered in the European bond market—has set up a green listing venue for bonds, whose listing rules may become the de facto European green bond standard (Luxembourg Stock Exchange, November). But even any European legislative or regulatory green bond standard is likely to remain fairly loose, and avoid discouraging issuers through being too hard to achieve. Nevertheless, some market advocates hope that eventually a green bond standard can feed into the regulation for banks, insurers and investors (European Banking Federation, 2017).

This would incentivize market players to purchase green bonds, by giving them better regulatory treatment, and therefore open any price gap between green bonds and conventional issues much wider.

Whereas this would supercharge the development of the market, as issuers rushed to take advantage, it would also encourage less scrupulous bond issuers to create green programes, which could see standards in the market getting diluted, given the minimal oversight which exists at present.

So, if there is to be any attempt to use regulation to encourage the green bond market, regulators will need to make sure there is a robust, verifiable way to trace carbon emissions through the chain of loans and bonds, using a transparent and secure data architecture. That, naturally, is where Blockchain could come in.

20.8 Impact Reporting

For the most forward-thinking issuers, waiting for regulation isn't quite good enough. Some issuers, mainly supranational or development banks with multi-billion dollar green bond issuing programes, have voluntarily tightened up their carbon reporting, going beyond the "second opinion" at issue, to continuous impact reporting for their green issues.

As discussed above, green bonds from corporates or utilities tend to finance environmental improvements directly—greener freezers for selling ice-cream at Unilever, an energy-efficient terminal building at Mexico City Airport, better and safer sewerage for

Washington DC's water board—but banks, whether public sector development banks or ordinary private banks, tend to lend their green bond proceeds on to other projects.

Lending money to a borrower's project is clearly different to running the project—so, it has a less distinct carbon audit trail associated with it. In July 2017, for example, The European Bank for Reconstruction and Development (EBRD) lent money to a Greek oil company to support a move away from hydrocarbons. This raises additionality issues—if EBRD hadn't lent the money, someone else would most likely have done, though perhaps more expensively—and makes the need for proper impact reporting all the more acute.

France, entering the green bond market with a €7 billion bang in January 2017 (by far the largest green bond ever issued, subsequently increased to €11 billion) includes a use of proceeds check–annual examinations–until all the cash from the bonds has been allocated.

German development bank KfW has a dual-track system, where its bond programe is verified, and the loans that back it are verified separately to come up with a fairly rigorous carbon reduction figure. Its green loans in recent years have primarily financed European wind farms, a set of assets for which it is relatively easy to demonstrate and calculate the carbon benefits. The EIB, a market pioneer through its "Climate Awareness Bond" programe, has its own impact reporting structure, with its environmental benefits audited by KPMG, and the World Bank and the International Finance Corporation (another part of the World Bank Group) offer extensive reports on which projects their green bonds finance and the corresponding carbon benefits.

Even these market leaders, though, are cautious about how firmly they can claim carbon cuts. The World Bank notes:

> *"reporting is based on 'ex-ante' estimates of climate and environmental impacts at the time of project appraisal and mostly for direct project effects. . .an important consideration in estimating impact indicators and projecting impacts is that they are based on assumptions. While technical experts aim to make sound and conservative assumptions that are reasonable based on the information available at the time, the actual environmental impact of the projects generally diverge from initial projections."*

The World Bank is leading a project to try to standardize green bond impact reporting across the development bank issuers that do it–a clear boon for investors that want to track their own carbon impact–but however standard the reporting format, it cannot compensate for weaknesses in the underlying data. It is difficult to track carbon reduction, and new projects will always require some amount of guesswork.

Blockchain, as a mechanism for transparently tracing carbon emission reduction, can help, but feed the system with poor data or loose predictions, and these, too, will be transmitted to investors.

20.9 The Future of Monitoring

Here is how, one day, the structure of green bonds could improve, with the help of Blockchain technology.

For a vision of how bonds can rigourously track the performance of loans which back them, it makes sense to consider the asset-backed securities market today. These are bonds structured by grouping together large numbers of smaller loans, diverting the cashflows from these loans into layers called tranches.

The technique received a bad reputation during the financial crisis, as it was used to transmit the risk from the collapsing US subprime mortgage market around the world and into the heart of the banking system. But there is nothing inherently wrong with the idea of bundling up small loans and using them to back bonds. There is even a subsection of the green bond market already taking this approach.

US issuers are bundling up Property Assessed Clean Energy (marketed as the rather snappier "PACE") loans, made to homeowners to fit solar panels or make other improvements, to back bonds. Elon Musk's solar installation company SolarCity does the same. Toyota has been structuring bonds backed by loans secured on its hybrid vehicles for more than three years, whereas the Dutch bank, Rabobank, has also issued bonds backed by residential mortgage loans on energy efficient homes.

The relevant thing about this market, though, is not the green issues which have taken place, but how tightly a specific group of loans can be tied to a specific set of bond issues.

This tight connection has the potential to solve some of the major issues in today's green bond market, such as the fungibility of cash borrowed in green and brown bond markets, and the difficulty of proving a bond issue's green purity.

The legal form used to tie the loans and bonds together is, today, a "special purpose vehicle"—an empty shell company holding only the benefit of loans and issuing only bonds.

Because the underlying loans are the only thing backing their bonds, bond investors usually require a great deal of information about them—if the loans are domestic mortgages, bond investors want income details. If the loans are used to buy cars, investors want the models, ages and depreciation rates. If the loans are against commercial property, investors want to see rent rolls and tenancy details.

Some issuers provide high-quality data, which plugs easily into investor tools for modeling risk and payments; some do not, which makes it harder to issue bonds, and a cottage industry of data specialists has grown up around the asset-backed securities market.

20.10 Bring on the Blockchain

Where there are data to be corralled, trusted, verified, and distributed, there is an application for Blockchain—indeed, the power of Blockchain technology to create trust and, perhaps, legal enforceability, raises the possibility of doing away with the shell company and its clumsy contractual set-up entirely.

As these structures work today, there are several real corporations clustered around the shell corporation, performing the functions it is unable to do for itself. The "trustee" is the legal representative of investor interests, and is charged with observing and ensuring the contracts are adhered to as written. The "cash manager," a bank, hosts accounts for the shell, making and receiving payments. The "servicer" collects money from borrowers—straightforward enough if the payments are made in full and on time, progressively more difficult if a borrower is struggling to pay.

The shell company functions on a one month or three month cycle, meaning payments are collected, stored, and then pumped through the structure according to the contracts.

Many of these functions, and the chapter contracts defining the different functions and cashflows, are amenable to disruption by a new data technology. Cashflows and data reporting could straightforwardly be switched to a trusted distributed ledger, allowing investors to receive cash and examine loan performance in real time, and giving the trustee a transparent audit trail.

For the moment, none of this has happened, and, right now, the incentives to strike out in this direction are small. The overheads of the shell company are tiny compared to the benefits of the financing technique, and the market is replete with incumbents who will fight hard to keep their place in it.

But like other areas of the financial system, Blockchain will come in by fits and starts. Loan servicers, who professionally track many millions of loans for hundreds or thousands of shell companies issuing asset-backed securities, will start to use distributed ledgers internally—not for their trust properties, but for their accessibility and immutability.

That will open the eyes of market participants, who typically used to manipulating large datasets as part of their investment process to the possibilities of applying the technology more widely. They will be pushing at an open door. As loan data migrate to Blockchain, in parallel, many of the banks, asset managers, and trading firms that populate the capital markets will have rolled out shiny Blockchain applications in other areas, such as clearing and settlement.

Nonethless, Blockchain is unlikely to take hold quickly. The asset-backed securities market has been in existence since the 1980s, and already has a highly developed set of digital tools to manage, model, and track loans. These tools could always be made to work better, but for most investors most of the time, they are good enough.

What might change the status quo is a radical shift to green underlying loans, and a serious attempt to accurately monitor their carbon characteristics and progress. This creates a whole new set of properties of loans to record and evaluate, new criteria against which to write contracts, new information to pass to investors, and new reasons to bundle up loans and use them to back securities.

That is roughly where the green bond market might find itself, a few years down the road. As investors ask for ever-tighter impact reporting, that implies closer and closer definitions of green bond use of proceeds—and a firmer segregation of assets funded by green bonds from general corporate balance sheets.

Blockchain provides a bulwark against double-counting, a secure and enforceable way to track carbon, and, if the market ever gets there, an effective way to price the carbon saved by green bonds in real time.

Start with asset-backed securities, and watch it take hold from there.

20.11 The Magic of Traceability

Blockchain technology, with its power to enforce traceability, has the power to manufacture green-tagged assets from other pools of potentially carbon cutting assets.

The most obvious target is "forest bonds"—finance aimed at stopping deforestation, likely through the REDD+ framework, a UN-led initiative which aims to resolve some of the formidable problems of proof, verification, and monitoring inherent in an initiative which pays countries *not* to do something.

Forest bonds have been on the agenda for a long time—Goldman Sachs, WWF, and the Climate Bonds Initiative ran a roundtable in 2011 looking at their potential—but, so far, they have failed to take off, being partly eclipsed by the emergence and flourishing of the green bond market, which kicked into high gear from 2014, and partly because of a lack of strong institutional sponsorship.

Rich countries that promised to pay toward REDD programes, with some exceptions, failed to honor their commitments and the carbon markets which were supposed to provide private sector demand for carbon credits, more or less failed. Forests, if left alone, do not produce much revenue—hence, the need to pay to protect them—meaning any finance instrument needs donors and development banks in the middle.

Blockchain can help, at least a little. It cannot resolve the high level political failures around REDD+, or around carbon markets. But Blockchain technology excels at helping solve problems of trust, traceability, and contractual enforcement. To the extent that a lack of trust and standardization has stopped payments to avoid deforestation, a forest Blockchain could help solve that. Creating secured and trusted assets, and there are the raw materials for bond investment, which can be transformed into contracts and hence financial assets.

Indeed, in September 2017, Ecosphere + , a consultancy working to put a price on forest services, announced a Blockchain initiative called Poseidon, which it says is "the creation of a carbon currency that will revolutionize the value we value nature, using Blockchain technology to improve access to, and visibility of, the carbon market." It's not a green bond on Blockchain—but it's only a short step away.

20.12 Blockchain in Theory

It is important to remember that Blockchain for finance generally, and bonds in particular, is still a theoretical proposition. Few innovations have attracted so much attention, and so much feverish investment and excitement from financial institutions desperate to trumpet their tech credentials. But the results, so far, are limited, and merit sober consideration.

Only a handful of bonds, at the time of writing, have been issued using Blockchain technology, and were proofs of concept, not real, distributed issues. Companies such as Digital Asset Holdings, run by former JP Morgan executive Blythe Masters and R3, backed by a broad consortium of major banks, provide rock star panelists at industry events, and attract breathless press coverage. But few opportunities exist to apply Blockchain at the sharp end of the bond markets.

Digital Asset Holding is focusing its early efforts on loan settlement and repo (repurchase agreements; essentially borrowing bonds for short periods). Loan settlement suits well because it is so laughably inefficient—settlement cycles might be 40 days—whereas, repo suits because of the instant traceability characteristics of Blockchain. The challenges of the repo market involve traceability and understanding who owns what in real time.

But the ordinary activity of the bond market is already fast, efficient, and, for the most part, highly traceable. Clearing and settlement of bond trading is a tiny fraction of the costs associated with running a bond portfolio or trading bonds—regulation costs, central bank purchases, and the cost of staff are vastly more important.

That is not to say that Blockchain will not eventually arrive. The main settlement companies, the Depository Trust and Clearing Company in the US, and Euroclear and Clearstream in Europe, have Blockchain labs and trials running; their equivalents in smaller markets, such as the ASX in Australia or SIX in Switzerland have projects in the pipeline as well.

But it will not arrive soon, and, to bond people, that does not matter too much. Contracts on Blockchain, issuance, trading, and settlement on Blockchain all sound exciting and promise improvements over existing practice. But the improvements are small, the cost savings to ordinary investors negligible, and the obstacles of coordination and cooperation formidable.

Until the bond market more widely becomes comfortable using Blockchain, and not just for niche, proof-of-concept issues but for mainstream bonds in large size, expect little in terms of Blockchain for green bonds. Investors do not seek out the bond market for its complexity or obscurity. They like regular cashflow and regular structures, and settlement mechanisms that are noted for their reliability rather than their innovation.

Green bonds already demand more from investors than ordinary issues, and the prudent issuer or banker will introduce innovations a step at a time.

If Blockchain enables higher trust, more verifiability, less risk of double counting and firmer carbon commitments, green-minded investors will appreciate the change, whether or not they understand the technology. But the first issuer to introduce this level of rigor should take time to make sure it has the maximum impact.

Similarly, the first issuer to bring Blockchain to other parts of the bond market will need to be ordinary to a fault. It is no accident that the Australian state of Queensland provided one of the Blockchain proof-of-concept issues so far—in the Australian market, Queensland Treasury Corporation is a regular issuer of unimpeachable solidity. There is no point rocking the boat in more than one direction at once.

20.13 Carbon Needs a Price

What will really drive both Blockchain and green bonds to new heights is the same solution that will fuel all manner of other green innovations—a meaningful, tradeable market in carbon. That does not mean handing out carbon permits to every polluter with political connections.

What a carbon price implies is a direct way for the powerful trust and traceability characteristics of Blockchain technology to connect to finance. Once there is a financial incentive to avoid double-counting carbon—but to count the carbon that has been saved carefully—then a sophisticated way to track it becomes vital, and that could plug straight into green bonds, rescuing the market from charges of greenwash, and rendering green bond verification and monitoring essential, rather than an optional extra to flatter the social responsibility pretensions of corporates.

A carbon price and market is a game-changer—and once it arrives, Blockchain and green bonds can make sure the awesome power of the institutional capital markets comes into its own for the planet.

References

ABN Amro. (2015). *ABN Amro investor relations. ABN Amro.* <https://www.abnamro.com/en/sustainable-banking/finance-and-investment/green-bonds/index.html>. Accessed on 30.11.2017.

Climate Bonds Association. (2017). <https://www.climatebonds.net/>. Accessed on 30.11.2017.

Climate Bonds Initiative. (2017). *Climate bonds initiative market history. Climate Bonds Initiative.* <https://www.climatebonds.net/market/history>. Accessed on 30.11.2017.

Davies, T. (2017). *QBE gears up for first insurance green bond. GlobalCapital.* <http://www.globalcapital.com/article/b12mrym76rtj4j/qbe-gears-up-for-first-insurance-green-bond>.

Dealogic. (2016). *Dealogic debt capital markets annual review 2016. http://www.dealogic.com/insight/key-trends-shaped-markets-2016/. Accessed on 30.11.2017.*

European Banking Federation. (2017). *Towards a green finance framework. European Banking Federation.* <https://www.ebf.eu/towards-a-green-finance-framework/>. Accessed on 30.11.2017.

European Commision HLEG on Green Finance. (2017). *European commission high level expert group on green finance interim report. European Commission.* <https://ec.europa.eu/info/sites/info/files/170713-sustainable-finance-report_en.pdf>. Accessed on 30.11.2017.

European Commission. (2016). *Staff impact assessment on energy efficiency directive. European Commission.* <https://ec.europa.eu/energy/sites/ener/files/documents/1_en_impact_assessment_part1_v4_0.pdf>. Accessed on 30.11.2017.

European Investment Bank. (2017). *European investment bank: 10 years of green bonds. European Investment Bank.* <http://www.eib.org/infocentre/ten-years-of-green-bonds.htm>. Accessed on 30.11.2017.

International Capital Market Association. (2017). <https://www.icmagroup.org/Regulatory-Policy-and-Market-Practice/green-social-and-sustainability-bonds/green-bond-principles-gbp/>. Accessed on 30.11.2017.

Lloyds Banking Group Investor Relations. (2014). *Lloyds bank ESG bond. Lloyds Bank Investor Relations.* <http://www.lloydsbankinggroup.com/globalassets/documents/investors/2014/esg_bond_lb_q3_2014_sept2014.pdf>. Accessed on 30.11.2017.

Luxembourg Stock Exchange. (2016). *Luxembourg stock exchange green bond exchange. Luxembourg Stock Exchange.* <https://www.bourse.lu/lgx>. Accessed on 30.11.2017.

McGlashan, C. (2017). *Roundtable: SSAs sowing the seeds of green pricing performance. GlobalCapital.* <http://www.globalcapital.com/article/b14xjyrnt982vf/ssas-sowing-the-seeds-of-green-pricing-performance>.

Packer, T.E. (2017). *Green finance and certification. Bank for International Settlements Quarterly Review.* <https://www.bis.org/publ/qtrpdf/r_qt1709h.htm>. Accessed on 30.11.2017.

People's Bank of China. (2015). <http://www.pbc.gov.cn/goutongjiaoliu/113456/113469/2993398/index.html>. Accessed on 30.11.2017.

Repsol. (2017). *Repsol green bond framework. Repsol Investor Relations.* <https://www.repsol.energy/imagenes/global/es/Repsol_Green_Bond_Framework_tcm13-71041.pdf>. Accessed on 30.11.2017.

Sanderson, O. (2016). *GlobalCapital: How can the worth of green investments be proved? GlobalCapital.* <http://www.globalcapital.com/article/zr086sbskbjq/how-can-the-worth-of-green-investments-be-proved>

Turner, M. (2017). *Repsol oils way for green bond first. GlobalCapital.* <http://www.globalcapital.com/article/b12y0rfsrgq078/repsol-oils-way-for-green-bond-first>.

Wong, L. (2017). *Monetary authority of Singapore: Keynote speech at investment management association of Singapore. Monetary Authority of Singapore.* <http://www.mas.gov.sg/News-and-Publications/Speeches-and-Monetary-Policy-Statements/Speeches/2017/Keynote-Address-at-the-Investment-Management-Association-of-Singapores-20th-Anniversary-Conference.aspx>. Accessed on 30.11.2017.

World Bank Group. (2017). *World Bank: MDBs and green bonds.* <http://treasury.worldbank.org/cmd/htm/Chapter-2-MDBs-and-Green-Bonds.html>. Accessed on 30.11.2017.

Further Reading

World Bank. *(n.d.).* <http://treasury.worldbank.org/cmd/htm/Chapter-2-MDBs-and-Green-Bonds.html>. Accessed on 30.11.2017.

CHAPTER 21

Utilizing Blockchain for Better Enforcement of Green Finance Law and Regulations[1]

Xiaochen Zhang[1], Matias Aranguiz[2], Duoqi Xu[3], Xing Zhang[4] and Xinran Xu[5]

[1]President, FinTech4Good, United States [2]PhD Candidate, Shanghai Jiaotong University, Shanghai Shi, China [3]Professor, Shanghai Jiaotong University, Shanghai Shi, China [4]Operations Officer, IFC, Washington, DC, United States [5]Fellow, FinTech4Good, United States

21.1 Why Blockchain Matters?

Integrated with Smart Contract, Blockchain technology realizes much wider applications through allowing automatic execution of obligations, reduction of costs, increase of security, and improvement of analytics in upscaling climate actions. A multitude of benefits of the use of Blockchain technology has been explained throughout Sections 1–4 of this book. In this chapter, we will review a few major international and domestic regulations addressing challenges of climate change, sustainable development, and in particular, green finance (including those for the banking sector), which sets the scene for green investment. On that basis, we will explore a number of practical applications of the Blockchain technology in the space of green investment. Lastly, the potential areas for further improvement to unleash the potential of Blockchain will be discussed.

Briefly, Blockchain is "a distributed database of records or public ledger of all transactions or digital events that have been executed and shared among participating parties"[2]. As Sebastien Meunier summarized in his chapter entitled "Blockchain 101," Blockchain or Distributed Ledgers Technology (DLT) allows users to record and share a common view of a system's state across a distributed network. This opens up unlimited possibilities for (1) peer-to-peer value transfers; (2) shared trusted registries as immutable source of truth (which guarantees the traceability and auditability of data); and (3) securely

[1] The authors would like to thank Professor Wenjun Chen and Li An from the Shanghai Algorithm Economic Research Center for their great contribution in reviewing the paper.

[2] Crosby, et al (2015) p. 3.

Transforming Climate Finance and Green Investment with Blockchains.
DOI: https://doi.org/10.1016/B978-0-12-814447-3.00021-5

executable agreements through the use of Smart Contracts. Blockchain, therefore, poses a serious challenge to our traditional models of regulatory compliance and legislation.

21.2 Blockchain as an Enabling Instrument for the International Legal Frameworks

International legal frameworks related to environmental and social protection benefit mankind in many aspects, from addressing global climate change to achieving sustainable economic development. In Section 21.2.1, we will review some existing mechanisms that incentivize investments in the environment and climate change actions, and explain why Blockchain can promote and ensure legal protections for stakeholders of green investment.

21.2.1 Implementation of the Paris Agreement—Internationally Transferred Mitigation Outcomes (Article 6)

One of the objectives of the Paris Agreement, which was reached at COP21 in 2015, was to mitigate global greenhouse gas emissions. Initiated under the United Nations Framework Convention on Climate Change (UNFCCC), 170 out of 197 countries have ratified the Paris Agreement as of mid-November, 2017[3]. Article 6.2 of the Paris Agreement promotes a system to allow Parties to work cooperatively (i.e., the Cooperative Approach): In certain circumstances, Parties can use mitigation outcomes transferred internationally toward achieving their nationally determined contributions (NDCs). The use of internationally transferred mitigation outcomes (ITMOs) may involve some forms of international carbon credit trading as in the case of the cean development mechanism.

Against this backdrop, from our viewpoint, one of the most important requirements of the Paris Agreement for the carbon market is "robust accounting." The main difficulties, aside from administering the systems of carbon credit issuance and pricing, in the market mechanism are the validation of information (e.g., compliance with relevant rules or regulations); the linkages among different emissions trading systems (e.g., domestic and international); and building a trusted transaction system (e.g., a carbon inventory and registry).

21.2.2 Implementation of the Paris Agreement—Global Climate Finance Transfers (Article 9)

Article 9 of the Paris Agreement encourages developed countries to collaborate on the mobilization of financial resources to help developing countries cope with climate change. As of today, there are still neither pre-set conditions nor roadmaps for such collaboration among developed countries to this end. Although developed countries have pledged to

[3] United Nations (2016).

mobilize US$ 100 billion a year through the Green Climate Fund set up at COP16 in 2010, it has been widely accepted that the goal of US$ 100 billion per year up to 2020 is far less than the finance needed for developing countries to cope with the increasingly tangible impact of climate change. The World Economic Forum projects that by 2020 about US$ 5.7 trillion will need to be invested annually in green infrastructure alone[4]. To fill such a tremendous gap, innovative technologies including Blockchain is undoubtedly worth exploring.

Throughout Sections 1–4 of the book, cases on how Blockchain technology may be deployed to enhance climate actions have been introduced. The UNFCCC has classified these use cases into four categories:[5] (1) improved carbon emission trading; (2) facilitated clean energy trading; (3) enhanced climate finance flows; and (4) better tracking and reporting of greenhouse gas emissions reduction and avoidance of double counting. Noteworthy is that Blockchain has been officially written into the Co-chair's note at COP23 in Bonn in 2017 during the discussions on potential guidance regarding the cooperative approaches stated in Article 6.2 of the Paris Agreement[6].

21.2.3 Achievement of Sustainable Development Goals

In 2015, the UN launched the Sustainable Development Goals (SDGs), which include 17 goals and 169 targets to be achieved by the year 2030[7]. The release of SDGs helps shape governments' development agendas and hence, private capital into new investment opportunities. However, most of the targets or criteria set forth under the SDGs are confined to the macro level and abstract, which may hinder the deployment of policy actions and their impact evaluation.

To make the high-level SDGs more operational, many Blockchain start-ups aim to address specific challenges around the implementation of SDGs. These new ideas focus on areas such as financial and education access, connectivity, legal identity, protection of natural resources, and social integration. But the key component underlying all these focused areas is the reduction of market entry barriers (including costs and time involved) with increased security and transparency of many big data sets.

4 World Resources Institute (2017).

5 United Nations Framework Convention on Climate Change (2017).

6 UNFCCC (2017).

7 No Poverty; Zero Hunger; Good Health and Well-being; Quality Education; Gender Equality; Clean Water and Sanitation; Affordable and Clean Energy; Decent Work and Economic Growth; Industry, Innovation, and Infrastructure; Reduced Inequality; Sustainable Cities and Communities; Responsible Consumption and Production; Climate Action; Life Below Water; Life on Land; Peace and Justice Strong Institutions; Partnerships to achieve the Goal.

Blockchain innovations featuring such increased data security and transparency are set to facilitate many difficult aspects of (international) law enforcement, which will, in turn, support international climate finance transfers. Some examples are:

- "ID2020"—A public-private partnership aiming to assign unique digital identities to people without any legal identity recognition. Target 16.9 of the SDGs is to provide legal identity to all people in the world, including birth registration, which should be achieved by 2030. Today, there are by estimate 1.1 billion people without official recognition of identity,[8] most of which are in developing countries. The undocumented population is more vulnerable to climatic shocks as they are mostly unable to receive any aid internationally and support themselves through undertaking proper job roles by law.
- Bitpesa[9], Everex[10], and Humaniq[11]—Companies providing unbanked households access to banking facilities and wider financial infrastructure for cross-border transactions and payments. Their services are faster and cheaper than traditional banking services. Their target market is people from lower socio-economic classes, especially in the least developed countries, who can benefit from swift and cheap transactions.
- IXO Foundation[12]—IXO Foundation created the "ixo protocol"—technology that harnesses Blockchain and next generation web standards to collect and validate high quality data and create "proof of impact" to optimize sustainable development impact. Proof is generated when an impact claim is made by a service provider and is verified by an evaluator or an oracle that the impact has been achieved. As a result, the risks of fraud and corruption in program delivery processes are reduced. Both the service provider and the funder can gain access to invaluable data to evaluate the authentic impact of their programs.

These Blockchain-based innovations are laying some solid foundations to unlock both public and private green financing to unserved people and underserved areas (in the least developed countries), which are the blind spots to the traditional centralized financial system.

21.2.4 Execution of Results-Based Financing by Multilateral Development Finance Institutions

As promising instruments for enabling green actions financing, Blockchain applications are envisioned to be widely adopted across the international climate finance architecture,

[8] ID2020 (2017).
[9] Bitpesa (2017).
[10] Everex (2017).
[11] Humaniq (2017).
[12] Ixo Foundation (2017).

especially multilateral development finance institutions, in the spirit of results-based financing as a "contractual requirement."

In 1944, the International Bank for Reconstruction and Development (i.e., the World Bank) was founded. It was the first institution to take an international approach to development through finance. The International Bank for Reconstruction and Development started with a specific goal, which is the reconstruction of Europe after the Second World War. But its objective has been extended into stimulating economic development and eliminating poverty since then.

Many development finance institutions have followed with similar objectives. Most of the new institutions have a regional scope but follow the same logic, including the Inter-American Development Bank (est. 1959), Africa Development Bank (1964), Asian Development Bank (1966), Caribbean Development Bank (1970), CAF-Development Bank of Latin America (1970), European Bank for Reconstruction and Development (1991), and New Development Bank (2012); and lately the Bank of the South (2009) and the Asian Infrastructure Investment Bank (2015). Many of those institutions have already recognized the principles underlying the Sustainable Development Goals and the Paris Agreement not only as a programatic goal but also as a methodology for planning or approving development projects.

As the most recent example, the newest Asian Infrastructure Investment Bank, at its second meeting of the Board of Governors held in June 2017, addressed several issues relevant to SDGs implementation. The President of the Bank stressed the idea about their roles is to facilitate the implementation of the Paris Agreement and the 2030 Agenda for Sustainable Development[13].

Since 2016, many multilateral development banks (MDBs) have coordinated their efforts into climate *mitigation* or *adaptation* investment[14]:

- Mitigation investment: MDBs committed a total of US$ 21.2 billion to mitigation investment which is defined as activities or investments that reduce or limit greenhouse gases.

  This is a very promising area in which Blockchain technology can play a critical role. For example, Blockchain-enabled initiatives such as DAO IPCI (introduced in Chapter 10: Blockchain—Powering and Empowering the Poor in Developing Countries), which tailors an Ethereum-based public Blockchain protocol (a set of digital smart-contracts and modules) to help eliminate the barriers limiting the growth of green finance markets by addressing their lack of transparency, trust, and unification, will

[13] Jungcurt (2017).
[14] Inter-American Development Bank (2016) p. 6.

potentially dramatically improve both efficiency and effectiveness of MDBs' mitigation investment.

- Adaptation investment: MDBs committed a total of US$ 6.2 billion to adaptation investment, which is defined as a process to reduce risks and vulnerabilities of communities to the impacts of climate change.

  Adaptation differs from mitigation in a way that it is often about factoring projected climate change impact (e.g., higher flood or drought risk and natural disaster preparedness) in the planning of infrastructure. Infrastructure is seldom a target of private investors and hence, reliant on public (aid) funding. To ensure the accountability of adaptation finance, the public sector, MDBs, or international climate funds, from which the money is drawn, increasingly adopt the principle of results-based finance (i.e., payments by results) for many climate change projects. Blockchain presents itself as a powerful instrument for tracking elements or data which help donor institutions evaluate the impact of the adaptation measures they funded and activate transfers of financial resources through smart contracts. For example, the potential application of IXO Protocol in adaptation program evaluation could increase public confidence toward the credibility of infrastructure planners and hence, incentivize more international or even private funding for adaptation initiatives. More adaptation programs at multiple levels mean more accumulation of precise climate adaptation project data, which improves adaptation policy-making for many developing countries.

21.3 Blockchain in Enforcing Domestic Green Finance Law: China's Experience

Over the past two decades, climate change has been escalated from being a sectoral issue to a top priority for China's national development strategy, including its "Five Year Plans." China pledged that its carbon emissions would peak by 2030, or earlier if possible. To achieve this, China must reduce its carbon intensity by about 3.6%–4.1% while add an annual capacity equivalent to the entire US power grid, all with nonfossil-fuel energy sources. It has been estimated that an additional US$600 billion each year of green finance is needed to achieve China's climate ambitions. Owing to the country's fiscal constraints, more than 85%of that amount will have to come directly from the private sector[15].

The interaction between green finance law and Blockchain innovations in China, specifically how Blockchain will facilitate the enforcement of China's green finance regulations, can be approached from the three pillars of Chinese green finance law: (1) establishing the green financial system; (2) integrating environmental risks consideration in investment (e.g., through green bonds); and (3) piloting emissions trading system.

[15] UNEP (2016).

21.3.1 Establishing the Green Financial System

In greening the financial system, China has made great efforts in introducing new policies, regulations, standards and market instruments over the past 5 years[16]. In 2014, the Green Finance Task Force, convened by the People's Bank of China, presented an ambitious framework of recommendations to establish the green financial system in China.

The recommendations cover broad areas such as specialized investment vehicles, fiscal and financial support measures, and financial and legal infrastructure. To advance the legal infrastructure to scale up green finance, the task force recommended the Central Government include provisions to advance lenders' liabilities to environmental sustainability, as well as mandate the purchase of environmental liability insurance and disclosure of environmental information[17].

Based on lessons learnt and experience gained from 2014 to 2016, the People's Bank of China, in coordination with six other financial and regulatory authorities, issued the "Guidelines for Establishing the Green Financial System" in 2016[18]. Under the guidelines, five provinces (including Guangdong, Guizhou, Jiangxi, and Zhejiang provinces and Xinjiang Uygur Autonomous Region)[19] were commanded to pilot "green financial system" and "innovations" locally from 2017 onwards.

Information Box 1: The adoption of Blockchain in China's financial system

A key component of establishing the green financial system in China is to advance technology use in the country's domestic financial services industry. After successfully completing a Blockchain trial run together with the Industrial and Commercial Bank of China (ICBC), Bank of China and WeBank in 2016, China's Central Bank (PBOC) made the development of new digital technologies such as Blockchain and artificial intelligence (AI) a priority in China's 13th Five-Year Plan for the Finance Sector. PBOC also established a "digital currency research institute" to prepare for a potential digital currency pilot. ICBC has not only piloted a Blockchain-based Financial Product Trading Platform but also employed Blockchain technology to support RMB300 billion worth of government grant deployment in Guizhou province. China Construction Bank is developing a distributed ledger system to sell third-party insurance products. Shanghai Stock Exchange, Shenzhen Stock Exchange, Shanghai Insurance Exchange, and many other major financial institutions are racing to research on use cases, develop proofs of concept and pilot various Blockchain-based financial transaction systems to better meet the requirements of financial regulations around lenders' environmental liability at much lower costs.

(*Continued*)

16 Arkin (2017).
17 UNEP (2015).
18 People's Bank of China (2016).
19 Jiangxi Government (2017).

Information Box 1: The adoption of Blockchain in China's financial system (Continued)

In addition to Blockchain's potential benefits in greening the existing financial instruments, the technology also offers the opportunity to create new green financial instruments in mobilizing resources for China's green transformation. Thanks to the technologies developed by the Greeneum Network, China is creating a decentralized sustainable energy market that enables smart monetization for real-time energy transactions. Through the launching of a Green Digital Exchange, the country can harness the power of hundreds of climate and SDGs-related coins and mobilize billions of additional private capital for green finance.

Although many of the current efforts were originated from a nongreen related purpose, the new system capacity developed and products tested are preparing the country for rolling out Blockchain technology to green China's financial system.

21.3.2 Integrating Environmental Risks in Assessing Investment Proposals

In China, one of the key barriers for businesses to invest in green projects or adopt green standards lies in their awareness. When environmental risks are not being considered in investment decision-making processes, or the carbon price has not been included in modeling operational costs, businesses will have neither incentives nor tools to green their operations.

To help address this challenge, the Ministry of Environmental Protection (MEP) and the China Securities Regulatory Commission (CSRC) signed an agreement to require mandatory disclosure of environmental information by listed companies[20]. The MEP also announced the "Guidelines for Environmental Protection in Foreign Investment and Cooperation" to green foreign investment and international cooperation initiatives. While the China Banking Regulatory Commission has integrated sustainability risk considerations into the regular bank supervision process, the CSRC has also publicly encouraged Chinese investors to follow the Principles of Responsible Investment (PRI).

Another barrier to scaling up green finance is monitoring and evaluating the effectiveness of each instrument. Without clear guidance, comprehensive indicators, and solid data, it is difficult to assess the effectiveness of green finance instruments.

Take green bond as an example. Green bonds, a debt instrument designed for financing green projects, are the most advanced financial instrument for investors to engage in climate finance[21]. The total cumulative volume of climate bonds issued globally was estimated at US$ 160 billion as of 2016, of which US$ 70 billion was issued in 2016 alone.

[20] Ministry of Environmental Protection (2017).
[21] World Bank Group (2017).

As more green bonds help mobilize private sector finance for clean development, sustainable investing is expected to become the standard of managing fixed-income portfolios among institutional investors. More and more portfolio managers would value long-term profits from sustainable growth rather than solely short-term gains.

Following the issuance of a domestic "green bond catalogue" and "green bond guidelines," China's green bond issuance reached US$34 billion in 2016, up from US$1 billion in 2015[22]. To further scale up green bonds, the CSRC released specific guidelines for green bonds issued by listed companies in March 2017. Though China's green bonds market has experienced a dramatic increase in volume, it accounts for less than 2% of all bonds in China. Green bond investors are still facing challenges in assessing the effectiveness of green bond projects. If investors lack the historic data and credit ratings of the newly issued green bonds, in the long run, they might tend to invest more on traditional bonds.

The use of Blockchain in green bond is still at the initial stage in China. FinTech4Good[23] is working with a few key partners to launch a Blockchain Bond Initiative to facilitate the adoption of Blockchain in bonds markets. The potential applicability of Blockchain in this respect is outlined in the information box below.

Information Box 2: Scaling up green bond issuance with Blockchain

Green bond is a promising instrument, however with challenges. The following questions need to be considered for its upscaling: (1) Are investors willing to accept lower yield of green bonds in exchange for demonstrable environmental benefits? (2) Do new buyers want to join the "usual suspects" in green bond market to secure and sustain a green bond premium? (3) Are there technologies available for cost-effective standardization, verification, certification, and reporting to ensure market integrity and an opportunity for investors to perform their own due diligence? (4) Are there technologies available to prove and disclose additionality without increasing transaction costs? (5) Are there effective methods to raise awareness and improve professional education in investment community on green bonds?

Blockchain technology is capable of addressing some of the above questions for scaling up green bonds. First of all, Blockchain has the potential to add greater transparency and security to the whole transaction process. While the immutability of transaction data stored on a Blockchain guarantees its security, Blockchain can also ensure the transparency of green bond proceedings being invested in qualified environmentally friendly projects by tracing the movement of every dollar along the value chain; and, in turn, publicizing the authentic environmental impact of the green bond for investors. Thus Blockchain technology could redefine the green bond certification process.

(*Continued*)

[22] United Nations Environment (2017).

[23] https://www.fintech4good.co/

Information Box 2: Scaling up green bond issuance with Blockchain (Continued)

Use case: DisLedger is a Blockchain-enabled system designed for green bonds. It is a fast, scalable, distributed ledger that can improve transaction efficiency and attain cost reduction by handling hundreds of thousands of security transactions per second with complete privacy. It provides the definitive, final settlement in milliseconds, which frees up cash deposits and collateral, and simplifies ownership recording and coupon payment processing. DisLedger is environmentally friendly in that it requires small amount of computing power, thus consumes minimal electricity in contrast to the case of Bitcoin Blockchain. In addition, not only can the use of smart contracts increase efficiency through automated execution when certain conditions are met, but also can it increase accuracy by avoiding human errors in documentation.

21.3.3 Emissions Trading as a Market-Based Instrument

In 2013, China introduced a pilot emissions cap-and-trade program which covers seven major cities and provinces—which was the world's second largest after the European Union's. In just 5 years, China's pilot carbon market program has become an important market mechanism to drive financial innovations and cross-sector investment collaboration[24]. Built on the experience gained from the pilot program, in late 2017, the country launched the nation-wide emissions trading system to accelerate the deployment of clean technologies and raising of capitals to invest in low-carbon transformation at a national level.

As more and more sectors are going to be covered in the national emissions trading market (which starts with the power sector), the private sector needs to get prepared for internalizing carbon price in its operations in order to maintain its market competitiveness. As the largest carbon market in the world, the market operator will be confronted by challenges related to the administration cost, as well as the security and stability of the transaction system. Moreover, whenever projects supplying emissions off-setting credits are admitted into the trading system, full project cycles associated with project origination, monitoring, and evaluation will add further complexity to the system, which would make compliance with emissions trading regulations increasingly difficult. In light of this conundrum, since September 2017, China has been piloting the use of a Blockchain-based carbon management platform in partnerships with IBM China and the Energy Blockchain Lab with a view to streamlining the system's operation. Further research about the effectiveness of this Blockchain-based carbon management platform in facilitating regulatory compliance is necessary after the pilot period lapses.

[24] BSR (2015).

21.3.4 Unleashing the Potential of Blockchain With a Better Regulatory Environment

Regulations governing Blockchain technologies worldwide can be described as excitement, suspicion, or indifference. They are highly dependent on the attitude of a particular government toward this disruptive technology. In the United States, the federal government has not exercised any power to regulate Blockchain technology but intend to let state government introduce their own rules and regulations. According to the *Brave New Coin*, in 2017, at least eight US states have worked on bills accepting or promoting the use of Bitcoin and the underlying Blockchain technology, while a couple of them such as in Arizona, Vermont, Illinois, and Delaware have already passed them into law[25]. In Europe, the EU is embracing Blockchain technology with warm welcome. Brussels has revealed that it is working on bills related to Blockchain to support Union-wide distributed ledger-based projects. Switzerland has an enabling environment for Blockchain development and adoption, where a Swiss nonprofit, Crypto Valley Association (CVA), is developing an "ICO Code of Conduct" in hopes that it will be integrated in future financial regulations and clear foreseeable regulatory hurdles.

In order to further unleash the potential of Blockchain in mobilizing and leveraging green finance, more enabling legal environment is essential. When it comes to legal or regulatory reform to suit the 'Blockchain era', there are at least two points regulators need to consider:

- For certain Blockchain applications, particularly those involving legacy from traditional banking operating systems, a clear set of industry standards will have to be developed. It could be particularly challenging for governments to regulate open, permissionless distributed ledger systems because no legal entity can be identified as the owner of the system. It is, therefore, necessary to develop legally binding protocols to vet the creators of blocks of transaction data on a Blockchain network.
- Concerns are raised on cross-border transactions with regard to the jurisdictions of the underlying transaction data and information. International finance law should be revisited to fit the Blockchain technology into the requirement of Know-Your-Customer (KYC) and Customer Due Diligence (CDD) in order for stakeholders to constantly comply with the existing antimoney laundering and combating the financing of terrorism standards.

To expedite global efforts toward the "Blockchainization" of the greening of the international financial systems, governments should first enhance their institutional capacity to receive the pros and cons of Blockchain technology. Capacity building shall be provided to finance sector practitioners and regulators. Like all the other emerging technologies, the key to successful adoption of Blockchain to fuel the growth of green finance markets never goes far from building up credible multistakeholder partnerships among governments, finance sector practitioners, and Blockchain innovators to take stock of lessons learnt so far and examine the opportunities and limitations ahead.

[25] Brave New Coin (2017).

21.4 Conclusion

To facilitate the compliance and enforcement of increasing number of climate change law and green finance regulations internationally and domestically, the world needs a technological breakthrough that brings stakeholders low-cost yet high-speed instrument to achieve a low-carbon economy. Blockchain and other emerging digital technologies such as the IoT have appeared at this precise moment to be able to solve many challenges facing the implementation of the Paris Agreement and the achievement of SDGs, etc. The challenges are huge, but Blockchain gives the society an opportunity to develop new approaches in the interest of environmental sustainability for our future generations.

In other chapters of this book, you have already learnt some examples of how Blockchain technology may be deployed in addressing various challenges that have been hindering climate finance and green investment. The case of China's green finance law being better enforced through Blockchain-based innovations, outlined in this chapter, sets an example for many countries considering alleviating the burden of regulatory compliance for stakeholders. To fill the green finance gap of US$600 billion per year, green financing definitely needs to be more efficient and effective. The emergence of Blockchain technology will help address the problems of high transaction costs, low efficiency, high security risks, and low accountability, which have been some of the major barriers for green investment.

In the space of green investment, reforms of relevant regulatory frameworks should go ahead to unleash the full potential of Blockchain technology for the markets. Awareness of governments needs to be increased on the workings of Blockchain technology. It may not be silver bullet to all the challenges being faced today, but governments shall keep an open mind toward Blockchain and other fintech that align innovative technological solutions with the wider sustainable finance agenda. With appropriate legal frameworks in place, Blockchain innovations will certainly assist the governments in mobilizing tremendous green finance that is imperative to combat climate change and bring a sustainable living environment for the people.

References

Arkin, Fatima. (2017). *Through China, a green financing leader emerges*. Retrieved from https://www.devex.com/news/through-china-a-green-financing-leader-emerges-90829.

Bitpesa. (2017). *Home*. Retrieved from https://www.bitpesa.co/.

Brave New Coin. (2017). *US States working on blockchain legislation in 2017*. Retrieved from https://bravenewcoin.com/news/us-states-working-on-blockchain-legislation-in-2017/.

BSR. (2015) China insight. Briefing Retrieved from https://www.theclimategroup.org/sites/default/files/archive/files/China-Insight-briefing.pdf.

Crosby, Michael, et al. (2015). *Blockchain technology, beyond bitcoin*. Sutardja Center for Entrepreneurship & Technology, Berkeley Engineering.

Everex. (2017). Home. Retrieved from https://www.everex.io/.

Humaniq. (2017). *Home*. Retrieved from https://humaniq.com/.

ID2020. (2017). *Home*. Retrieved from http://id2020.org/.

Jiangxi Government. (2017). *China to set up pilot zones for green finance*. Retrieved from http://english.jiangxi.gov.cn/business/investmentenvironment/201706/t20170615_1341323.htm.

Jungcurt, Stephan. (2017). *AIIB to step up support for renewable energy, sustainable infrastructure development*. SDG Knowledge hub.

Ministry of Environmental Protection. (2017). *Authorities promote mandatory environmental disclosure as listed companies fail test*. Retrieved from http://english.mep.gov.cn/News_service/media_news/201706/t20170614_415970.shtml.

People's Bank of China. (2016). *The People's Bank of China and six other agencies jointly issue "Guidelines for Establishing the Green Financial System"*. Retrieved from http://www.pbc.gov.cn/english/130721/3131759/index.html.

UNEP. (2015). Establishing China's green financial system. Retrieved from http://unepinquiry.org/publication/establishing-chinas-green-financial-system/.

UNEP. (2016). *G20 financial leaders commit to exploring green finance options*. Retrieved from http://www.unep.org/northamerica/news/2016/g20-financial-leaders-commit-exploring-green-finance-options.

United Nations. (2016). *Paris Agreement – Status of ratification*. Retrieved from http://unfccc.int/paris_agreement/items/9444.php.

United Nations Environment. (2017). *Green finance progress*. Retrieved from Report http://unepinquiry.org/wp-content/uploads/2017/07/Green_Finance_Progress_Report_2017.pdf.

United Nations Framework Convention on Climate Change. (2017). *Blockchain technology can boost climate action*. Retrieved from http://newsroom.unfccc.int/climate-action/how-blockchain-technology-could-boost-climate-action/.

World Bank Group. (2017). *Russia green finance for climate and environmental protection*. Retrieved from https://www.mnr.gov.ru/upload/files/docs/1_vs_bank.pdf.

World Resources Institute. (2017). *Climate finance*. Retrieved from http://www.wri.org/our-work/project/climate-finance.

Further Reading

Barbour, Emily. (2010). International agreements on climate change: Selected legal questions. *Congressional Research Service*. R41175.

Deloitte. (2017). *Key characteristics of Blockchain*. Retrieved from https://www2.deloitte.com/content/dam/Deloitte/in/Documents/industries/in-convergence-blockchain-key-characteristics-noexp.pdf.

Green Climate Fund. (2017). *About the fund*. Retrieved from http://www.greenclimate.fund/who-we-are/about-the-fund.

Interamerican Development Bank, World Bank, et al. (2016). Joint report on multilateral development banks'. *Climate finance*.

Ixo Fundation. (2017). *Home*. Retrieved from http://ixo.foundation/.

United Nations Climate Change. (2017). *Guidance in cooperative approaches referred to in Article 6, paragraph 2, of the Paris Agreement*. Retrieved from http://unfccc.int/files/meetings/bonn_nov_2017/in-session/application/pdf/sbsta47_11a_third_informal_note_.pdf.

CHAPTER 22

Blockchain and Smart Contracts: Complementing Climate Finance, Legislative Frameworks, and Renewable Energy Projects

James Duchenne

Managing Partner of blockchain advisory firm, Sutton Stone, Houston, United States; Honorary Representative for The USA of the Board of Investment, Republic of Mauritius; Chief Operating Officer of Crowd Machine, San Jose, California; Legal practitioner of the High Court of Australia

> *"Blockchain could contribute to greater stakeholder involvement, transparency and engagement and help bring trust and further innovative solutions in the fight against Climate Change."*
>
> **– Alexandre G. Paris, UNFCCC Secretariat, Paris Agreement, in UNFCCC Article (2017).**

22.1 Introduction

The word "Blockchain," and more recently "smart contract," can be overhyped buzzwords (Irrera & Caspani, 2017) used to describe the impending transformation of entire industry verticals, or to some, the doom of those that do not pay attention to the disruption it may bring.

This often leads new Blockchain converts to equate the technology with a messianic solution to most problems and great wealth. In reality, it exposes a hard truth: Most feel that Blockchain is a big deal though beyond that, it still feels a little like magic. Without a well-rounded understanding of the subject, properly executing Blockchain-related projects can be tricky.

We have seen a move by governments and respected businesses to participate in investigatory studies to wrap their heads around this phenomenon, and the likes of the United Nations Framework Convention on Climate Change (UNFCCC) looking at how

Transforming Climate Finance and Green Investment with Blockchains.
DOI: https://doi.org/10.1016/B978-0-12-814447-3.00022-7

Blockchain should be leveraged to remove friction in the fight against climate change. This cautious approach is wise since rushing to satisfy a "fear of missing out" has led to costly lessons in this nascent ecosystem.

Blockchain is not business as usual—not by a long shot.

Thus, the business and legal implications of introducing Blockchain technology to underpin climate finance interactions requires the setting of a clear context. Further, by describing a landscape, rather than the specificity of individual elements therein, could spur novel solutions that have not been thought of yet. Otherwise, this would be like trying to describe Facebook in the early 1990s.

It is only then that pertinent information regarding peer-to-peer transactions can be offered, including the evolution of renewable energy trading, improvement of the system of carbon asset transactions, the tracking and reporting of emission reductions, as well as double counting issues. Especially as efforts to reduce carbon emissions are long-term goals that draw from the complex participation of numerous actors.

22.1.1 A New Paradigm

In 2013, Bill Gates mentioned that Bitcoin (the first commercial application of a Blockchain) *"is a technological tour de force"* (Chen, 2014). Focusing solely on the technological aspects of the Blockchain misses the point that it engenders highly disruptive transformations.

In the coming years, Blockchain will likely change how we communicate and transfer value in almost all industries. More than a business model evolution and the legal framework supporting the same, this means social re-engineering on a grand scale. This could be the opportunity that stakeholders were looking for to improve dealings between parties in the climate industry.

22.1.1.1 A Bird's Eye View of Blockchain

There is nothing exciting about describing Blockchain[1]. It is only a shared database, distributed to a number of participants and kept synchronized by an agreed-upon set of codified rules. For the layman, this mouthful is of little or no interest.

Like peeling layers off an onion, that bland façade is quickly replaced by the discovery of a new world, a leap where technology intertwines with human productivity like little else before. This is because the representation of value and its transfer can now be independent of a hierarchical central third party or even personhood. Consider that an autonomous agent today can now earn money and spend it on its own.

[1] The word "Blockchain" refers to an accounting ledger made up of pages (or blocks) of transactions linked (or chained) to each other.

22.1.1.1.1 A Brief, but Relevant History

As commercial databases became available in the 1960s, the information was stored in siloes and controlled by a centralized party (Danielsen, 1998) (under a hierarchical structure). With an increase in demand for the sharing of information across organizations, this model evolved to provide for permitted access to third parties[2]. This then spun into decentralized and distributed structures of storing and serving information (e.g., bitTorrent).

However, the reliability of the information was still under the purview of centralized actors. To ensure the sanctity of the shared information, the use of various auditors, clearing houses, document control specialists, firewalls, and the like were required. The result (although much better than the then alternative slower means of communication) was significant friction, checkpoints, time delays, costs, and more importantly, the introduction of single points of failure that could compromise the security of the information.

Further, as financial settlement was still done by third party institutions, the communication of information was decoupled from the passing of consideration. In complex structures, such as in climate finance, this led to political and vested interest interference and eventually, systemic risk in achieving the stated aim to combat climate change. As a result, a complex web of agreements, legislative provisions, and other rules were required to keep parties honest and in line. A by-product was to keep small players from meaningful active participation.

Blockchain technology has reinvented the interoperability of databases. It has pushed the verification, reliability, interactions, and security of data to a number of actors in a distributed network. In addition, by agreeing to the same set of rules and incentives structures, this also effectively created the possibility for data immutability, digital scarcity, and a solution to the double spend problem (or duplication of digital data)[3].

The first application of the Blockchain was, unsurprisingly, for use as a currency. Satoshi Nakamoto named this currency, Bitcoin, in the white paper: "Bitcoin: A Peer-to-Peer Electronic Cash System" (Nakamato, 2008). The world changed the day that Bitcoin achieved price discovery [a pizza was sold for Bitcoin in 2010 (Hinchlifee, 2017)] and the free market determined what value to associate to it without the involvement of financial institutions or governments. The incentives for actors to participate achieved a peculiar balance and a new digital economy was born around Bitcoin. Over time, the Blockchain on which Bitcoin ran became resilient (or antifragile).

---

[2] For example, Application Programming Interface (APIs) allows third party sites that have the right credentials to query databases and information from another.

[3] The double spend problem has been a computer science problem since the 1970s and called the Byzantine general's problem or the two generals problem. For more on this (accessed on 30 July 2017 at): https://en.wikipedia.org/wiki/Two_Generals%27_Problem.

Today, many variants of achieving consensus via Blockchain technology have emerged. However, a distinctive split exists between the public networks-type Blockchain (like Bitcoin and Ethereum), and permissioned or private Blockchains (such as Hyperledger, Corda, and Quorum[4]) favored by large organizations[5]. An analogy is the difference between an intranet and the Internet.

It should be noted that Ethereum has catered for the permissioned Blockchain space. This was done under the Enterprise Ethereum Alliance (or EEA[6]), and counts Fortune 500 companies as part of that alliance.

22.1.1.2 Relevant Features of Blockchain Applicable to Climate Finance

22.1.1.2.1 Designing the Right Incentive Model

A misunderstood aspect of Blockchain technology is how it uses the concepts of game mechanics and behavioral influence. This requires a deep understanding of how humans interact with machines in a carefully choreographed symbiosis. These incentives must reward honest behaviors and prevent malicious ones. Without this feature, at least in public Blockchains (and to a lesser degree, for permissioned Blockchains), it could affect the proper operation of the consensus algorithms and collapse the effectiveness of Blockchains[7].

However, there are many levers baked into Blockchain software that relevant actors run within a network, such as to dynamically adjust incentives to acceptable levels. Today, these levers are responsible for securing approximately US$42 billion of value in the Bitcoin network, while Ethereum secures approximately US$19 billion of value[8]. In private Blockchains, this is leading to the reinventing of the rails of settlement. For example, the Linq platform (from Nasdaq) could be used to record, audit, and settle significantly more value (Nordrum, 2017).

While it is obvious that the right incentives for actors must exist in a commercial transaction, in Blockchain, this must be baked into the software by carefully crafting what

[4] For an appreciation of the growth of Blockchain, a list of tokens secured by same can be found at Coinmarket Cap. Accessed on 20 July 2017: https://coinmarketcap.com/.

[5] Public Blockchains have significant mindshare, expertise, and innovation potential that are under public scrutiny as they benefit from open-sourced networks and voluntary participation on a global scale. Permissioned Blockchains are more like walled gardens for particular applications, generally under a consortium business models.

[6] For more information on Enterprise Ethereum Alliance (accessed 30 July 2017 at): https://entethalliance.org/.

[7] In public Blockchains, actors that maintain and audit transactions (typically called "miners"), receive rewards in exchange for their work or stake in sustaining a healthy network. For permissioned Blockchain the benefit of banding together with competitors or other organizations, in terms of costs savings, faster transactions and the like, far outweigh the risk of not doing so.

[8] Consult: https://coinmarketcap.com/.

is to be included in the software's protocol. This is no simple feat, especially for the streamlining of transactions toward a greener future that involves several parties with specific interests.

For example, we can now imagine a global Blockchain network for versioning control and approvals of local laws linked to international agreements. This can also be used to give access to third parties, from finance and administration to tracking and enforcement, which can only act in relation to codified rules. Using the incentives against the gaming of the system is critical.

22.1.1.2.2 Cryptography

Our current financial system operates on the concept of permission and personhood transactions. In other words, to deal with financial transactions, the identity of a person must be known and ticked against certain criteria. In this process, a financial institution can proceed to issue an account number by cross-checking its uniqueness with the current account database to prevent account collisions. Further, transactions can be reversed, blocked or otherwise dealt with. The downside is that bottlenecks can exist in a particular financial institution leading to abuse of the system.

A Blockchain-based system relies on cryptographic key pairs. This is interesting since "accounts" can be generated independently without registration on the Blockchain network (or under anyone's control). This is like picking an atom out of all the atoms in existence in the known universe and using one as a unique account. The likelihood of collision is virtually zero. The account is only registered on the Blockchain once it transacts on the network.

This is important because it is possible for a decentralized autonomous agent to automatically fund leading efforts on climate change based on a set of agreed rules and incentives. Such decentralization may not sit well with the way things are presently done, but it opens the door for participation of "the crowd" under the weight of their own efforts, and independently of any traditional state-based institutions or central authorities. This may mean a significant increase in considerations given to international cooperation at the grass roots level.

22.1.1.2.3 The Importance of Tokens

Bitcoin is a token, and so are Ether, Dash, Litecoin, and a thousand more. Some have value while some are representations of activities and unique fingerprints. However, given that the Blockchain records transactions from the first block, it should not be used to store large amounts of data, such as files, videos, music, and photos alike. These are stored in the Cloud and that the unique fingerprint of data is anchored into the Blockchain and tracked by tokens.

Any representation of activities within the climate finance field must, therefore, take note of this attribute of tokens to design an optimum system. For example, thanks to the immutable nature of the records, a token could represent notarized data [such as the origination of power, renewable energy certificates (RECs), and carbon credits] linked to an underlying financial transaction at a given time. The conditions attached to these tokens can then be self-enforceable.

22.1.1.2.4 Distributing the Role of the Third Party

By removing the friction for the interoperability of databases and its verification with the relevant consensus rules, it means that the need of a trusted third party is removed. Rather, that trust is distributed to all relevant participants. In fact, it is not that Blockchain is a trustless system or a trust in mathematics, it is all the above and the trust in the honest operation of those that secure and maintain the network.

A permissioned system may be more relevant for regional governmental or highly regulated actors since the trust in an open network that they do not control may be too risky. Thus, a decision on a suitable model is essential.

Here is a simple SWOT (Strength-Weakness-Opportunity-Threat) analysis for a suitable solution in the fight against climate change.

Permissioned (Private) Blockchains

Strengths	*Weaknesses*
1. Coming together of vetted actors under one umbrella with more control and *proof of authority* 2. Possibility of reversal of transactions when needed 3. Smaller pool of trusted actors to agree on change of software (consensus)	1. Consensus rules being changeable, threatening the immutability of the data where the pool of participants is small
Opportunities	*Threats*
1. Global unified system with the potential to remove friction and accelerate activities toward achieving climate change goals 2. The creation of regional Blockchains linked together to form a global climate Blockchain as a middle ground solution	1. Reintroducing single points of failure that may compromise the underlying data 2. An improper design may lead to favoritism and the undermining of the consensus algorithms

Public Blockchains

Strengths	*Weaknesses*
1. Leveraging the power of the crowd globally and aligning with the common value 2. The ability to have immutable data by distributing the consensus algorithms on a larger scale	1. At the mercy of potentially unknown actors 2. Changes in consensus not possible depending on the architecture of the system 3. The need to convince participants to participate

(*Continued*)

(Continued)

Opportunities	*Threats*
1. Global unified system with the potential to remove friction and accelerate activities toward achieving climate change goals 2. Linking regional Blockchains together to form a global climate Blockchain 3. Access to productivity and funding not previously available	1. A breakdown of the consensus algorithm or other compromise possibly leading to chaos 2. Blockchain technology still in an experimental phase and being premature to push control out to the crowd 3. The appetite for organizations to have their transactions recorded in a transparent public database probably resulting in unintended competitive advantage (it does not provide full anonymity although transactions are pseudonymous)

22.1.1.2.5 A Technical Challenge—Forks

Recently, a number of news articles have addressed the potential fork of the Bitcoin network in August 2017 (Condliffe, 2017). Since the Blockchain relies on every node running the same software, upgrades to the Blockchain are required to have the approval of all those using the software, failing which may cause a mismatch in case the upgrade is not compatible with earlier versions.

Where a nonbackwards-compatible upgrade is proposed without consensus of all network participants, the Blockchain may record and use that information differently, and this can lead to a split of the Blockchain, generally termed a "fork."

In the field of climate finance or the fight against climate change, the study of the effects of forks and the impact of consensus change must, therefore, be very well understood. Moreover, this understanding must trickle down to actors in the current industry that may take a long time to come to terms with this new paradigm. It is important that Blockchain does not introduce further risks that can destabilize any global initiative and capital markets.

22.1.1.2.6 Smart Contracts

A smart contract allows a piece of code to execute given particular conditions that need automated verifications for a particular purpose. On the Ethereum network, each smart contract running a computer cycle is required to pay a small proportion of Ether, the native currency of the Ethereum platform. This is so as not to flood the Blockchain with spam transactions. As more applications are built, this leads to more use of the Ether token, which should then appreciate over time.

Smart contracts have led to a new way of funding ventures within the Blockchain ecosystem. This is generally termed ICOs (initial coin offerings) or ITOs (initial token offerings). ICOs have been made more relevant with smart contracts as they can be used to generate a unit of account.

Further, they have been shown to leverage wealth made in the Blockchain ecosystem to fuel new ventures or ideas. In the climate finance field, this could lead to new business models to fund renewable energy projects by issuing tokens to record, track, and settle trades within the industry in return for funding in Ether, Bitcoin, or other digital Blockchain native currencies.

22.1.1.2.7 Multisignatures

While multisignatures were possible since the beginning of the decade, it was not until c. 2012 that Bitcoin Improvement Protocol (BIP) 16 and 17 made it easier to use advanced scripts for multisignatures (BIP11)[9].

Multisignatures are often an overlooked part of digital security by distributing private keys among several participants such that approval from multiple parties is obtained before executing a transaction. This is like having a requirement of two or more signatories on a check.

In both public and permissioned (private) Blockchains, multisignatures can empower the distribution of authoritative decisions with escrow accounts, reversible consumer trade, and boost the control and security of Blockchain transactions. This is critical with larger-scale projects.

22.1.1.3 Savings From Disintermediation

The implications of Blockchain technology should be seen as being in the same line as other human innovations (and maybe even more subversive since it deals with the immutability, or quasi-immutability, of communication of value) such as the invention of the Internet.

The potential disintermediation of the trusted third party in the financial sector (for investment banks alone) has been estimated to provide savings of US$12 billion annually (Irrera & Kelly, 2017). By considering all of the other inefficiencies in other sectors, including climate finance, this could boost projects by unlocking orders of magnitude more toward meeting climate change goals.

It is, indeed, a big deal to be able to put a bank as powerful as the largest banks today, or an open notarization service, land registry platform, or climate finance exchange, on a mobile device without requiring the traditional infrastructure that comes along with it.

[9] https://github.com/bitcoin/bips/blob/master/bip-0011.mediawiki.

22.2 Smart Contracts for Self-Executing Obligations

It is not correct to say that the law can be mathematically codified and run on top of the Blockchain, no matter how disruptive and significant it is. Or that legislature can devolve their responsibilities to programmers as some contend (Civalleri, 2017). Code is not law. However, it is possible for Blockchain (or smart contract) technology to support the applicability and enforceability of certain legislative provisions. However, to do so, a deep appreciation of some technical limitations of smart contracts must be understood by working side by side with legal professionals, industry experts and the community[10].

22.2.1 Leveraging Smart Contracts

Smart contracts allow the locking-in of consideration (tradable value), pursuant to rules that operate within the confines of the Blockchain itself. Thus, the transfer of the information can accompany the financial execution it requires with all in one motion.

Smart contracts that operate in this manner must be properly reviewed to avoid malfunctions and financial losses. For example, in May 2017, a Canadian exchange that locked in their revenues within a smart contract overlooked a mistake in their codebase and so could not retrieve their funds amounting to c. US$14 million (Higgins, 2017).

Through many experiments, smart contract architects are now learning the hard way that simple smart contracts are not only preferred, but critical. Hence, complexity is pushed off the Blockchain environment to an external environment, but one that relies on the immutability of the data sets recorded in the Blockchain[11].

To properly consider the smart contract bandwagon, those in the climate change field will need to bring together different expertise from legal, technology, marketing, and business modeling working in close collaboration.

22.2.2 Using Smart Contracts in Efforts Against Climate Change

Blockchain and smart contracts provide an elegant way to link renewable energy and carbon accounting, reporting, and tracking on a micro and macro economic level. These can be

[10] For example, it is possible that letters of credit can be backed in real time by proof of reserves, so the issuers know that the a particular business is financially healthy. This in turn may lead to lower overheads for the issuers and reliance by other parties.

[11] Unlike agile development and rapid iterations in a production environment, smart contracts or Blockchain-based applications are designed to be unstoppable and lasts as long as the Blockchain is operational. For this reason, the work involved in building, testing, and checking the code must be meticulous. The architecture of a smart contract must be airtight, and therefore a clear vision of what is to be done is required. Properly implemented smart contracts should take longer to release than expected.

connected to funding initiatives, trade settlements, lock in value for taxes and penalties, without the friction introduced by traditional intermediaries. However, doing this on a global scale is daunting. This is because climate finance involves national, regional, and international entities that participate in mitigation and adaptation of projects and programs. In designing such a system, the complexity of various participants must be pushed out of the smart contracts to create a simple solid base on which extensible feature sets can be built and developed.

Therefore, three questions that should be asked are:

1. Which is a solution for this industry, a public or private chain?
2. Which platform should be used?
3. What should be recorded and tracked into the Blockchain?

Using same question references, below are some possible answers.

For questions (1), (2), it is thought that a public network, like Ethereum, should be preferred because:

a. The traction and mindshare of the Ethereum network is, by far, the largest globally. It means that the technology used should be at the cutting edge for a significant period of time and reduces the risk of building within a network that is isolated;
b. The tools for development on Ethereum have been built to be user-friendly to accelerate adoption;
c. The tests being run on Ethereum right now in terms of real use cases and businesses are generating a lot of data for lesson learning. The failures of others have generated a roadmap of best practices;
d. Through the standardization of tokens issued on top of Ethereum (called 'ERC 20'), it is possible to tap into significant funding for a project that can then be traded globally and act as Foreign Direct Cryptocurrency Investments (or "FDCI");
e. Through the EEA, it is possible to use a familiar environment to bolt private Blockchains later on to the public Ethereum network, thus creating a network of private and public Blockchains (together forming a global Blockchain network for the climate change vertical); and
f. Ethereum is considered one of the most secure information technology networks in the world[12].

As indicated in the Section 22.2.1, the downside of using the Ethereum network is that although transactions do not disclose the entity making the transactions, the details of transacting entities are nonetheless recorded on the public network. Therefore, it does not provide complete anonymity and if the public addresses are linked with a particular participant, this could offer an advantage to a competitor. This is the cost of transparency.

[12] It should be noted that we haven't yet seen the susceptibility of Blockchains to viruses.

However, several solutions such as the zero knowledge proofs are being worked on to mitigate this problem.

(3) The question of what should be recorded in the Blockchain and how the smart contract should behave at the most basic level requires an introspection of the lifecycle of power attributes and typical legislative provisions.

In the United States, power generators (originators) are at the start of that lifecycle. Once the power is sold to the relevant participants, it goes through a process of consumption. The source of the power production identifies renewable sources. Attributes are attached to the renewable sources that are then certified as RECs (for mandated markets). These are then traded and retired as required. In this process, depending on the US state in which it is produced and the technology source from which the power originates, the pricing of each REC differs.

Therefore, the obvious start for a Blockchain solution is by tokenizing a unit of measure from the originators (e.g., a kWh equals one token). This can then be issued by a smart contract and tracked on a Blockchain. Since it is possible to track the type, location, and the relevant particulars of each power source, this can then be handled by a REC certifier, who can issue an REC certificate via smart contract functionality. Essentially, the smart contract would issue a new token representing each REC certified. Trading power and RECs can then occur in tandem and be burnt upon consumption or retirement.

The design is similar for carbon accounting and the issuance of carbon credits, although it is more complex owing to a variety of emission sources. However, in a system where reporting information for carbon accounting is reported on an industry-wide basis, an authorized actor can register the same via a smart contract and anchor it into the Blockchain. This could then be tracked on a per-organization basis, and aggregated at state level. It is not hard then to calculate offsets or shortfalls that can immediately be relied upon and dealt with.

The accounting of the entire industry vertical can then be audited in real time and accessed by auditors, regulators, and other relevant actors based on "read only" access levels.

Reliance on legislative licensing to approve authorized persons to certify documents or approve an origination is still required. Ultimately, where the data introduced in tokenized form is incorrect, this can lead to "garbage out issues." While new technologies such as the Internet of Things (IOT) can be used to automate part of the origination process, the problem of trusting the information must first be addressed, which take some years. Some ventures already working to tackle this problem are Volt Markets, L03, and Grid + .

22.2.3 Financing Flows

Smart contracts can help unscramble and inspect climate finance flows from government budgets and capital markets. For example, it is possible to create a new asset class to conduct settlement of trades such as "crypto-securities" or peg the creation of a USD token to the US dollar that is redeemable for US dollars via a liquid global market. Doing so would remove the need for the involvement of individual financial institutions, clearing houses, and other parties to verify information. It would, however, require a properly designed fund with proof of reserves that would act to back each USD token issued.

With the above, funds from governments to development cooperation agencies, bilateral finance institutions, multilateral finance institutions, and the UNFCCC could be tracked and reported in real time, globally, with minimal infrastructure cost. Likewise, funds from capital markets could be tracked for the private sector and the flow from carbon markets aiding in the funding of projects would be managed in a transparent and open manner. This should lead to less conflicts and a greater clarity about the state of this industry.

It also signifies that climate policy administrators can implement the relevant agreed-upon conditions within their jurisdictions together with layers of reporting, in granular details, such that, in relation to the settlement of trading-in assets, retirement is verified beyond doubt, without synchronization issues.

Imagine a world where Blockchain-based cooperation uses the same set of data in real time. This could lead to less time being spent on lengthy protectionist international agreements and; more focus on directing the flow of funds to projects that matter in order to achieve the stated goals of a cleaner, greener future.

22.3 The Dawn of a New Economy

One of the most exciting developments for climate finance and renewable energy trading with smart contracts is the explosion of global funding through token sales. This has accelerated in 2017 because of a number of reasons:

1. Early cryptocurrency adopters rode an appreciation wave whereby many new millionaires were created. As the growth of this industry is orders of magnitude more than what is experienced with stocks or commodities, they tend not to leave this booming digital economy, rather investing their earnings in Blockchain ventures via token sales;
2. As Blockchain technology proves itself, it is attracting a number of investors to invest in tokens such as Ether and Bitcoin, in addition to ERC20 tokens, demonstrable through the rapidly increasing liquidity in trading those assets; and
3. The Fat-Protocol model is now being more understood (Monegro, 2016). This postulates that during the Internet era, the protocol carrying information had virtually no way to capture value at the protocol layer, but applications built on top of that infrastructure are where most of the value was soaked up. In the Blockchain space, we are seeing a reversal of this, for instance: What if the Internet had tokens and anyone was required to have tokens to use it—how much would this token be worth today?

In this manner, the protocol layer is expected to produce most of the value in the Blockchain era and the applications running on top of it significantly less.

In this new economy, several conclusions can be drawn:

1. If a token represents a tailored approach to the climate change-related industries, then its successful deployment could mean unimaginable returns for those investing in the tokens;
2. The way in which contributions to climate finance is made can be highly creative although some patterns are starting to emerge in the way that institutional investors participate in token sales;
3. The way in which the community's participation is leveraged could mean a change in the current governance models with the input of the crowd through various voting mechanisms; and
4. The tokens are liquid from the day they are released on global exchanges, although this may change significantly in the coming years as more clarity is available from regulators.

22.3.1 A Novel Way to Funding Renewable Energy Projects

A number of funding models have been used to fund projects, such as the Bitcoin Investment Trust model, Coinlist, Crypto Assets Fund, and Blockchain Capital. Other models are being tested that have a number of different flavors as this field is evolving rapidly.

Another model—the Sutton Stone[13] model—can be used to address the climate-related industries through a utility token (where the token itself represents access to the platform, recordings, trades, or rewards). The model segregates funding for institutional-type investors (e.g., venture capitals) and public (crowd) sales. In addition, the funding period is over a long period of time to discourage major participants (or speculators) from influencing the price of the token (e.g., by dumping tokens to move the market).The opportunities previously only available to wealthy investors are now being commingled with the crowd.

These models are being tailored to include consumer protection initiatives, including "Know Your Customer" (KYC) and antimoney laundering (AML) policies and obligations, and the relevant incentives for multistakeholder participation.

The utility token sales model highlights a distinct shift from the way in which modern-day corporations operate. We have had hundreds of years of experience with the concept of the company and many corporate structures have been subject to adaptation over the years. However, this new way of funding means a rethinking of the concepts of shareholders

[13] https://suttonstone.com/

equity, officers' duties, compliance and planning, and more. For example, the same token can represent ownership or access to a platform, in addition to cash for cash-flow requirements, all under different conditions.

In the renewable energy sector, this funding model could be linked to assessments predicting the likely production of energy over an average year, or the average annual energy yield predicted for a site. These evaluations are often termed P50-P90 evaluations, which are interpretations of the simulation of energy productivity over several years. It is possible to link that simulation results to the release of funding for projects and complement traditional funding models. This democratization of for renewable energy financing can lead to significant upside toward the goals of a cleaner, low-carbon future.

By combining financing and enterprise activities, the cost of financing renewable projects such as administrative overheads can also be significantly sidestepped. Further, compliance is a matter of examining the Blockchain records with the relevant checkpoints, in real time. Penalties or taxes can be automatically charged to organizations or nation–states for under-performance; or grants can be linked to provable events that all participants agree on.

22.3.2 Appreciation of Token Values

Between January and July 2017, approximately US$1.2 billion has been raised through token sales (Soklin, 2017). While this may look small for capital markets, the rate of growth is significant. It means that significant deliberation should be given to how token sales can boost renewable project finance, plus other mitigation and adaptation projects against climate change.

Why could the crowd not participate in owning the tokens required to drive trades and finance climate investments? The more trading increases and its uses grow, the more likely that the tokens will appreciate in value and the less friction there will be in trades. The meshing of these activities with various international agreements (from which various national legislations, regulations, and sectoral standards are transposed) must, therefore, be studied as soon as possible. It is recommended that an ad hoc experts' committee be created under the UNFCCC to address just that. Blockchain allows us to live in a society of a different paradigm, which should be better understood and used to incentivize all parties to address climate change together.

22.4 Concluding Remarks

As indicated in this chapter, it is impossible to describe a narrative for accelerated climate finance and wider efforts against climate change without the prerequisite of understanding the nuances of Blockchain and smart contracts. Hence, by describing some of the major elements of the interaction among relevant stakeholders, the legislative environments in which they evolve and the new possibilities fetched by Blockchain technology, readers can identify (investment) opportunities in a practical manner and, as such, streamline it to their specific roles. A goal of this chapter and also the entire book is to generate ideas and drive for innovation that truly help various professionals working on climate change businesses toward developing real, executable (renewable energy) projects.

The power of this technology and the paradigm shift it is bringing is so real that it has the power to fundamentally transform climate finance and the global battle against climate change. But this new economy for the crowd will not happen unless it is complemented with a new paradigm for legislative and regulatory frameworks in the Blockchain era.

References

Chen, C. (2014). *Bill Gates: "Bitcoin Technology is Key"*, CryptoCoin News (CCN). Accessed on 30 July 2017 at https://www.cryptocoinsnews.com/bill-gates-bitcoin-technology-key/.

Civalleri J. (2017). *When "Code Is Law" meets "Law Is Law" – Inside the American Bar Association's Biggest Blockchain Event of the Year* in Hackernoon. Accessed on 30 July 2017. https://hackernoon.com/when-code-is-law-meets-law-is-law-inside-the-american-bar-associations-biggest-Blockchain-c76e3bafbb9a.

Condliffe J. (2017). *Bitcoin has avoided Tearing itself apart (for Now)*. Accessed on 30 July 2017: https://www.technologyreview.com/s/608332/bitcoin-has-avoided-tearing-itself-apart-for-now/.

Danielsen, A (1998). *The evolution of data models and approaches to persistence in database systems*. Accessed on 30 July 2017at https://www.fing.edu.uy/inco/grupos/csi/esp/Cursos/cursos_act/2000/DAP_DisAvDB/documentacion/OO/Evol_DataModels.html.

Higgins S. (2017). *Ethereum client update issue costs cryptocurrency exchange $14M* in coindesk. Accessed on 30 July 2017 at https://www.coindesk.com/ethereum-client-exchange-14-million/.

Hinchlifee E. (2017). *10,000 Bitcoin bought 2 pizzas in 2010 – and now it'd be worth $20 million* in Mashable.com. Accessed on 30 July 2017 at: http://mashable.com/2017/05/23/bitcoin-pizza-day-20-million/#A3FR3Ka_Lmqo.

Irrera A and Caspani, M. (2017). *Fintech "Lingo" Explained*, the New York Times. Accessed on 30 July 2017 at https://www.nytimes.com/reuters/2017/07/24/business/24reuters-usa-fintech-explainer.html.

Irrera A. and Kelly J. (2017). *Blockchain could save investment banks up to $12 billion a year: Accenture in Reuters*. Accessed on 30 July 2017 at http://www.reuters.com/article/us-banks-Blockchain-accenture-idUSKBN1511OU.

Monegro J. (2016). *Fat Protocols* in Union Square Ventures blog. Accessed on 30 July 2017 at http://www.usv.com/blog/fat-protocols.

Nakamato, S. (2008). Accessed on 30 July 2017 at: https://bitcoin.org/bitcoin.pdf.

Nordrum A. (2017). *Citi launches Blockchain-based payments service with Nasdaq or private equity in spectrum*. Accessed on 30 July 2017 at: http://spectrum.ieee.org/tech-talk/telecom/internet/citi-launches-Blockchainbased-payments-service-with-nasdaq-for-private-equity.

Paris Agreement, in UNFCCC Article (2017). *Blockchain technology can boost climate action*. Accessed on 30 July 2017 at newsroom.unfccc.int/climate-action/how-Blockchain-technology-could-boost-climate-action/.

Soklin L. (2017). Autonomous Research LLP. *#Token Mania – The new funding mechanism using distributed ledger technology that displaces both public markets (IPOs) and private investment (venture capital) with a billion of USD equivalent in Cryptocurrency*.

Editor's Epilogue

Alastair Marke

The objective of this book has not been to give readers any model answer to such profound question that how the international community can deploy Blockchain technology to reach or even surpass climate finance goals. But rather it is to ignite extensive discussions among all of us in search of even better answers, though they could be evolving constantly alongside technological advancement. One year of activities in the field of Blockchain may be equal to 10 years in other fields.

Now that you may have gained some more ideas about the potential applicability of Blockchain technology in some key areas of global climate change governance: renewable energy trading, international climate finance transfers, emissions management, and green finance law enforcement; you may have developed some views around this topic. I do not expect all readers to agree to all—or even most of—the insights shared by our contributors, but it is time for everybody to, at least, start thinking about it. It is because we have spent no less than 25 years on climate negotiations, the progress of which just cannot keep pace with the speed of climate change.

Blockchain might be a new "thing" to many climate change policy makers. Yet, an old quote from a French poet in the 19th century, Victor Hugo, should be still meaningful today.

> There is one thing more powerful than all the armies in the world, and that is an idea whose time has come.

Over the last two centuries, that time came to technology development which has created climate change. Today, in 2017, the "idea whose time has come" may be a technology which can turn it around—Blockchain. No matter their professional backgrounds, all our contributors share a common vision: Blockchain, Artificial Intelligence, the Internet of Things, and other emerging digital technologies are set to revolutionize global climate change governance. They would start with facilitating the implementation of the Paris Agreement, for examples, Articles 6 and 9 on emissions reduction and climate finance, respectively, as a representative of the UNFCCC Secretariat presented at a side event at COP23 in Bonn, Germany.

Time Has Come to Blockchain as a "Trust Machine"

Why Blockchain? As uttered in the first section of this book, one of the biggest challenges to fighting climate change effectively is a sheer lack of trust. It has been such lack of trust that cultivates so much "anti-tamper" bureaucracy in many international climate funds, through which some least developed countries cannot navigate easily to get the funding they desperately need to cope with the looming impact of climate change. An unintended consequence has been lengthy delays in the disbursement of climate finance to the climate-vulnerable. It becomes increasingly a norm. I would imagine that by the time the aggrieved receive the much-needed funding, the water level will have already soaked their eyebrows! The current climate finance system has got to change to break the deadlock.

The Economist has rightly said that Blockchain is a "trust machine." Recent research by Edelman found that only 15% of people trust the systems in their countries to work for them, and that trust across every sector declined for the first time ever in 2016 while the trend continued in 2017. Blockchain will help build or rebuild trust among nations, without which the successful conclusion, and effective implementation of climate agreements will become a mission impossible. We can expect that trust is bolstered by the ability to see whether rich countries are in fact living up to their commitment to emissions reductions and climate finance transfers. With such key features as distributed ledgers and smart contracts, Blockchain ensures that developed country governments or private investors actually provide the funds they promise: and that the funds, once committed, actually are spent on areas that need them badly. The buzzword we use to refer to this paradigm of climate finance governance today is MRV (measurement, reporting, and verification), but MRV should be done in a "smarter" and "faster" way to keep pace with climate change.

Blockchain as an Enabler of New Climate Solutions

I have suggested that Blockchain might be a new "thing" to climate policy makers. It is because "Blockchain is not the 'thing'. It's the 'thing' that enables the 'things'," as Mark Buitenhek, ING's Global Head of Transaction Services, frames it. This book has presented many "things" to our readers.

Let me recapitulate some key ideas in previous chapters. In the second section, we presented how Blockchain-based innovations can increase the efficiency of the energy markets by re-engineering the renewable energy smart grids for peer-to-peer energy trading among prosumers. It will involve new cryptocurrency systems that reward green energy prosumers fairly and replace some traditional policy instruments such as feed-in tariff and net metering which have not taken current grid designs into account. In the third section, we introduced some concepts about how Blockchain-supported algorithms can "speed-matche" funders or investors with climate-vulnerable communities by fading out intermediaries in the current climate finance architecture, while its smart contracts ensure the disbursement of all climate finance being results-based. In the fourth section, we discussed how Blockchain-based transaction platforms can disrupt the emissions trading markets by redefining greenhouse gas inventories management, as well as the approach to carbon pricing, accounting, and reporting. In the last section, we examined how Blockchain-based validation of green bonds impact can improve the enforceability of many green finance-related regulations. To explore the potential interaction between Blockchain and relevant legal frameworks, we discussed how regulatory frameworks should be tuned and pruned to govern disruptive technologies like Blockchain in an effective fashion.

Having said it, I would also stress that Blockchain is not a panacea to all the problems in climate change governance. I am not saying the current governance systems being superfluous just because of the complexity of procedures to follow, but I reckon that we need Blockchain to "automate" these governance systems to administer climate projects smarter and faster. As of today, there is no definitive or universal Blockchain system built in this space. If we are to fully roll out Blockchain as an enabler to various new climate solutions, there are still a plethora of challenges ahead we have to weather together. They include the huge amount of energy needed for data processing, and the handling of privacy-sensitive data such as consumption, locations, or financial transactions. It may take at least half a decade more for us to combine jurisdiction and timestamp on a Blockchain. It may even take a little longer for stakeholders to unanimously accept it and integrate relevant technological tools together to address the needs of everywhere and everyone.

Next Steps...

This book is just the beginning of this Blockchain climate movement. Together with all the experts who have contributed to this piece of hard work, we are upgrading the International Core Group on Blockchain Climate Finance to an incorporated not-for-profit—**Climate Blockchain Institute**, http://blockchainclimateinstitute.org—to continue the momentum and welcome new members. It is our intention to translate at least some of the concepts presented in this book into reality in collaboration with various institutions. If you are interested in our work, we can continue the discussions through our Readers eForum https://www.linkedin.com/groups/13580160.

Let me conclude the book with some words of wisdom from Professor Clayton Christensen of Harvard Business School in *The Innovator's Dilemma* (1997):

> By and large, a disruptive technology is initially embraced by the least profitable customers in a market. Hence, most companies with a practiced discipline of listening to their best customers and identifying new products that promise greater profitability and growth are rarely able to building a case for investing in disruptive technologies until it is too late.

I am sure we have learnt a lot of lessons from business history. After thinking about it, the next step for us may be to start investing our time or even money to embrace believably the most exciting part of the ongoing digital revolution before it is too late—when the impact of climate change becomes unmanageable with our current system. The potential uses of Blockchain against climate change are unlimited. There is nothing more difficult than to take the lead in the introduction of a new order of "things" in climate change governance. We have to create this turning point together.

Index

Note: Page numbers followed by "*f*," "*t*," and "*b*" refer to figures, tables, and boxes, respectively.

C

9780128144473